AQA Physics

A LEVEL
YEAR 2

2nd Edition

D0921479

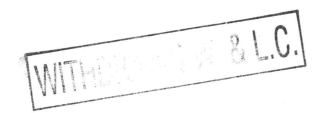

WITH... & L.C.

Jim Breithaupt

OXFORD
UNIVERSITY PRESS

856178

OXFORD
UNIVERSITY PRESS

Great Clarendon Street, Oxford, OX2 6DP, United Kingdom

Oxford University Press is a department of the University of Oxford.
It furthers the University's objective of excellence in research, scholarship, and
education by publishing worldwide. Oxford is a registered trade mark of Oxford
University Press in the UK and in
certain other countries

© Jim Breithaupt 2015

The moral rights of the authors have been asserted

First published in 2015

All rights reserved. No part of this publication may be reproduced, stored in a
retrieval system, or transmitted, in any form or by any means, without the prior
permission in writing of Oxford University Press, or as expressly permitted by
law, by licence or under terms agreed with the appropriate reprographics rights
organization. Enquiries concerning reproduction outside the scope of the above
should be sent to the Rights Department, Oxford University Press,
at the address above.

You must not circulate this work in any other form and you must impose this same
condition on any acquirer

British Library Cataloguing in Publication Data
Data available

978-0-19-835772-8

10 9 8 7 6 5

Paper used in the production of this book is a natural, recyclable product made from
wood grown in sustainable forests.
The manufacturing process conforms to the environmental regulations of the
country of origin.

Printed in China by Leo Paper Products Ltd

Message from AQA

This textbook has been approved by AQA for use with our qualification. This
means that we have checked that it broadly covers the specification and we
are satisfied with the overall quality. Full details of our approval process can be
found on our website.

We approve textbooks because we know how important it is for teachers and
students to have the right resources to support their teaching and learning.
However, the publisher is ultimately responsible for the editorial control and
quality of this book.

Please note that when teaching the *AQA AS or A-Level Physics* course, you must
refer to AQA's specification as your definitive source of information. While this
book has been written to match the specification, it cannot provide complete
coverage of every aspect of the course.

A wide range of other useful resources can be found on the relevant subject
pages of our website: www.aqa.org.uk.

WIGAN & LEIGH COLLEGE
LOCATION: LC RWSTO
ORDER NO:
COLLECTION: STANDARD
CLASS MARK: 530 BRE
BARCODE: 856178
DATE: 9/11/17 PRICE 28.50

AS/A Level course structure

This book has been written to support students studying for AQA A Level Physics. It covers the A Level Year 2 only content from the specification. The sections covered are shown in the contents list, which also shows you the page numbers for the main topics within each section. There is also an index at the back to help you find what you are looking for. This student book contains summaries of the options. The full options are available on Kerboodle with supporting material.

AS exam

Year 1 content

1 Particles and radiation
2 Waves
3 Mechanics and energy
4 Electricity

Year 2 content

6 Further mechanics and thermal physics
7 Fields
8 Nuclear physics

Plus one option from the following:
- Astrophysics
- Medical physics
- Engineering physics
- Turning points in physics
- Electronics

A level exam

A Level exams will cover content from Year 1 and Year 2 and will be at a higher demand than the AS exams. You will also carry out practical activities throughout your course.

Contents

Please note, Sections 1–5 of this course can be found in the equivalent AS/Year 1 book.

How to use this book

Learning objectives

→ At the beginning of each topic, there is a list of learning objectives.

→ These are matched to the specification and allow you to monitor your progress.

→ A specification reference is also included.

Specification reference: 3.1.1

This book contains many different features. Each feature is designed to foster and stimulate your interest in Physics, as well as supporting and developing the skills you will need for your examinations.

Terms that you will need to be able to define and understand are highlighted in **bold orange text**. You can look these words up in the glossary.

Sometimes a word appears in bold. These are words that are useful to know but are not used on the specification. They therefore do not have to be learnt for examination purposes.

Synoptic link

These highlight how the sections relate to each other. Linking different areas of physics together is important, and you will need to be able to do this.

There are also links to the maths section to support the development of these skills.

 Application features

These features contain important and interesting applications of chemistry in order to emphasise how scientists and engineers have used their scientific knowledge and understanding to develop new applications and technologies. There are also application features to develop your maths skills, with the icon $\sqrt{x}$, and to develop your practical skills, with the icon ⚗

Extension features

These features contain material that is beyond the specification, designed to stretch and provide you with a broader knowledge and understanding and lead the way into the types of thinking and areas you might study in further education. As such, neither the detail nor the depth of questioning will be required for the examinations. But this book is about more than getting through the examinations.

> 1 Extension and application features have questions that link the material with concepts that are covered in the specification. Again, short answers are inverted at the bottom of the feature, whilst longer answers can be found in the answers section at the back of the book.

Study Tips

Study tips contain prompts to help you with your revision. They can also support the development of your practical skills (with the practical symbol ⚗) and your mathematical skills (with the maths symbol $\sqrt{x}$)

Hint

Hint features give other information or ways of thinking about a concept to support your understanding. They can also relate to practical or mathematical skills and use the symbols ⚗ and $\sqrt{x}$

Summary Questions

1 These are short questions that test your understanding of the topic and allow you to apply the knowledge and skills you have aquired. The questions are ramped in order of difficulty.

2 $\sqrt{x}$ Questions that will test and develop your mathematical and practical skills are labelled with the mathematical symbol ($\sqrt{x}$) and the practial symbol (⚗).

Introduction at the opening of each section summarises what you need to know.

Section 6

...d thermal physics

...tion

...nsists of two main areas: further mechanics and thermal ...areas depend on a good understanding of A Level ...studies on force and energy.

17 Motion in a circle
18 Simple harmonic motion
19 Thermal physics
20 Gases

Further mechanics develops the A Level Year 1 link between force and acceleration in the context of objects in circular motion at constant speed and objects oscillating in simple harmonic motion. You will also develop an awareness of the wide application of the conservation of energy. The first two chapters explore the topics in a range of real-world contexts. In addition, the study of mechanics in this section covers effects such as damping and resonance that are found not just in mechanical systems but also in acoustic, atomic, and electrical systems. By studying the further mechanics topics in this section, you should establish a solid basis for further studies in both physics and engineering. You will also gain an appreciation of the historical importance of Newton's laws in providing the key principles behind the design, construction, and use of the machines and engines that powered the Industrial Revolution and those we rely on in our own scientific age.

Thermal physics is the study of the thermal properties of materials, including the relationships between energy, temperature, and the physical state of a substance. You will learn to explain the physical properties of an ideal gas by applying the kinetic theory of matter and the laws of mechanics to a simple molecular model of a gas. This model is a classic example of the successful use of a theory to explain a law founded on experimental observations.

Working scientifically

In this part of the course, you will develop and broaden your practical skills by, for example, measuring the time period of oscillating objects and using data loggers to measure and record temperature changes. You will also deepen your understanding of the physical properties of gases and of the importance of experiments in developing and testing physical theories and applications.

The maths skills you will develop in this section include arithmetical skills involving radians and trigonometry functions; algebraic skills involving rearranging non-linear equations; skills in using sines and cosines to represent oscillating motion; skills in using vectors to understand circular motion; modelling skills to understand the properties of a gas; and graphical skills. These maths skills will improve your confidence for the final parts of the A Level Physics course.

What you already know

From your A Level Physics Year 1 studies on mechanics and materials, you should know that:

- ☐ A vector quantity has magnitude and direction, whilst a scalar quantity has magnitude only.
- ☐ Displacement, velocity, acceleration, and force are vector quantities.
- ☐ A force F can be resolved parallel and perpendicular to a line at angle θ to the direction of the force into a parallel component $F\cos\theta$ and a perpendicular component $F\sin\theta$.
- ☐ $\text{Speed} = \dfrac{\text{distance}}{\text{time}}$ for constant speed.
- ☐ Velocity = rate of change of displacement = the gradient of a displacement–time graph.
- ☐ Acceleration = rate of change of velocity = the gradient of a velocity–time graph.
- ☐ An object continues at rest or at constant velocity unless acted on by a resultant force.
- ☐ The resultant force on an object = its mass × its acceleration, and its weight is its mass × g.
- ☐ The symbols s, u, v, t, a, and F represent displacement, initial velocity, velocity at time t, acceleration, and force, respectively.
- ☐ Momentum = mass × velocity, and resultant force on an object = rate of change of momentum of the object.
- ☐ The equations for uniform acceleration, $v = u + at$, $s = \frac{1}{2}(u + v)t$, $s = ut + \frac{1}{2}at^2$ and $v^2 = u^2 + 2as$, and $F = ma = \dfrac{mv - mu}{t}$ can be used to predict the motion of an object moving along a straight line.
- ☐ The work done W on an object is given by $W = Fs\cos\theta$, where θ is the angle between the direction of the force and the displacement of the object.
- ☐ For an object of mass m, its kinetic energy at speed v is $\frac{1}{2}mv^2$ (if $v \ll c$, the speed of light in free space) and its change of gravitational potential energy = $mg\Delta h$ when moved up or down through a vertical distance Δh.
- ☐ Power = rate of transfer of energy.
- ☐ $\text{Density} = \dfrac{\text{mass}}{\text{volume}}$.
- ☐ The force needed to stretch a spring is directly proportional to the extension of the spring from its natural length.

A checklist to help you assess your knowledge from KS4 or Year 1 of this course, before starting work on the section.

2

Visual summaries of each section show how some of the key concepts of that section interlink with other sections.

Section 9 Engineering physics

In this option, you can reinforce skills that you developed in the compulsory topics in your A Level Physics course. By studying this option, you will strengthen your competence in the practical and mathematical skills listed below. In addition, you will develop a deep insight into the application of mathematical knowledge gained in this option to a range of engineering contexts associated with rotational dynamics and thermodynamics. The extension task is intended to boost your skills of written communication, and also lets you work within a group by creating presentations on different topics.

Practical skills

In this option you meet the following skills:

- use of a stopwatch or stopclock to measure the period of rotation of a rotating object
- use of metre rules and vernier callipers to measure distances (e.g. diameters of shafts and wheels)
- use of thermometers or temperature sensors connected safely and accurately

to a data logger, in demonstrations and experiments involving heating or cooling (e.g. work done on a gas)
- use of equipment safely and correctly (e.g. in the measurement of the moment of inertia of a flywheel)
- use of a manometer and a pressure gauge to measure the pressure of a gas (e.g. in an adiabatic change).

Maths skills

In this option you meet the following skills:

- use of the radian and its conversion to and from degrees (e.g. angular displacement calculations)
- calculation of a mean from a set of repeat readings (e.g. time period of a rotating object)
- interpret graphs in terms of relevant physics variables (e.g. pressure–volume changes for an ideal gas)

- interpret diagrams such as heat engines in terms of work done and energy transfers
- use a calculator or a spreadsheet to carry out calculations using data and appropriate formulae, expressing answers to an appropriate number of significant figures and with appropriate units
- use appropriate formulae and experimental data from experiments (e.g. moment of inertia of a flywheel).

Extension task

Use the internet to find out more about an application of a device you have studied in this option. For example, you could find out more about flywheels used for energy

storage. Or you could find out about the principles of operation of a heat pump or heat engine, ... engine, ... to give t...

A synoptic extension task to bring everything in the section together and start leading you towards higher study at university.

Summaries of the key practical and maths skills of the section.

242

vii

Practical skills section with skills for the suggested practicals on the specification. Remember, 15% of your exam will be based on practical skills.

Mathematical section to support and develop the mathematical skills required for your course. Remember, 40% of your exam will involve mathematical skills.

Practice questions at the end of each chapter and each section, including questions that cover practical and maths skills. There are also additional practice questions at the end of Section 8.

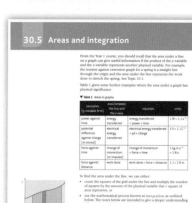

Kerboodle

This book is supported by next generation Kerboodle, offering unrivalled digital support for independent study, differentiation, assessment, and the new practical endorsement.

If your school subscribes to Kerboodle, you will also find a wealth of additional resources to help you with your studies and with revision.

- Study guides
- Maths skills boosters and calculation worksheets
- On your marks activities to help you achieve your best
- Practicals and follow up activities to support the practical endorsement
- Interactive objective tests that give question-by-question feedback
- Animations and revision podcasts
- Self-assessment checklists

Test your knowledge with the progress quizzes, and learn from your mistakes with the detailed explanations given for each answer.

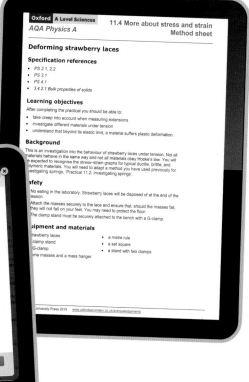

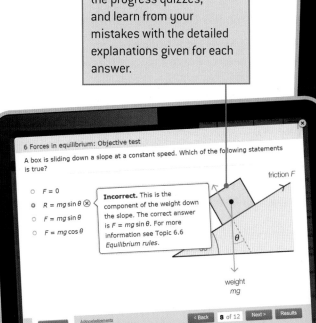

If you are a teacher reading this, Kerboodle also has plenty of further assessment resources, answers to the questions in the book, and a digital markbook along with full teacher support for practicals and the worksheets, which include suggestions on how to support and stretch your students. All of the resources are pulled together into teacher guides that suggest a route through each chapter.

Skills for starting A level Physics

1. Using a calculator

Practice makes perfect when it comes to using a calculator. For A level physics, you need no more than a scientific calculator. You need to make sure you master the technicalities of using a scientific calculator as early as possible in your physics course. At this stage, you should be able to use a calculator to add, subtract, multiply, divide, find squares and square roots and calculate sines, cosines and tangents of angles. But remember when using a calculator, it's all too easy to make a mistake, for example pressing the wrong button. So when carrying out a calculation using a calculator, check the answer by making an order-of-magnitude **estimate** of the answer in your head.

Maths link

Further important calculator functions are described in 16.1, in the Year 1 book.

2. Making measurements

You should know at this stage how to make measurements using basic equipment such as metre rules, protractors, stopwatches, thermometers, balances (for weighing an object) and ammeters and voltmeters. During the course, you will also be expected to use equipment such as micrometers, verniers, oscilloscopes and data loggers.

Practical link

The use of these items is described for reference in 14.3, in the Year 1 book.

Here are some useful reminders about making measurements:

- check the zero reading when you use an instrument to make a measurement, for example, a metre ruler worn away at one end might give a zero error
- when a multi-range instrument is used, start with the highest range and switch to a lower range if the reading is too small to measure accurately

Practical link

See 14.2 in the Year 1 book for anomalous measurements.

- make sure you record all your measurements in a logical order, stating the correct unit of each measured quantity
- don't pack equipment away until you are sure you have enough measurements or you have checked unexpected readings

3. Using measurements in calculations

whenever you make a record of a measurement, you must always note the correct unit as well as the numerical value of the measurements.

The scientific system of units is called the S.I. system. This is described in more detail in 16.1 and 16.3. The base units of the S.I. system you need to remember are listed below. All other units are derived from the S.I. base units.

1   the metre (m) is the S.I. unit of length. Note also that
    $1\,m = 100\,cm = 1000\,mm$.
2   the kilogram (kg) is the S.I. unit of mass. Note that
    $1\,kg = 1000$ grams.
3   the second (s) is the S.I. unit of time.
4   the ampere (A) is the S.I. unit of current.

Powers of ten and numerical prefixes are used to avoid unwieldy numerical values. For example:

- $1\,000\,000 = 10^6$ which is 10 raised to the power 6 (usually stated as '10 to the 6').
- $0.000\,000\,1 = 10^{-7}$ which is 10 raised to the power -7 (usually stated as '10 to the -7').

Prefixes are used with units as abbreviations for powers of ten. For example, a distance of 1 kilometre may be written as $1000\,\mathrm{m}$ or $10^3\,\mathrm{m}$ or $1\,\mathrm{km}$. The most common prefixes are shown in Table 1.

▼ **Table 1** *Prefixes*

Prefix	pico-	nano-	micro-	milli-	kilo-	mega-	giga-	tera-
Value	10^{-12}	10^{-9}	10^{-6}	10^{-3}	10^{3}	10^{6}	10^{9}	10^{12}
Prefix symbol	p	n	μ	m	k	M	G	T

Note that the cubic centimetre (cm^3) and the gram (g) are in common use and are therefore allowed as exceptions to the prefixes shown.

Standard form is usually used for numerical values smaller than 0.001 or larger than 1000.

- The numerical value is written as a number between 1 and 10 multiplied by the appropriate power of ten. For example,

  $64\,000\,\mathrm{m} = 6.4 \times 10^4\,\mathrm{m}$

  $0.000\,005\,1\,\mathrm{s} = 5.1 \times 10^{-6}\,\mathrm{s}$

- A prefix may be used instead of some or all of the powers of ten. For example:

  $35\,000\,\mathrm{m} = 35 \times 10^3\,\mathrm{m} = 35\,\mathrm{km}$

  $0.000\,000\,59\,\mathrm{m} = 5.9 \times 10^{-7}\,\mathrm{m} = 590\,\mathrm{nm}$

To convert a number to standard form, count how many places the decimal point must be moved to make the number between 1 and 10. The number of places moved is the power of ten that must accompany the number between 1 and 10. Moving the decimal place:

- to the left gives a positive power of ten (for example, $64\,000 = 6.4 \times 10^4$)
- to the right gives a negative power of ten (for example, $0.000\,005\,1 = 5.1 \times 10^{-6}$)

4. Using the right-angled triangle

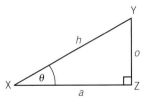

▲ **Figure 1** *The right-angled triangle*

The sine, cosine and tangent of an angle are defined from the right-angled triangle. Figure 1 on the previous page shows a right-angled triangle XYZ in which side XY is the hypotenuse (i.e. the side opposite the right angle), side YZ is opposite angle θ and side XZ is adjacent to angle θ.

$$\sin \theta = \frac{YZ}{XY} = \frac{o}{h}$$

where o = YZ, the side opposite angle θ

$$\cos \theta = \frac{XZ}{XY} = \frac{a}{h}$$

h = XY, the hypotenuse

$$\tan \theta = \frac{YZ}{XZ} = \frac{o}{a}$$

and a = XZ, the side adjacent to angle θ

To remember these formulas, recall SOHCAHTOA.

Pythagoras' theorem states that for any right angle triangle,

the square of the hypotenuse = the sum of the squares of the other two sides

Applying Pythagoras' theorem to the right angle triangle XYZ in Figure 1 gives:

$$(\mathbf{XY})^2 = (\mathbf{XZ})^2 + (\mathbf{YZ})^2$$

5. Using equations

Symbols are used in equations and formulas to represent physical variables. In your GCSE course, you may have used equations with words instead of symbols to represent physical variables. For example, you will have met the equation 'distance moved = speed × time'. Perhaps you were introduced to the same equation in the symbolic form '$s = v\,t$' where s is the symbol used to represent distance, v is the symbol used to represent speed and t is the symbol used to represent time. The equation in symbolic form is easier to use because the rules of algebra are more easily applied to it than to a word equation. In addition, writing words in equations wastes valuable time. However, you need to remember the agreed symbol for each physical quantity.

Equations often need to be rearranged. This can be confusing if you don't learn the following basic rules at an early stage:

1 Learn to read an equation properly. For example, the equation $v = 3\,t + 2$ is not the same as the equation $v = 3\,(\,t + 2\,)$. If you forget the brackets when you use the second equation to calculate v when $t = 1$, then you will get $v = 5$ instead of the correct answer $v = 9$. The first equation tells you to multiply t by 3 then add 2. The second equation tells you to add t and 2 then multiply the sum by 3.
2 Learn to rearrange an equation properly. In simple terms, always make the same change to both sides of an equation.

For example, to make t the subject of the equation $v = 3t + 2$

Step 1: Subtract 2 from both sides of the equation so
$$v - 2 = 3t + 2 - 2 = 3t$$

Step 2: The equation is now $v - 2 = 3t$ and can be written $3t = v - 2$

Step 3: Divide both sides of the equation by 3 so $\dfrac{3t}{3} = \dfrac{v-2}{3}$

Step 4: Cancel 3 on the top and the bottom of the left hand side to finish with $t = \dfrac{v-2}{3}$

To use an equation as part of a calculation:

- start by making the quantity to be calculated the subject of the equation
- write the equation out with the numerical values in place of the symbols
- carry out the calculation and make sure you give the answer with the correct unit

AND remember to check your answer by making an order-of-magnitude estimate.

Unless the equation is simple (for example, $V = IR$), don't insert numerical values then rearrange the equation. Rearrange then insert the numerical values as you are less likely to make an error if the numbers are inserted later in the process rather than earlier.

Maths link

See 16.1 to 16.3 in the Year 1 book for more about data handling, trigonometry and algebra.

Section 6
Further mechanics and thermal physics

Introduction

This section consists of two main areas: further mechanics and thermal physics. Both areas depend on a good understanding of A Level Physics Year 1 studies on force and energy.

Further mechanics develops the A Level Year 1 link between force and acceleration in the context of objects in circular motion at constant speed and objects oscillating in simple harmonic motion. You will also develop an awareness of the wide application of the conservation of energy. The first two chapters explore the topics in a range of real-world contexts. In addition, the study of mechanics in this section covers effects such as damping and resonance that are found not just in mechanical systems but also in acoustic, atomic, and electrical systems. By studying the further mechanics topics in this section, you should establish a solid basis for further studies in both physics and engineering. You will also gain an appreciation of the historical importance of Newton's laws in providing the key principles behind the design, construction, and use of the machines and engines that powered the Industrial Revolution and those we rely on in our own scientific age.

Thermal physics is the study of the thermal properties of materials, including the relationships between energy, temperature, and the physical state of a substance. You will learn to explain the physical properties of an ideal gas by applying the kinetic theory of matter and the laws of mechanics to a simple molecular model of a gas. This model is a classic example of the successful use of a theory to explain a law founded on experimental observations.

Working scientifically

In this part of the course, you will develop and broaden your practical skills by, for example, measuring the time period of oscillating objects and using data loggers to measure and record temperature changes. You will also deepen your understanding of the physical properties of gases and of the importance of experiments in developing and testing physical theories and applications.

The maths skills you will develop in this section include arithmetical skills involving radians and trigonometry functions; algebraic skills involving rearranging non-linear equations; skills in using sines and cosines to represent oscillating motion; skills in using vectors to understand circular motion; modelling skills to understand the properties of a gas; and graphical skills. These maths skills will improve your confidence for the final parts of the A Level Physics course.

What you already know

From your A Level Physics Year 1 studies on mechanics and materials, you should know that:

- ○ A vector quantity has magnitude and direction, whilst a scalar quantity has magnitude only.

- ○ Displacement, velocity, acceleration, and force are vector quantities.

- ○ A force F can be resolved parallel and perpendicular to a line at angle θ to the direction of the force into a parallel component $F\cos\theta$ and a perpendicular component $F\sin\theta$.

- ○ Speed $= \dfrac{\text{distance}}{\text{time}}$ for constant speed.

- ○ Velocity = rate of change of displacement = the gradient of a displacement–time graph.

- ○ Acceleration = rate of change of velocity = the gradient of a velocity–time graph.

- ○ An object continues at rest or at constant velocity unless acted on by a resultant force.

- ○ The resultant force on an object = its mass × its acceleration, and its weight = its mass × g.

- ○ The symbols s, u, v, t, a, and F represent displacement, initial velocity, velocity at time t, acceleration, and force, respectively.

- ○ Momentum = mass × velocity, and resultant force on an object = rate of change of momentum of the object.

- ○ The equations for uniform acceleration, $v = u + at$, $s = \dfrac{1}{2}(u + v)t$, $s = ut + \dfrac{1}{2}at^2$ and $v^2 = u^2 + 2as$, and $F = ma = \dfrac{mv - mu}{t}$ can be used to predict the motion of an object moving along a straight line.

- ○ The work done W on an object is given by $W = Fs\cos\theta$, where θ is the angle between the direction of the force and the displacement of the object.

- ○ For an object of mass m, its kinetic energy at speed v is $\dfrac{1}{2}mv^2$ (if $v \ll c$, the speed of light in free space) and its change of gravitational potential energy = $mg\Delta h$ when moved up or down through a vertical distance Δh.

- ○ Power = rate of transfer of energy.

- ○ Density $= \dfrac{\text{mass}}{\text{volume}}$.

- ○ The force needed to stretch a spring is directly proportional to the extension of the spring from its natural length.

17 Motion in a circle
17.1 Uniform circular motion

Learning objectives:

→ Recognise uniform motion in a circle.

→ Describe what you need to measure to find the speed of an object moving in uniform circular motion.

→ Define angular displacement and angular speed.

Specification reference: 3.6.1.1

▲ **Figure 1** *In uniform circular motion*

▲ **Figure 2** *The London Eye*

Hint

Remember that $360° = 2\pi$ radians.
$180° = \pi$ radians
$90° = \dfrac{\pi}{2}$ radians.
Prove for yourself that 1 rad = 57.3°.

In a cycle race, the cyclists pedal furiously at top speed. The speed of the perimeter of each wheel is the same as the cyclist's speed, provided the wheels do not slip on the ground. If the cyclist's speed is constant, the wheels must turn at a steady rate. An object rotating at a steady rate is said to be in **uniform circular motion**.

Consider a point on the perimeter of a wheel of radius r rotating at a steady speed.

- The circumference of the wheel = $2\pi r$
- The frequency of rotation $f = \dfrac{1}{T}$, where T is the time for one rotation.

The speed v of a point on the perimeter = $\dfrac{\text{circumference}}{\text{time for one rotation}}$

$= \dfrac{2\pi r}{T} = 2\pi r f$

$$v = \dfrac{2\pi r}{T}$$

Worked example

A cyclist is travelling at a speed of $25\,\text{m s}^{-1}$ on a bicycle that has wheels of radius 750 mm. Calculate:

a the time for one rotation of the wheel,
b i the frequency of rotation of the wheel,
ii the number of rotations of the wheel in one minute.

Solution

a Rearranging speed $v = \dfrac{2\pi r}{T}$ gives the time for one rotation,
$T = \dfrac{2\pi r}{v}$
Therefore, $T = \dfrac{2\pi \times 0.75}{25} = 0.19\,\text{s}$

b i Frequency $f = \dfrac{1}{T} = \dfrac{1}{0.19} = 5.3\,\text{Hz}$

ii Number of rotations in 1 min = $60 \times 5.3 = 320$ (2 s.f.)

Angular displacement and angular speed

The London Eye is a very popular tourist attraction. The wheel has a diameter of 130 m and takes passengers high above the surrounding buildings, giving a glorious view on a clear day. Each full rotation of the wheel takes 30 minutes. So each capsule takes its passengers through an angle of 0.2° per second (= $\dfrac{\pi}{900}$ radians per second). This means that each capsule turns through an angle of

- 2° in 10 s,
- 20° (= $\dfrac{2\pi}{18}$ radians) in 100 s,
- 90° (= $\dfrac{\pi}{2}$ radians) in 450 s.

For any object in uniform circular motion, the object turns through an angle of $\dfrac{2\pi}{T}$ radians per second, where T is the time taken for one complete rotation. In other words, the angular displacement of the object per second is $\frac{2\pi}{T}$.

The **angular displacement**, θ (i.e., the angle in radians), of the object in time t is $\frac{2\pi t}{T} = 2\pi ft$, where T is the time for one rotation and $f\,(= \frac{1}{T})$ is the frequency of rotation.

The **angular speed**, ω, is defined as the angular displacement per second.

So, $\omega = \dfrac{\text{angular displacement}}{\text{time taken}} = \dfrac{\theta}{t} = \dfrac{2\pi}{T} = 2\pi f$

The unit of ω is the *radian per second* (rad s^{-1}).

Worked example

A cyclist travels at a speed of $12\,\text{m s}^{-1}$ on a bicycle that has wheels of radius $0.40\,\text{m}$. Calculate:

a the frequency of rotation of each wheel,

b the angular speed of each wheel,

c the angle the wheel turns through in $0.10\,\text{s}$ in
  i radians,   **ii** degrees.

Solution

a Circumference of wheel $= 2\pi r = 2\pi \times 0.4 = 2.5\,\text{m}$

Time for one wheel rotation, $T = \dfrac{\text{circumference}}{\text{speed}} = \dfrac{2.5}{12} = 0.21\,\text{s}$

Frequency $f = \dfrac{1}{T} = \dfrac{1}{0.21} = 4.8\,\text{Hz}$

b Angular speed $\omega = \dfrac{2\pi}{T} = 30\,\text{rad s}^{-1}$

c i Angle the wheel turns through in $0.10\,\text{s}$,

$\theta = \dfrac{2\pi t}{T} = \dfrac{2\pi \times 0.10}{0.21} = 3.0\,\text{rad}$

ii $\theta = 3.0 \times \dfrac{360}{2\pi} = 170°$ (2 s.f.)

Hint

1 In time t, an object in uniform circular motion at speed v moves along the arc of the circle through a distance s of
$$s = vt = \frac{2\pi rt}{T} = \theta r$$

2 Speed $v = \dfrac{2\pi r}{T} = \omega r$

(because $\omega = \dfrac{2\pi}{T}$)

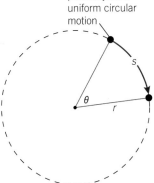

point object in uniform circular motion

▲ **Figure 3** *Arcs and angles*

Study tip

The unit symbol for the radian is 'rad'. Make sure you know how to convert angles from degrees to radians, and remember to use rad for θ and rad s^{-1} for ω.

Summary questions

1 Calculate the angular displacement in radians of the tip of the minute hand of a clock in:

  a 1 second,

  b 1 minute,

  c 1 hour.

2 An electric motor turns at a frequency of 50 Hz. Calculate:

  a its time period,

  b the angle it turns through in radians in
    i 1 ms,   **ii** 1 s.

3 The Earth takes exactly 24 h for one full rotation. Calculate:

  a the speed of rotation of a point on the equator,

  b the angle the Earth turns through in 1 s in
    i degrees,   **ii** radians.

  The radius of the Earth $= 6400\,\text{km}$.

4 A satellite in a circular orbit of radius 8000 km takes 120 minutes per orbit. Calculate:

  a its speed,

  b its angular displacement in 1.0 s in
    i degrees,   **ii** radians.

17.2 Centripetal acceleration

Learning objectives:

→ Explain why velocity is not constant when an object is travelling uniformly in a circle.

→ Determine the direction of the acceleration.

→ Calculate the centripetal force.

Specification reference: 3.6.1.1

When an object is moving in a circle at a constant speed, the **velocity** of the object is continually changing direction. Because its velocity is changing, the object is *accelerating*. If this seems odd because the object's speed always stays constant (uniform), remember that acceleration is the change of velocity per second. Passengers on the London Eye might not notice they are being accelerated, but if the wheel rotated at a bigger speed than usual, they certainly would!

The velocity of an object in uniform circular motion at any point along its path is along the tangent to the circle at that point (see Figure 1). The direction of the velocity changes continually as the object moves along its circular path. This change in the direction of the velocity is towards the centre of the circle. So, the acceleration of the object is towards the centre of the circle and is called **centripetal acceleration**. Centripetal means *towards the centre of the circle*.

For an object moving at constant speed v in a circle of radius r,

its centripetal acceleration $a = \dfrac{v^2}{r}$

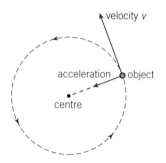

▲ **Figure 1** *Centripetal acceleration*

Proof of $a = \dfrac{v^2}{r}$

- Consider an object in uniform circular motion at speed v moving in a short time interval δt from position A to position B along the perimeter of a circle of radius r. The distance AB along the circle is $\delta s = v\delta t$. You can see this in Figure 2.

- The line from the object to the centre of the circle at C turns through angle $\delta\theta$ when the object moves from A to B. The velocity direction of the object turns through the same angle $\delta\theta$, as you can see in Figure 2.

- The change of velocity is δv = velocity at B − velocity at A. You can see this in Figure 2 in the velocity vector triangle.

- The triangles ABC and the velocity vector triangle have the same shape because they both have two sides of equal length with the same angle, $\delta\theta$, between the two sides.

If $\delta\theta$ is small, then $\quad \dfrac{\delta v}{v} = \dfrac{\delta s}{r}$

Because $\delta s = v\delta t$, then $\quad \dfrac{\delta v}{v} = \dfrac{v\delta t}{r}$

So, acceleration $a = \dfrac{\text{change of velocity}}{\text{time taken}} = \dfrac{\delta v}{\delta t} = \dfrac{v^2}{r}$ towards the centre

Study tip

You *don't* need to know how to prove this equation at A Level. The proof is given here just to give you a better understanding of the idea of centripetal acceleration. Because speed $v = \omega r$, then

$$a = \frac{v^2}{r} = \frac{(\omega r)^2}{r} = \omega^2 r$$

The equation for centripetal acceleration can then be written as $a = \omega^2 r$

Centripetal force

To make an object move around on a circular path, it must be acted on by a resultant force that changes its direction of motion.

The resultant force on an object moving around a circle at constant speed is called the **centripetal force** because it acts in the same direction as the centripetal acceleration, which is towards the centre of the circle.

- For an object whirling around rapidly on the end of a string, the tension in the string provides most of the centripetal force.

- For a satellite moving around the Earth, the force of gravity between the satellite and the Earth is the centripetal force.

Synoptic link

Remember that acceleration and velocity are vectors and that the direction of acceleration of an object is always the same as the direction of the resultant force on the object. See Topic 8.1, Force and acceleration.

- For a planet moving around the Sun, the force of gravity on the planet due to the Sun is the centripetal force.
- For a capsule on the London Eye, the centripetal force is the resultant of the support force on the capsule and the force of gravity on it.
- In Chapter 24, you will meet the use of a magnetic field to bend a beam of charged particles (e.g., electrons) in a circular path. The magnetic force on the moving charged particles is the centripetal force.

Any object that moves in circular motion is acted on by a resultant force that always acts towards the centre of the circle. The resultant force is the centripetal force and so causes centripetal acceleration.

Equation for centripetal force

For an object moving at constant speed v along a circular path of radius r, its centripetal acceleration $a = \dfrac{v^2}{r} = \omega^2 r$ (where $\omega = \dfrac{v}{r}$)

Therefore, applying Newton's second law for constant mass in the form $F = ma$ gives

$$\textbf{centripetal force } F = \frac{mv^2}{r} = m\omega^2 r$$

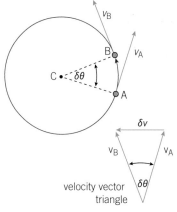

▲ **Figure 2** Proving $a = \dfrac{v^2}{r}$

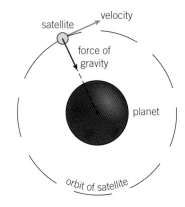

▲ **Figure 3** A satellite in uniform circular motion

Summary questions

1 The wheel of the London Eye has a diameter of 130 m and takes 30 minutes for one full rotation. Calculate:

   a the speed of a capsule,

   b i the centripetal acceleration of a capsule,

   ii the centripetal force on a person of mass 65 kg in a capsule.

2 An object of mass 0.15 kg moves around a circular path of radius 0.42 m at a steady rate once every 5.0 s. Calculate:

   a the speed and acceleration of the object,

   b the centripetal force on the object.

3 a The Earth moves around the Sun on a circular orbit of radius 1.5×10^{11} m, taking $365\frac{1}{4}$ days for each complete orbit. Calculate:

   i the speed,

   ii the centripetal acceleration of the Earth on its orbit around the Sun.

   b A satellite is in orbit just above the surface of a spherical planet which has the same radius as the Earth and the same acceleration of free fall at its surface. Calculate:

   i the speed,   ii the time for one complete orbit of this satellite.

   Radius of the Earth = 6400 km   Acceleration of free fall = 9.8 m s^{-2}

4 A hammer thrower spins a 2.0 kg hammer on the end of a rope in a circle of radius 0.80 m. The hammer took 0.60 s to make one full rotation just before it was released. Calculate:

   a the speed of the hammer just before it was released,

   b its centripetal acceleration,

   c the centripetal force on the hammer just before it was released.

Hint

1 If the object is acted on by a single force only (e.g., a satellite in orbit around the Earth), that force is the centripetal force and causes the centripetal acceleration.

2 The centripetal force is at right angles to the direction of the object's velocity. Therefore, no work is done by the centripetal force on the object because there is no displacement in the direction of the force. The kinetic energy of the object is therefore constant, so its speed is unchanged.

Learning objectives:

→ Explain why a passenger in a car seems to be thrown outwards if the car rounds a bend too quickly.

→ Describe what happens to the force between the passenger and his seat when travelling over a curved bridge.

→ Identify the forces that provide the centripetal force on a banked track.

Specification reference: 3.6.1.1

Even on a very short journey, the effects of circular motion can be important. For example, a vehicle that turns a corner too fast could skid or topple over. A vehicle that goes over a curved bridge too fast might even lose contact briefly with the road surface. To make any object move on a circular path, the object must be acted on by a resultant force that is always towards the centre of curvature of its path.

Over the top of a hill

Consider a vehicle of mass m moving at speed v along a road that passes over the top of a hill or over the top of a curved bridge.

At the top of the hill, the support force S from the road on the vehicle is directly upwards in the opposite direction to its weight, mg. The resultant force on the vehicle is the difference between the weight and the support force. This difference acts towards the centre of curvature of the hill as the centripetal force. In other words,

$$mg - S = \frac{mv^2}{r}$$

where r is the radius of curvature of the hill.

The vehicle would lose contact with the road if its speed is equal to or greater than a particular speed, v_0. If this happens, then the support force is zero (i.e., $S = 0$), so $mg = \frac{mv_0^2}{r}$.

Therefore, the vehicle speed should not exceed v_0, where $v_0^2 = gr$, otherwise the vehicle will lose contact with the road surface at the top of the hill. Prove for yourself that a vehicle travelling over a curved bridge of radius of curvature 5 m would lose contact with the road surface if its speed exceeded $7\,\text{m s}^{-1}$.

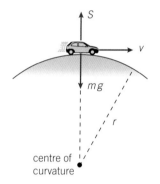

▲ **Figure 1** *Over the top*

On a roundabout

Consider a vehicle of mass m moving at speed v in a circle of radius r as it moves around a roundabout on a level road. The centripetal force is provided by the sideways force of friction between the vehicle's tyres and the road surface. In other words,

$$\text{force of friction } F = \frac{mv^2}{r}$$

For no skidding or slipping to occur, the force of friction between the tyres and the road surface must be less than a limiting value, F_0, that is proportional to the vehicle's weight.

Therefore, for no slipping to occur, the speed of the vehicle must be less than a particular value v_0 that is given in the equation

$$\text{limiting force of friction } F_0 = \frac{mv_0^2}{r}$$

Note: Because F_0 is proportional to the vehicle's weight, then $F_0 = \mu mg$, where μ is the coefficient of friction.

Therefore, $\mu mg = \frac{mv_0^2}{r}$

The maximum speed for no slipping $v_0 = (\mu gr)^{\frac{1}{2}}$

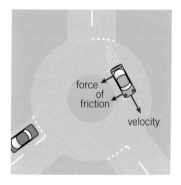

▲ **Figure 2** *On a roundabout*

Study tip

μ is not needed for A Level Physics, but you might meet it if you are studying A Level Mathematics.

On a banked track

A race track is often banked where it curves. Motorway slip roads often bend in a tight curve. Such a road is usually banked to enable vehicles to drive around without any sideways friction acting on the tyres. Rail tracks on curves are usually banked to enable trains to move around the curve without slowing down too much. Imagine you are an engineer and you are asked to design a banked track for a horizontal motorway curve.

- Without any banking, the centripetal force on a road vehicle is provided only by sideways friction between the vehicle wheels and the road surface. As you saw in the example of a roundabout, a vehicle on a bend slips outwards if its speed is too high.
- On a banked track, the speed can be higher. To understand why, consider Figure 3, which represents the front-view of a racing car of mass m on a banked track, where θ = the angle of the track to the horizontal. For there to be no sideways friction on the tyres from the road, the horizontal component of the support forces N_1 and N_2 must act as the centripetal force.

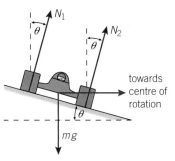

▲ **Figure 3** *A racing car taking a bend*

Resolving these forces into horizontal components $(= (N_1 + N_2) \sin \theta)$ and a vertical component $(= (N_1 + N_2) \cos \theta)$, then

- because $(N_1 + N_2) \sin \theta$ acts as the centripetal force,
$$(N_1 + N_2) \sin \theta = \frac{mv^2}{r},$$
- because $(N_1 + N_2) \cos \theta$ balances the weight (mg),
$$(N_1 + N_2) \cos \theta = mg$$

Therefore, $\tan \theta = \dfrac{(N_1 + N_2) \sin \theta}{(N_1 + N_2) \cos \theta} = \dfrac{mv^2}{mgr}$

Simplifying this equation gives the condition for no sideways friction:
$$\tan \theta = \frac{v^2}{gr}$$

In other words, there is no sideways friction if the speed v is such that
$$v^2 = gr \tan \theta$$

Synoptic link

The horizontal and vertical components of a force F that is at an angle θ to the vertical must be $F \sin \theta$ and $F \cos \theta$, respectively. See Topic 6.1, Vectors and scalars.

Study tip

Prove for yourself that if the banking angle θ is not to exceed $5°$ and the radius of curvature is $360\,m$, the speed for zero sideways friction is $18\,m\,s^{-1}$.

Summary questions

$g = 9.8\,m\,s^{-2}$.

1 A vehicle of mass 1200 kg passes over a bridge of radius of curvature 15 m at a speed of $10\,m\,s^{-1}$. Calculate:

   a the centripetal acceleration of the vehicle on the bridge,

   b the support force on the vehicle when it was at the top.

2 The maximum speed for no skidding of a vehicle of mass 750 kg on a roundabout of radius 20 m is $9.0\,m\,s^{-1}$. Calculate:

   a the centripetal acceleration,

   b the centripetal force on the vehicle when moving at this speed.

3 Explain why a circular athletics track is banked for sprinters but not for marathon runners.

4 At a racing car circuit, the track is banked at an angle of $25°$ to the horizontal on a bend that has a radius of curvature of 350 m.

   a Use the equation $v^2 = gr \tan \theta$ to calculate the speed of a vehicle on the bend if there is to be no sideways friction on its tyres.

   b Discuss and explain what could happen to a vehicle that took the bend too fast.

Learning objectives:

→ Describe when the contact force on a passenger on a 'big dipper' ride is the greatest.

→ Describe the condition that applies when a passenger just fails to keep in contact with her seat.

Specification reference: 3.6.1.1

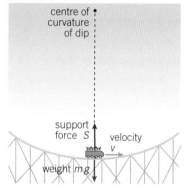

▲ **Figure 1** *In a dip*

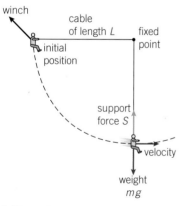

▲ **Figure 2** *The very long swing*

Many of the rides at a fairground or amusement park take people around in circles. Some examples are analysed below. It is worth remembering that centripetal acceleration values of more than 2 to 3g can be dangerous to the average person.

The Big Dipper

A ride that takes you at high speed through a big dip pushes you into your seat as you pass through the dip. The difference between the support force on you (acting upwards) and your weight acts as the centripetal force.

At the bottom of the dip, the support force S on you is vertically upwards, as shown in Figure 1.

Therefore, for a speed v at the bottom of a dip of radius of curvature r,

$$S - mg = \frac{mv^2}{r}$$

So the support force $S = mg + \frac{mv^2}{r}$.

The extra force you experience due to circular motion is therefore $\frac{mv^2}{r}$.

The very long swing

In another ride, a person of mass m on a very long swing of length L is released from height h above the equilibrium position. The maximum speed occurs when the swing passes through the lowest point. You can work this out by equating the gain of kinetic energy to the loss of potential energy:

$$\frac{1}{2}mv^2 = mgh$$

where v is the swing's speed as it passes through the lowest point.

$$\text{Therefore, } v^2 = 2gh$$

The person on the swing is on a circular path of radius L. At the lowest point, the support force S on the person from the rope is in the opposite direction to the person's weight, mg. The difference, $S - mg$, acts towards the centre of the circular path and provides the centripetal force. Therefore,

$$S - mg = \frac{mv^2}{L}$$

Because $v^2 = 2gh$, so $S - mg = \frac{2mgh}{L}$

In other words, $\frac{2mgh}{L}$ represents the extra support force the person experiences because of circular motion. Prove for yourself that for $h = L$ (i.e., a 90° swing), the extra support force is equal to twice the person's weight.

The Big Wheel

This ride takes its passengers around in a vertical circle on the inside of the circumference of a very large wheel. The wheel turns fast enough to stop the passengers falling out as they pass through the highest position.

At maximum height, the reaction R from the wheel on each person acts downwards. So, the resultant force at this position is equal to $mg + R$. This reaction force and the weight provide the centripetal force. Therefore, at the highest position, if the wheel speed is v,

$$mg + R = \frac{mv^2}{r},$$ where r is the radius of the wheel,

therefore, $R = \frac{mv^2}{r} - mg$

At a particular speed v_0 such that $v_0^2 = gr$, then $R = 0$, so there would be no force on the person due to the wheel.

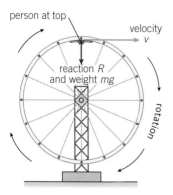

▲ **Figure 3** *The Big Wheel*

 Safe rides

Amusement rides are checked regularly to ensure they are safe. Accidents on such rides in the United Kingdom have to be investigated by the Health and Safety Executive (HSE). A passenger on a ride may experience 'g-forces' in different directions that may be back and forth, side to side, or normal to the track. HSE have found that the g-forces in past accidents were within acceptable limits for amusement rides, and that other factors such as passenger behaviour and passenger height in relation to compartment design were more likely to have caused the accidents. In particular, passengers should be able to sit back in their seats with their feet on the foot rests or floor and be able to comfortably reach the hand holds.

▲ **Figure 4** *g-forces at work*

Q: Calculate the minimum speed of the train in Figure 4 at the top of the loop if it is to stay in contact with the rails at the top. Assume that the loop radius is 10 m.

A: 10 m s⁻¹

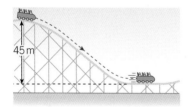

45 m

▲ **Figure 5**

Summary questions

$g = 9.8 \, \text{m s}^{-2}$.

1 A train on a fairground ride is initially stationary before it descends through a height of 45 m into a dip that has a radius of curvature of 78 m, as shown in Figure 5.

   a Calculate the speed of the train at the bottom of the dip, assuming air resistance and friction are negligible.

   b Calculate
      i   the centripetal acceleration of the train at the bottom of the dip,
      ii the extra support force on a person of weight 600 N in the train.

2 A very long swing at a fairground is 32 m in length. A person of mass 69 kg on the swing descends from a position when the swing is horizontal. Calculate:

   a the speed of the person at the lowest point,

   b the centripetal acceleration at the lowest point,

   c the support force on the person at the lowest point.

3 The Big Wheel at a fairground has a radius of 12.0 m and rotates once every 6.0 s. Calculate:

   a the speed of rotation of the perimeter of the wheel,

   b the centripetal acceleration of a person on the perimeter,

   c the support force on a person of mass 72 kg at the highest point.

4 The wheel of the London Eye has a diameter of 130 m and takes 30 minutes to complete one revolution. Calculate the change due to rotation of the wheel of the support force on a person of weight 500 N in a capsule at the top of the wheel.

1 **Figure 1** shows a cross-section of an automatic brake fitted to a rotating shaft. The brake pads are held on the shaft by springs.

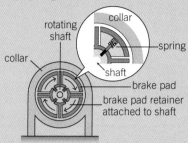

▲ Figure 1

   (a) Explain why the brake pads press against the inner surface of the stationary collar if the shaft rotates too fast. *(3 marks)*

   (b) Each brake pad and its retainer has a mass of 0.30 kg and its centre of mass is 60 mm from the centre of the shaft. The tension in the spring attached to each pad is 250 N. Calculate the maximum frequency of rotation of the shaft for no braking. *(4 marks)*

   (c) Automatic brakes of the type described above are used on ships to prevent lifeboats falling freely when they are lowered on cables onto the water. Discuss how the performance of the brake would be affected if the springs gradually became weaker. *(2 marks)*

2 (a) A particle that moves uniformly in a circular path is accelerating yet moving at a constant speed.
      Explain this statement by reference to the physical principles involved. *(3 marks)*

   (b) A 0.10 kg mass is to be placed on a horizontal turntable that is then rotated at a fixed rate of 78 revolutions per minute. The mass may be placed on the table at any distance, r, from the axis of rotation, as shown in **Figure 2**.

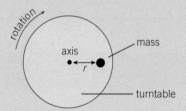

▲ Figure 2

If the maximum frictional force between the mass and the turntable is 0.50 N, calculate the maximum value of the distance r at which the mass would stay on the turntable at this rate of rotation. *(4 marks)*

AQA, 2007

3 **Figure 3** shows a dust particle at position **D** on a rotating vinyl disc. A combination of electrostatic and frictional forces act on the dust particle to keep it in the same position.

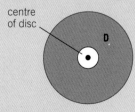

▲ Figure 3

The dust particle is at a distance of 0.125 m from the centre of the disc. The disc rotates at 45 revolutions per minute.

(a) Calculate the linear speed of the dust particle at D. *(3 marks)*

(b) (i) Copy the diagram and mark an arrow to show the direction of the resultant horizontal force on the dust particle.

    (ii) Calculate the centripetal acceleration at position D. *(3 marks)*

(c) On looking closely at the rotating disc, it can be seen that there is more dust concentrated on the inner part of the disc than the outer part. Suggest why this should be so. *(3 marks)*

AQA, 2002

4   A strimmer is a tool for cutting long grass. A strimmer head such as that shown in **Figure 4a** is driven by a motor. This makes the plastic line rotate, causing it to cut the grass. To simplify analysis, the strimmer line is modelled as the arrangement shown in **Figure 4b**. In this model the effective mass of the line is considered to rotate at the end of the line.

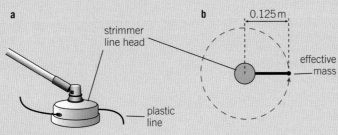

▲ **Figure 4**

In one strimmer, the effective mass of 0.80 g rotates in a circle of radius 0.125 m at 9000 revolutions per minute.

(a) Show that the angular speed of the line is approximately $9.4 \times 10^2 \, \text{rad s}^{-1}$. *(2 marks)*

(b) (i) Explain how the centripetal force is applied to the effective mass.

    (ii) Calculate the centripetal force acting on the effective mass. *(4 marks)*

(c) The line strikes a pebble of mass 1.2 g, making contact for a time of 0.68 ms. This causes the pebble to fly off at a speed of $15 \, \text{m s}^{-1}$.
Calculate the average force applied to the pebble. *(3 marks)*

AQA, 2007

5   **Figure 5** shows a toy engine moving with a constant speed on a circular track of constant radius.

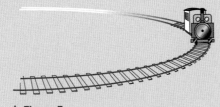

▲ **Figure 5**

(a) (i) Explain why the engine is accelerating even though its speed remains constant.

    (ii) Mark on a copy of **Figure 5** the direction of the centripetal force acting on the engine. *(3 marks)*

(b) The total mass of the toy engine is 0.14 kg and it travels with a speed of $0.17 \, \text{m s}^{-1}$. The radius of the track is 0.80 m. Calculate the centripetal force acting on the engine. *(2 marks)*

Figure 6 shows a close up of a pair of wheels as the engine moves towards you in the forward direction shown in **Figure 5**.

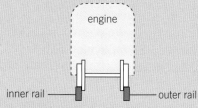

engine

inner rail ———————— outer rail

▲ Figure 6

(c) (i) State and explain on which wheel the centripetal force acts at the instant shown. You may use **Figure 6** to help your explanation.

(ii) For the toy engine going round a curved track, state and explain the two factors which determine the stress on each wheel. (*5 marks*)

AQA, 2004

6 A mass of 30 g is attached to a thread and spun in a circle of radius 45 cm. The circle is in a horizontal plane. The tension in the thread is 0.35 N.

(a) Calculate:
(i) the speed of the mass,
(ii) the period of rotation of the mass. (*4 marks*)

(b) The mass M is now spun in a circle in a vertical plane as shown in **Figure 7**.

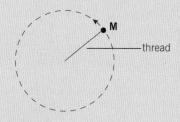

M

thread

▲ Figure 7

(i) On a copy of **Figure 7**, label the forces acting on the mass, and use arrows to show their direction.

(ii) Without performing calculations, state and explain the difference between the tension in the thread when M is at the top of the circle and when it is at the bottom. (*6 marks*)

AQA, 2007

7 **Figure 8** shows the initial path taken by an electron when it is produced as a result of a collision in a cloud chamber.

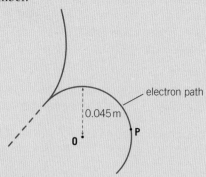

electron path

0.045 m

O

P

▲ Figure 8

The path is the arc of a circle of radius 0.045 m with centre O.

The speed of the electron is $4.2 \times 10^7 \, \text{m s}^{-1}$. The mass of an electron is $9.1 \times 10^{-31} \, \text{kg}$.

(a) Calculate the momentum of the electron. *(2 marks)*
(b) Calculate the magnitude of the force acting on the electron that makes it follow the curved path. *(2 marks)*
(c) Show on a copy of **Figure 8** the direction of this force when an electron is at point P. *(1 mark)*

AQA, 2002

8 **Figure 9** shows a simple accelerometer designed to measure the centripetal acceleration of a car going around a bend following a circular path.

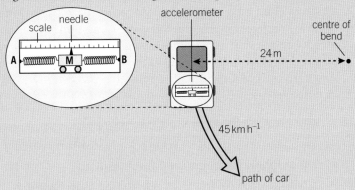

▲ **Figure 9**

The two ends A and B are fixed to the car. The mass M is free to move between the two springs.
The needle attached to the mass moves along a scale to indicate the acceleration.
In one instant a car travels around a bend of radius 24 m in the direction shown in **Figure 9**. The speed of the car is 45 km h^{-1}.

(a) State and explain the direction in which the pointer moves from its equilibrium position. *(3 marks)*
(b) (i) Calculate the acceleration that would be recorded by the accelerometer.
  (ii) The mass M between the springs in the accelerometer is 0.35 kg. A test shows that a force of 0.75 N moves the pointer 27 mm.
    Calculate the displacement of the needle from the equilibrium position when the car is travelling with the acceleration in part (i). *(4 marks)*

AQA, 2003

Learning objectives:

→ Explain what is meant by one complete oscillation.

→ Define amplitude, frequency, and period.

→ Describe the phase difference between two oscillators that are out of step.

Specification reference: 3.6.1.2

Synoptic link

The displacement of an object from a fixed point is its distance from that point in a particular direction. See Topic 7.1, Speed and velocity.

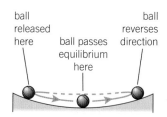

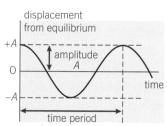

▲ **Figure 1** *Oscillating motion*

Study tip

Practise applying these definitions.

Measuring oscillations

There are many examples of oscillations in everyday life. A car that travels over a bump bounces up and down for a short time afterwards. Every microcomputer has an electronic oscillator to drive its internal clock. A child on a swing moves forwards then backwards repeatedly. In this simple example, one full cycle of motion is from maximum height at one side to maximum height on the other side and then back again. The lowest point is called the **equilibrium** position, because it is where the child eventually comes to a standstill. The child in motion is described as *oscillating* about equilibrium.

Other examples of oscillating motion include:

- an object on a spring moving up and down repeatedly
- a pendulum moving to and fro repeatedly
- a ball-bearing rolling from side to side as in Figure 1
- a small boat rocking from side to side.

An oscillating object moves repeatedly one way then in the opposite direction through its equilibrium position. The **displacement** of the object (i.e., distance and direction) from equilibrium continually changes during the motion. In one full cycle after being released from a non-equilibrium position, the displacement of the object:

- decreases as it returns to equilibrium, then
- reverses and increases as it moves away from equilibrium in the opposite direction, then
- decreases as it returns to equilibrium, then
- increases as it moves away from equilibrium towards its starting position.

The **amplitude** of the oscillations is the maximum displacement of the oscillating object from equilibrium. If the amplitude is constant and no frictional forces are present, the oscillations are described as **free vibrations**. See Topic 18.5, Energy and simple harmonic motion.

The **time period**, *T*, of the oscillating motion is the time for one complete cycle of oscillation. One full cycle after passing through any position, the object passes through that same position in the same direction.

The **frequency** of oscillations is the number of cycles per second made by an oscillating object.

The unit of frequency is the hertz (Hz), which is one cycle per second.

For oscillations of frequency *f*, the time period $T = \dfrac{1}{f}$

Note: The **angular frequency** ω of the oscillating motion is defined as $\dfrac{2\pi}{T}$

$(= 2\pi f)$. The unit of ω is the radian per second (rad s^{-1}).

Phase difference

Imagine two children on adjacent identical swings. The time period, T, of the oscillating motion is the same for both, because the swings are identical. If one child reaches maximum displacement on one side a particular time, Δt, later than the other child, then the children oscillate out of phase. Their **phase difference** stays the same as they oscillate, always corresponding to a fraction of a cycle equal to $\dfrac{\Delta t}{T}$. For example, if the time period is 2.4 s and one child reaches maximum displacement on one side 0.6 s later than the other child, then the later child will always be a quarter of a cycle $\left(= \dfrac{0.6\,\text{s}}{2.4\,\text{s}} \right)$ behind the other child. Their phase difference, in radians, is therefore $0.5\pi \left(= \dfrac{2\pi \Delta t}{T} \right)$.

In general, for two objects oscillating at the same frequency,

$$\textbf{their phase difference in radians} = \frac{2\pi \Delta t}{T}$$

where Δt is the time between successive instants when the two objects are at maximum displacement in the same direction.

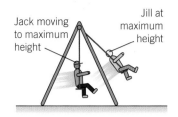

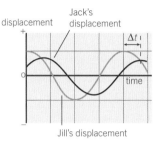

▲ **Figure 2** *Phase difference*

Summary questions

1 Describe how the velocity of a bungee jumper changes during a jump from the moment he jumps off the starting platform to the next time he reaches the highest point of his jump.

2 **a** Define free vibrations (free oscillations).

   b A metre rule is clamped to a table so that part of its length projects at right angles from the edge of the table, as shown in Figure 3. A 100 g mass is attached to the free end of the rule. When the free end of the rule is depressed downwards then released, the mass oscillates. Describe how you would find out if the oscillations of the mass are free oscillations.

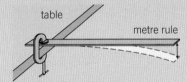

▲ **Figure 3**

3 An object suspended from the lower end of a vertical spring is displaced downwards from equilibrium. It takes 9.6 s to undergo 20 complete cycles of oscillation. Calculate:

   a its time period,

   b its frequency of oscillation.

4 Two identical pendulums X and Y each consist of a small metal sphere attached to a thread of a certain length. Each pendulum makes 20 complete cycles of oscillation in 16 s. State the phase difference, in radians, between the motion of X and that of Y if

   a X passes through equilibrium 0.2 s after Y passes through equilibrium in the same direction,

   b X reaches maximum displacement at the same time as Y reaches maximum displacement in the opposite direction.

Hint

1 2π radians = 360°, so the phase difference in degrees is $360 \times \dfrac{\Delta t}{T}$

2 The two objects oscillate in phase if $\Delta t = T$. The phase difference of 2π is therefore equivalent to zero.

3 Table of phase differences

Δt	phase difference in radians	phase difference in degrees
0	0	0
0.25T	$\dfrac{\pi}{2}$	90
0.50T	π	180
0.75T	$\dfrac{3\pi}{2}$	270
T	2π	360 or 0

Learning objectives:

→ State the two fundamental conditions about acceleration that apply to simple harmonic motion.

→ Describe how displacement, velocity, and acceleration vary with time.

→ Describe the phase difference between displacement and (i) velocity, (ii) acceleration.

Specification reference: 3.6.1.2

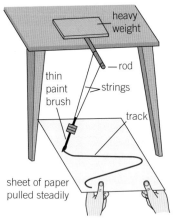

▲ **Figure 1** *Investigating oscillations*

Synoptic link

The gradient of a displacement–time graph gives velocity. The gradient of a velocity–time graph gives acceleration. See Topic 7.5, Motion graphs.

An oscillating object speeds up as it returns to equilibrium and it slows down as it moves away from equilibrium. Figure 1 shows one way to record the displacement of an oscillating pendulum.

The variation of displacement with time is shown in Figure 2(i). As long as friction is negligible, the amplitude of the oscillations is constant.

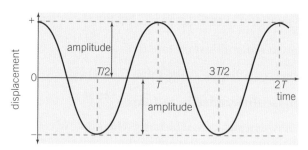

▲ **Figure 2(i)** *Displacement against time*

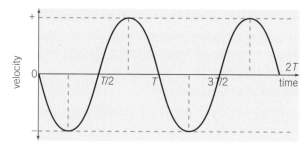

▲ **Figure 2(ii)** *Velocity against time*

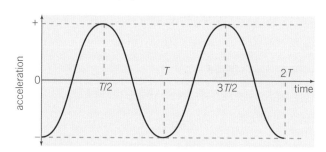

▲ **Figure 2(iii)** *Acceleration against time*

The variation of velocity with time is given by the gradient of the displacement–time graph, as shown by Figure 2(ii).

- The magnitude of the velocity is greatest when the gradient of the displacement–time graph is greatest (i.e., at zero displacement when the object passes through equilibrium).

- The velocity is zero when the gradient of the displacement–time graph is zero (i.e., at maximum displacement in either direction).

The variation of acceleration with time is given by the gradient of the velocity–time graph, as shown by Figure 2(iii).

- The acceleration is greatest when the gradient of the velocity–time graph is greatest. This is when the velocity is zero and occurs at maximum displacement in the opposite direction.

- The acceleration is zero when the gradient of the velocity–time graph is zero. This is when the displacement is zero and the velocity is a maximum.

By comparing Figures 2(i) and 2(iii) directly, it can be seen that

the acceleration is always in the opposite direction to the displacement

In other words, if you call one direction the positive direction and the other direction the negative direction, the acceleration direction is always the opposite sign to the displacement direction.

Simple harmonic motion is defined as oscillating motion in which the acceleration is

1 proportional to the displacement, and

2 always in the opposite direction to the displacement.

$$a \propto -x$$

In other words, acceleration $a = -$constant $\times$ displacement x.

The minus sign in both equations tells you that the acceleration is in the opposite direction to the displacement. The constant of proportionality depends on the time period T of the oscillations. The shorter the time period, the faster the oscillations, which means the larger the acceleration at any given displacement. So the constant is greater, the shorter the time period. As you will see in Topic 18.3, the constant in this equation is ω^2, where ω is the angular frequency $= \frac{2\pi}{T}$.

Therefore, the defining equation for simple harmonic motion is

acceleration $a = -\omega^2 x$

where $x =$ displacement and $\omega =$ angular frequency.

> **Hint**
>
> 1 The time period is independent of the amplitude of the oscillations.
>
> 2 Maximum displacement $x_{max} = \pm A$, where A is the amplitude of the oscillations. Therefore,
>
>   - when $x_{max} = +A$, the acceleration $a = -\omega^2 A$, and
>
>   - when $x_{max} = -A$, the acceleration $a = +\omega^2 A$.
>
> 3 The acceleration equation may also be written as $a = -(2\pi f)^2 x$, where frequency $f = \frac{1}{T}$.

Summary questions

1 A small object attached to the end of a vertical spring oscillates with an amplitude of 25 mm and a time period of 2.0 s. The object passes through equilibrium moving upwards at time $t = 0$. What is the displacement and direction of motion of the object:

   a $\frac{1}{4}$ cycle later,   b $\frac{1}{2}$ cycle later,

   c $\frac{3}{4}$ cycle later,   d 1 cycle later?

2 For the oscillations in Q1, calculate:

   a the frequency,

   b the acceleration of the object when its displacement is

      i +25 mm,   ii 0,   iii −25 mm.

3 A simple pendulum consists of a small weight on the end of a thread. The weight is displaced from equilibrium and released. It oscillates with an amplitude of 32 mm, taking 20 s to execute 10 oscillations. Calculate:

   a its frequency,   b its initial acceleration.

4 For the oscillations in Q3, the object is released at time $t = 0$. State the displacement and calculate the acceleration when:

   a $t = 1.0$ s,   b $t = 1.5$ s.

Learning objectives:

→ State the equation that relates displacement to time for a body moving with simple harmonic motion.

→ State the point at which the oscillations must start for this equation to apply.

→ Calculate the velocity for a given displacement.

Specification reference: 3.6.1.2

Circles and waves

Consider a small object P in uniform circular motion, as shown in Figure 1. Measured from the centre of the circle at O, the coordinates of P are therefore $x = r\cos\theta$ and $y = r\sin\theta$, where θ is the angle between the x-axis and the radial line OP. The graph shows how the x-coordinate changes as angle θ changes. The curve is a cosine wave. It has the same shape as the simple harmonic motion curves in Figure 2 in Topic 18.2.

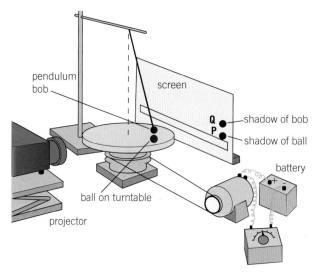

▲ **Figure 2** *Comparing simple harmonic motion with circular motion*

To see directly the link between simple harmonic motion and sine curves, consider the motion of the ball and the pendulum bob in Figure 2. A projector is used to cast a shadow of the ball (P) in uniform circular motion onto a screen alongside the shadow of the oscillating bob (Q). The two shadows keep up with each other exactly when their time periods are matched. In other words, P and Q at any instant have the same horizontal motion. So the acceleration of Q is the same as the acceleration of P's shadow on the screen.

- Because the ball (P) is in uniform circular motion, its acceleration $a = -\dfrac{v^2}{r}$, where v is its speed and r is the radius of the circle. Note that the minus sign indicates that its direction is towards the centre.

Because speed $v = \omega r$ (see Topic 17.1), then $a = -\omega^2 r$.

- The component of acceleration of the ball parallel to the screen is $a_x = a\cos\theta$, so the acceleration of the ball's shadow is $a_x = -\omega^2 r\cos\theta = -\omega^2 x$, where $x = r\cos\theta$ is the displacement of the shadow from the midpoint of the oscillations.

- Because the bob's motion is the same as the motion of the ball's shadow,

the acceleration of the bob (Q) $a_x = -\omega^2 x$

This is the defining equation for simple harmonic motion, and it shows why the constant of proportionality is ω^2.

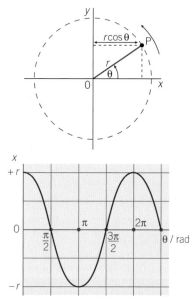

▲ **Figure 1** *Circles and waves*

Sine wave solutions

For any object oscillating at frequency f in simple harmonic motion, its acceleration a at displacement x is given by

$$a = -\omega^2 x$$

where $\omega = 2\pi f$. The variation of displacement with time depends on the initial displacement and the initial velocity (i.e., the displacement and velocity at time $t = 0$).

As shown by the bottom graph in Figure 3, if $x = +A$ when $t = 0$ and the object has zero velocity at that instant, then its displacement at a later time t is given by

$$x = A \cos (\omega t)$$

Notes:

1   In the above equation, x is the displacement of the bob in Figure 2 from its equilibrium position. Its value changes from $-r$ to $+r$ and back again as the bob oscillates. Therefore, the amplitude of oscillation of the bob is $A = r$.

2   The displacement of the bob from equilibrium is $x = A \cos \theta$, where θ is the angle the ball moves through from its position when $x = A$.

At time t after the ball passes through this position,

$$\theta \text{ (in radians)} = \frac{2\pi t}{T} = \omega t$$

Therefore, the displacement of the bob at time t is given by

$$x = A \cos (\omega t)$$

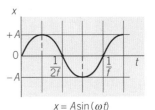

$x = A\sin(\omega t)$

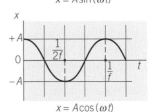

$x = A\cos(\omega t)$

▲ **Figure 3** *Graphical solutions*

More on sine waves

1   The shape of the curves in Figure 3, described as sinusoidal curves, is the same as the shape of the simple harmonic motion curves in Figure 2 in Topic 18.2.

2   The general solution of $a = -\omega^2 x$ is $x = A \sin (\omega t + \phi)$, where ϕ is the phase difference between the instants when $t = 0$ and when $x = 0$. If timing were to start at the centre with oscillations moving positive, the x against t equation would become $x = A \sin \omega t$ as $\phi = 0$. You *don't* need to know this general solution.

3   The time period T does *not* depend on the amplitude of the oscillating motion. For example, the time period of an object oscillating on a spring is the same, regardless of whether the amplitude is large or small.

Summary questions

1   An object oscillates in simple harmonic motion with a time period of 3.0 s and an amplitude of 58 mm. Calculate:

   a   its frequency,   **b**   its maximum acceleration.

2   The displacement of an object oscillating in simple harmonic motion varies with time according to the equation x (mm) $= 12 \cos 10t$, where t is the time in seconds after the object's displacement was at its maximum positive value.

   a   Determine:

   i   the amplitude,   **ii**   the time period.

   b   Calculate the displacement of the object at $t = 0.1$ s.

3   An object on a spring oscillates with a time period of 0.48 s and a maximum acceleration of 9.8 m s^{-2}. Calculate:

   a   its frequency,   **b**   its amplitude.

4   An object oscillates in simple harmonic motion with an amplitude of 12 mm and a time period of 0.27 s. Calculate:

   a   its frequency,

   b   its displacement and its direction of motion

   i   0.10 s,   **ii**   0.20 s after its displacement was +12 mm.

Study tip

Remember that ωt must be in radians. So make sure your calculator is in radian mode when you use the equation $x = A \cos (\omega t)$.

18.4 Applications of simple harmonic motion

Learning objectives:

→ State the conditions that must be satisfied for a mass–spring system or simple pendulum to oscillate with simple harmonic motion.

→ Describe how the period of a mass–spring system depends on the mass.

→ Describe how the period of a simple pendulum depends on its length.

Specification reference: 3.6.1.3

For any oscillating object, the resultant force acting on the object acts towards the equilibrium position. The resultant force is described as a *restoring force* because it always acts towards equilibrium. As long as the restoring force is proportional to the displacement from equilibrium, the acceleration is proportional to the displacement and always acts towards equilibrium. Therefore, the object oscillates with simple harmonic motion.

The oscillations of a mass–spring system

Use two stretched springs and a trolley, as shown in Figure 1. When the trolley is displaced then released, it oscillates backwards and forwards.

- The first half-cycle of the trolley's motion can be recorded using a length of ticker tape attached at one end to the trolley. When the trolley is released, the tape is pulled through a ticker timer that prints dots on the tape at a rate of 50 dots per second.

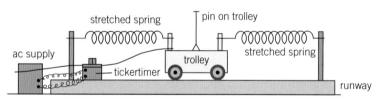

▲ **Figure 1** *Investigating oscillations*

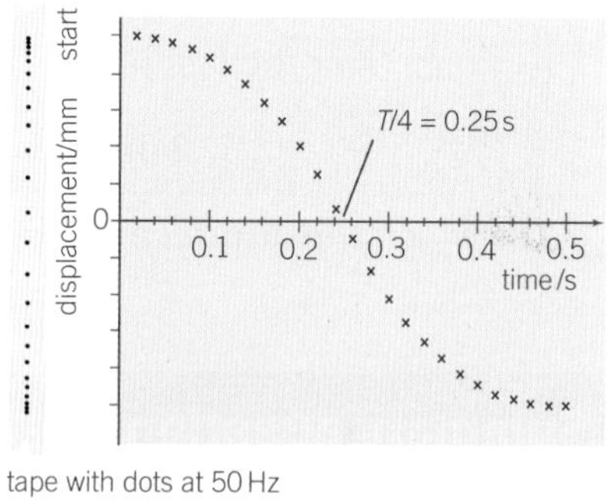

- A graph of displacement against time for the first half-cycle can be drawn using the tape, as shown in Figure 2. The graph can be used to measure the time period, which can be checked (see the Hint box on the next page) if the trolley mass *m* and the combined spring constant *k* are known.

tape with dots at 50 Hz

▲ **Figure 2** *Displacement–time curve from a ticker tape*

- A motion sensor linked to a computer can also be used to record the oscillating motion of the trolley. See Topic 9.3.

Study tip

When timing oscillations, use a fiduciary marker at the centre of oscillations and start counting with 0 as the oscillating object passes the mark in a particular direction.

What determines the frequency of oscillation of a loaded spring?

In the above investigation, the frequency of oscillation of the trolley can be changed by loading the trolley with extra mass or by replacing the springs with springs of different stiffness. The frequency is reduced by:

1 **Adding extra mass**. This is because the extra mass increases the inertia of the system. At a given displacement, the trolley would therefore be slower than if the extra mass had not been added. Each cycle of oscillation would therefore take longer.

2 **Using weaker springs**. The restoring force on the trolley at any given displacement would be less, so the trolley's acceleration and speed at any given displacement would be less. Each cycle of oscillation would therefore take longer.

To see exactly how the mass and the spring constant affect the frequency, consider a small object of mass m attached to a spring.

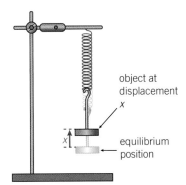

object at displacement x

equilibrium position

▲ **Figure 3** *The oscillations of a loaded spring*

- Assuming the spring obeys Hooke's law, the tension T_s in the spring is proportional to its extension ΔL from its unstretched length. This relationship can be expressed using the equation $T_s = k\Delta L$, where k is the spring constant.

- When the object is oscillating and is at displacement x from its equilibrium position, the change of tension in the spring provides the restoring force on the object. Using the equation $T_s = k\Delta L$, the change of tension ΔT_s from equilibrium is therefore given by $\Delta T_s = -kx$, where the minus sign represents the fact that the change of tension always tries to restore the object to its equilibrium position.

- So the restoring force on the object is equal to $-kx$

- Therefore, the acceleration $a = \dfrac{\text{restoring force}}{\text{mass}} = \dfrac{-kx}{m}$

This equation may be written in the form $a = -\omega^2 x$, where $\omega^2 = \dfrac{k}{m}$.

The object therefore oscillates in simple harmonic motion because its acceleration $a = -\omega^2 x$.

Investigating the oscillations of a loaded spring

Measure the oscillations of a loaded spring for different masses and verify the equation $T = 2\pi\sqrt{\dfrac{m}{k}}$.

Synoptic link

Because $T^2 = \dfrac{4\pi^2 m}{k}$, plotting a graph of T^2 on the y-axis against m on the x-axis should give a straight line through the origin with a gradient of $\dfrac{4\pi^2}{k}$ in accordance with the general equation for a straight line graph $y = mx + c$.

See Topic 16.4, Straight line graphs.

Hint

1. Because $\omega = 2\pi f$, then $\omega^2 = \dfrac{k}{m}$ can be written as $(2\pi f)^2 = \dfrac{k}{m}$. Rearranging this equation gives $f = \dfrac{1}{2\pi}\sqrt{\dfrac{k}{m}}$. You can use this equation to calculate f if k and m are known. The equation also shows that the frequency is increased if k is increased or if m is reduced.

2. The time period of the oscillations $T = \dfrac{1}{f} = 2\pi\sqrt{\dfrac{m}{k}}$.

   The time period does *not* depend on g. A mass–spring system on the Moon would have the same time period as it would on Earth.

3. The tension in the spring varies from $mg + kA$ to $mg - kA$, where A = amplitude.

   - Maximum tension is when the spring is stretched as much as possible (i.e., $x = -A$ = maximum displacement downwards)

   - Minimum tension is when the spring is stretched as little as possible (i.e., $x = +A$ = maximum displacement upwards)

Worked example

$g = 9.8\,\text{m s}^{-2}$

A spring of natural length 300 mm hangs vertically with its upper end attached to a fixed point. When a small object of mass 0.20 kg is suspended from the lower end of the spring in equilibrium, the spring is stretched to a length of 379 mm. Calculate:

a **i** the extension of the spring at equilibrium,
   ii the spring constant,

b the time period of oscillations that the mass on the spring would have if the mass was to be displaced downwards slightly then released.

Solution

a **i** Extension of spring at equilibrium is $\Delta L_0 = 79\,\text{mm} = 0.079\,\text{m}$

   ii Spring constant $k = \dfrac{mg}{\Delta L_0} = \dfrac{0.20 \times 9.8}{0.079} = 25\,\text{N m}^{-1}$

b $T = 2\pi\sqrt{\dfrac{0.20}{25}} = 0.56\,\text{s}$

Hint

1 Because $\omega = 2\pi f$, then $\omega^2 = \dfrac{g}{L}$ can be written as $(2\pi f)^2 = \dfrac{g}{L}$. Rearranging this equation gives $f = \dfrac{1}{2\pi}\sqrt{\dfrac{g}{L}}$.

So the time period

$$T = \dfrac{1}{f} = 2\pi\sqrt{\dfrac{L}{g}}$$

as long as the angle of the thread to the vertical does not exceed about 10°.

2 The time period T can be increased by increasing the length L of the pendulum. The length of the pendulum is the distance from the point of support to the centre of the bob.

3 As the bob passes through equilibrium, the tension T_s acts directly upwards. Therefore, the resultant force on the bob at this instant

$$T_s - mg = \dfrac{mv^2}{L}$$

where v is the speed as it passes through equilibrium.

The theory of the simple pendulum

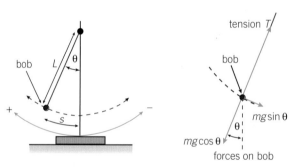

▲ **Figure 4** *The simple pendulum*

Consider a simple pendulum that consists of a bob of mass m attached to a thread of length L, as shown in Figure 4. If the bob is displaced from equilibrium then released, it oscillates about the lowest point. At displacement s from the lowest point, when the thread is at angle θ to the vertical, the weight mg has components

- $mg\cos\theta$ perpendicular to the path of the bob, and
- $mg\sin\theta$ along the path towards the equilibrium position.

The restoring force $F = -mg\sin\theta$, so the acceleration

$$a = \dfrac{F}{m} = \dfrac{-mg\sin\theta}{m} = -g\sin\theta$$

As long as θ does not exceed approximately 10° then $\sin\theta = \dfrac{s}{L}$, therefore the acceleration $a = -\dfrac{g}{L}s = -\omega^2 s$, where $\omega^2 = \dfrac{g}{L}$.

So the object oscillates with simple harmonic motion because its acceleration is proportional to the displacement from equilibrium and always acts towards equilibrium.

Summary questions

$g = 9.8\,\text{m s}^{-2}$

1 An object was suspended from the end of a vertical spring and set into oscillating motion along a vertical line. The amplitude of its oscillations was 20 mm, and it took 6.5 s to perform 20 oscillations. Calculate:

   a **i** its time period,   **ii** its frequency,

   b its acceleration when its displacement was

     i 0 mm,   **ii** 10 mm,   **iii** 20 mm.

2 In the arrangement described in Q1, the object was replaced by an object of different mass. When the second object was oscillating vertically, its acceleration a at displacement x was given by $a = -360x$.

   a Calculate:

     i the frequency,

     ii the time period of the oscillations.

   b By comparing the frequency of the oscillations of the second object with that of the first, discuss whether the mass of the second object is greater than or less than the mass of the first object.

3 The upper end of a vertical spring of natural length 250 mm is attached to a fixed point. When a small object of mass 0.15 kg is attached to the lower end of the spring, the spring stretches to an equilibrium length of 320 mm.

   a Calculate:

     i the extension of the spring at equilibrium,

     ii the spring constant.

   b The object is displaced vertically from its equilibrium position and released. Show that it oscillates at a frequency of 1.9 Hz and calculate its period of oscillation.

4 A mass of 0.50 kg is attached to the lower end of a vertical spring that has a spring constant of 25 N m^{-1}. The mass is displaced downwards by a distance of 50 mm then released.

   a Calculate:

     i the force on the object at a displacement of 50 mm,

     ii the acceleration of the object at the instant it was released.

   b **i** Show that the acceleration a at displacement x is given by $a = -50x$.

     ii Calculate the frequency of the oscillations and the displacement of the mass 0.050 s after it was released.

5 Calculate the time period of a simple pendulum:

   a of length

     i 1.0 m,   **ii** 0.25 m.

   b of length 1.0 m on the surface of the Moon, where $g = 1.6\,\text{m s}^{-2}$.

6 A simple pendulum and a mass suspended on a vertical spring have equal time periods on the Earth. Discuss whether or not they would have the same time periods on the surface of the Moon, where $g = 1.6\,\text{m s}^{-2}$.

Investigating the simple pendulum

Measure the oscillations of a simple pendulum for different lengths and verify the equation

$$T = 2\pi\sqrt{\frac{L}{g}}.$$

Synoptic link

Plotting a graph of T^2 on the y-axis against L on the x-axis should give a straight line through the origin with a gradient of $\dfrac{4\pi^2}{g}$.

See Topic 16.4, Straight line graphs.

Learning objectives:

→ Describe how, in simple harmonic motion, kinetic energy and potential energy vary with displacement.

→ Describe how these energies vary with time if damping is negligible.

→ Describe the effects of damping on the characteristics of oscillations.

Specification reference: 3.6.1.3

Free oscillations

A freely oscillating object oscillates with a constant amplitude because there is no friction acting on it. The only forces acting on it combine to provide the restoring force. If friction was present, the amplitude of oscillations would gradually decrease, and the oscillations would eventually cease.

Observe the oscillations of a simple pendulum over many cycles, and you should find that the decrease of amplitude from one cycle to the next is scarcely measurable. Nevertheless, over many cycles, the amplitude does decrease noticeably. So friction is present, even if its effect is insignificant over a single cycle.

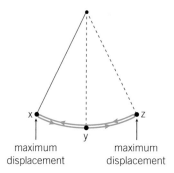

position	E_p	E_k
x	E_{TOTAL}	0
y	0	E_{TOTAL}
z	E_{TOTAL}	0

▲ **Figure 1** *The energy changes of a simple pendulum*

Consider the example of a small object of mass m oscillating on a spring. The energy of the system changes from kinetic energy to potential energy and back again every half-cycle after passing though equilibrium. As long as friction is absent, the total energy of the system is constant and is equal to its maximum potential energy.

- The potential energy E_P changes with displacement x from equilibrium, in accordance with the equation $E_P = \frac{1}{2}kx^2$, where k is the spring constant of the spring.

- The total energy, E_T, of the system is therefore $\frac{1}{2}kA^2$, where A is the amplitude of the oscillations.

- Because the total energy is $E_T = E_K + E_P$, where E_K is the kinetic energy of the oscillating mass, then $E_K = E_T - E_P = \frac{1}{2}k(A^2 - x^2)$.

The simple harmonic motion speed equation

Using $E_K = \frac{1}{2}mv^2$ gives $\frac{1}{2}mv^2 = \frac{1}{2}k(A^2 - x^2)$, where v is the speed of the object at displacement x.

Because $\omega^2 = \frac{k}{m}$, the above equation can be written as $v^2 = \omega^2(A^2 - x^2)$.

So, $v = \pm\omega\sqrt{(A^2 - X^2)}$.

Note that making $x = 0$ in this equation gives the maximum speed $= \omega A$.

Synoptic link

The energy stored in a spring stretched by extension x from its equilibrium length is $\frac{1}{2}kx^2$, where k is the spring constant. See Topic 11.2, Springs.

Energy–displacement graphs

1 The potential energy curve is parabolic in shape, given by $E_P = \frac{1}{2}kx^2$

2 The kinetic energy curve is an inverted parabola, given by

$$E_K = E_T - E_P = \frac{1}{2}k(A^2 - x^2)$$

The sum of the kinetic energy and the potential energy is always equal to $\frac{1}{2}kA^2$, which is the potential energy at maximum displacement. This is the same as the kinetic energy at zero displacement. So the two curves add together to give a horizontal line for the total energy.

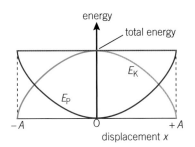

▲ **Figure 2** *Energy variation with displacement*

Damped oscillations

The oscillations of a simple pendulum gradually die away because air resistance gradually reduces the total energy of the system. In any oscillating system where friction or air resistance is present, the amplitude decreases. The forces causing the amplitude to decrease are described as *dissipative forces* because they dissipate the energy of the system to the surroundings as thermal energy. The motion is said to be damped if dissipative forces are present.

> **Study tip**
>
> Take the total energy to be equal to E_K at $x = 0$ (i.e., $E_P = 0$ when $x = 0$).

- **Light damping** occurs when the time period is independent of the amplitude so each cycle takes the same length of time as the oscillations die away. Figure 3 shows how the displacement of a lightly damped oscillating system decreases with time. The amplitude gradually decreases, reducing by the same fraction each cycle.

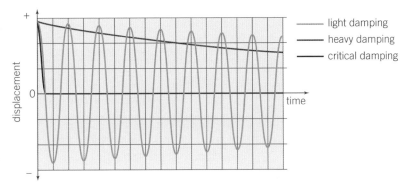

light damping
heavy damping
critical damping

▲ **Figure 3** *Damping*

- **Critical damping** is just enough to stop the system oscillating after it has been displaced from equilibrium and released. The oscillating system returns to equilibrium in the shortest possible time without overshooting if the damping is critical. Such damping is important in mass–spring systems such as a vehicle suspension system where you would have an uncomfortable ride if the damping was too light or too heavy. Figure 3 also shows how the displacement changes with time when the damping is critical.

- **Heavy damping** occurs when the damping is so strong that the displaced object returns to equilibrium much more slowly than if the system is critically damped. No oscillating motion occurs. For example, a mass on a spring in thick oil would return to equilibrium very slowly after being displaced and released.

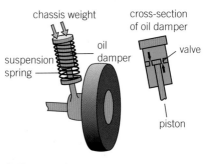

▲ **Figure 4** *Car suspension*

A car suspension system

The suspension system of a car includes a coiled spring near each wheel between the wheel axle and the car chassis. When the wheel is jolted, for example on a bumpy road, the spring smooths out the force of the jolts. An oil damper fitted with each spring prevents the chassis from bouncing up and down too much.

Without oil dampers, the occupants of the car would continue to be thrown up and down until the oscillations died away. The flow of oil through valves in the piston of each damper provides a frictional force that damps the oscillating motion of the chassis. The dampers are designed to ensure the chassis returns to its equilibrium position in the shortest possible time after each jolt with little or no oscillations. The suspension system is therefore at or close to critical damping.

Summary questions

$g = 9.81 \, \text{m s}^{-2}$

1 A simple pendulum consists of a small metal sphere of mass 0.30 kg attached to a thread. The sphere is displaced through a height of 10 mm with the thread taut then released. It takes 15.0 s to make 10 complete cycles of oscillation.

   a Calculate:

   i the time period of the pendulum,

   ii the length of the pendulum,

   iii the initial potential energy of the pendulum relative to its equilibrium position.

   b Sketch graphs on the same axes to show how the potential energy and the kinetic energy of the pendulum vary with its displacement from equilibrium.

2 A glider of mass 0.45 kg on a frictionless air track is attached to two stretched springs at either end, as shown in Figure 5. A force of 3.0 N is needed to displace the glider from equilibrium and hold it at a displacement of 50 mm. The glider is then released, and it oscillates freely on the air track.

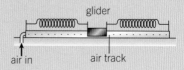

▲ **Figure 5**

Calculate:

   a i the spring constant k for the system,

   ii the time period of the oscillations,

   b i the initial potential energy of the system when the glider is held at a displacement of 50 mm,

   ii the maximum kinetic energy of the glider,

   iii the speed of the glider at a displacement of 25 mm.

3 a State whether the damping in each of the following examples is light, critical, or heavy:

   i a child on a swing displaced from equilibrium then released,

   ii oil in a U-shaped tube displaced from equilibrium then released.

   b Discuss how effective a car suspension damper would be, if the oil in the damper was replaced by oil that was much more viscous.

4 The amplitude of an oscillating mass on a spring decreases by 4% each cycle from an initial amplitude of 100 mm. Calculate the amplitude after:

   a 5 cycles of oscillation,

   b 20 cycles of oscillation.

18.6 Forced vibrations and resonance

Forced vibrations

Imagine pushing someone on a swing at regular intervals. If each push is timed suitably, the swing goes higher and higher. These pushes are a simple example of a **periodic force**, which is a force applied at regular intervals.

- When the system oscillates without a periodic force being applied to it, the system's frequency is called its *natural frequency*.
- When a periodic force is applied to an oscillating system, the response depends on the frequency of the periodic force. The system undergoes **forced vibrations**, or *forced oscillations*, when a periodic force is applied to it.

Figure 1 shows how a periodic force can be applied to an oscillating system consisting of a small object of fixed mass attached to two stretched springs.

The bottom end of the lower spring is attached to a mechanical oscillator, which is connected to a signal generator. The top end of the upper spring is fixed. The mechanical oscillator pulls repeatedly on the lower spring at a frequency that can be changed by adjusting the signal generator. The frequency of the oscillator is the *applied frequency*. The response of the system is measured from the amplitude of oscillations of the system. The variation of the response with the applied frequency is shown in Figure 2.

Consider the effect of increasing the applied frequency from zero:

As the applied frequency increases,

- the amplitude of oscillations of the system increases until it reaches a maximum amplitude at a particular frequency, and then the amplitude decreases again,
- the phase difference between the displacement and the periodic force increases from zero to $\frac{1}{2}\pi$ at the maximum amplitude, then from $\frac{1}{2}\pi$ to π as the frequency increases further.

Resonance

When the system is oscillating at the maximum amplitude, the phase difference between the displacement and the periodic force is $\frac{1}{2}\pi$.
The periodic force is then exactly in phase with the velocity of the oscillating system, and the system is in **resonance**. The frequency at the maximum amplitude is called the **resonant frequency**.

The lighter the damping,

- the larger the maximum amplitude becomes at resonance, and
- the closer the resonant frequency is to the natural frequency of the system.

So, when the damping is lighter, the resonance curve that you see in Figure 2 is sharper.

Learning objectives:

→ State the circumstances in which resonance occurs.

→ Distinguish between free vibrations and forced vibrations.

→ Explain why a resonant system reaches a maximum amplitude of vibration.

Specification reference: 3.6.1.4

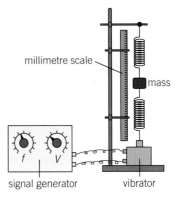

millimetre scale

mass

signal generator

f V

vibrator

▲ **Figure 1** *Forced vibrations*

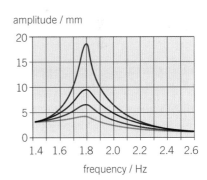

amplitude / mm

frequency / Hz

▲ **Figure 2** *Resonance curves*

As the applied frequency becomes increasingly larger than the resonant frequency of the mass–spring system,

- the amplitude of oscillations decreases more and more,
- the phase difference between the displacement and the periodic force increases from $\frac{1}{2}\pi$ until the displacement is π radians out of phase with the periodic force.

For an oscillating system with little or no damping, at resonance,

**the applied frequency of the periodic force
= the natural frequency of the system**

More examples of resonance

Barton's pendulums

Figure 3 shows five simple pendulums, P, Q, R, S, and T, of different lengths hanging from a supporting thread that is stretched between two fixed points. A single driver pendulum D of the same length as one of the other pendulums is also tied to the thread.

The driver pendulum D is displaced and released so that it oscillates in a plane perpendicular to the plane of the pendulums at rest. The effect of the oscillating motion of D is transmitted along the support thread, subjecting each of the other pendulums to forced oscillations. Pendulum R responds much more than any other pendulum. This is because it has the same length and therefore the same time period as D. So its natural frequency is the same as the natural frequency of D. Therefore, R oscillates in resonance with D because it is subjected to forced oscillations of the same frequency as its own natural frequency of oscillations. The response of each of the other pendulums depends on how close its length is to the length of D and whether it is shorter or longer than D.

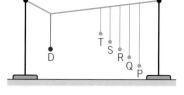

▲ **Figure 3** *Barton's pendulums*

Bridge oscillations

A bridge span can oscillate because of its springiness and its mass. If a bridge span is not fitted with dampers, it can be made to oscillate at resonance if the bridge span is subjected to a suitable periodic force.

1 A crosswind can cause a periodic force on the bridge span. If the wind speed is such that the periodic force is equal to the natural frequency of the bridge span, resonance can occur in the absence of damping. The dramatic collapse of the Tacoma Narrows Bridge in the United States in 1940 was due to such resonance.

2 A steady trail of people in step with each other walking across a footbridge can cause resonant oscillations of the bridge span if there is not enough damping. Soldiers marching in columns are taught to break out of step with each other when they cross a footbridge, to avoid causing resonance. Soon after it was opened, the Millennium Bridge in London had to be closed and fitted with a more suitable damping system because it swayed in resonance when people first walked across it.

Synoptic link

When **stationary waves** are formed on a stretched string, the string may be said to be in resonance. The string resonates and forms stationary wave patterns called harmonics. The stationary wave formed at the lowest frequency is called the first harmonic. Stationary waves are formed at higher frequencies that are multiples of the first-harmonic frequency. See Topic 4.5, Stationary and progressive waves, and Topic 4.6, More about stationary waves on strings.

Notes

1 At resonance, the periodic force acts on the system at the same point in each cycle, causing the amplitude to increase to a maximum value limited only by damping. At maximum amplitude, energy supplied by the periodic force is lost at the same rate because of the effects of damping.

2 The applied frequency at resonance (i.e., the resonant frequency) is equal to the natural frequency only when there is little or no damping. Resonance occurs at a slightly lower frequency than the natural frequency if the damping is not light. The lighter the damping, the closer the resonant frequency is to the natural frequency.

▲ **Figure 4** *The collapse of the Tacoma Narrows Bridge*

Summary questions

1 **a** A mass suspended on a vertical spring is made to oscillate by applying a periodic force of natural frequency f_0.

  i Define resonance.

  ii Explain why the frequency of the periodic force needs to be f_0 to cause resonance.

  b With reference to the mass–spring system shown in Figure 1, state and explain what the effect would be on the resonant frequency of

  i increasing the mass,

  ii replacing the springs with stiffer springs.

2 A 0.12 kg mass suspended on a vertical spring is made to resonate by applying a periodic force of frequency 2.4 Hz to it. Calculate:

  a the spring constant of the system,

  b the frequency at which the system would resonate if the mass were doubled.

3 The panel of a washing machine vibrates loudly when the drum rotates at a particular frequency. Explain why this happens only when the drum rotates at this frequency.

4 A vehicle of mass 850 kg has a suspension system that is lightly damped. When it is driven without extra load by a driver of mass 50 kg over speed bumps spaced 15 m apart at a speed of $3.0\,\mathrm{m\,s^{-1}}$, the vehicle resonates.

  a Explain why this effect happens.

  b Calculate the speed that resonance would occur at over the same speed bumps if the vehicle had also been carrying an extra load of 130 kg.

▲ **Figure 5** *The Millennium Bridge, London*

1   In an investigation, a small object was suspended from the lower end of a vertical steel spring which was fixed at its upper end, as shown in **Figure 1**.

A horizontal marker pin P was attached to the object. The vertical position, x, of the pin was measured against the millimetre scale of a metre rule clamped vertically in a fixed position. The measurement was made three times without, then with, the small object suspended from the spring.

(a) The readings obtained are shown in Table 1.

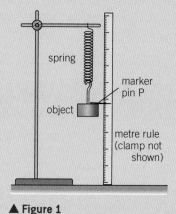

spring

marker pin P

object

metre rule (clamp not shown)

▲ **Figure 1**

▼ **Table 1**

	x / mm			mean x / mm	extension e / mm
without the object on the spring	2	2	2	2.0	0
with the object on the spring	71	72	73		

(ii) Copy and complete Table 1 by calculating the mean vertical position of P and the extension of the spring when the object was placed on it.

(iii) The readings were taken to a precision of 0.5 mm using a millimetre ruler. Estimate the percentage 'uncertainty' in the extension.   *(2 marks)*

(b) The time period, T, of small vertical oscillations of the object on the spring was also measured by timing 20 oscillations three times.

The timing readings for 20 oscillations were 10.98 s, 11.11 s, and 10.97 s.

(i) Calculate the time period T.

(ii) Use the readings to estimate the percentage 'uncertainty' in T.   *(2 marks)*

(c) (i) Give an expression for the extension e of the spring in terms of the mass m of the object and the spring constant k of the spring.

(ii) Hence show that $T = 2\pi\sqrt{\dfrac{e}{g}}$.   *(3 marks)*

(d) The experiment was repeated with objects of different mass suspended from the spring. The measurements obtained are given in Table 2.

▼ **Table 2**

object	e / mm	T / s
1	70	0.551
2	139	0.761
3	205	0.923
4	271	1.062
5	341	1.187
6	409	1.291

Plot a suitable graph using the above measurements to confirm the equation
and to determine g. (9 marks)

(e) Discuss the accuracy of your determination of g. (4 marks)

2 The tuning fork shown in **Figure 2** is labelled 512 Hz and has the tip of each
of its two prongs vibrating with simple harmonic motion of amplitude 0.85 mm.

▲ Figure 2

(a) (i) **Figure 2** shows the extreme positions of the prongs. How is the
distance marked d related to the amplitude of the prongs?

(ii) Sketch a graph to show how the displacement of one tip of the tuning
fork changes with time. Mark each axis with an appropriate scale. (4 marks)

(b) (i) Calculate the maximum speed of the tip of a prong.

(ii) Calculate the maximum acceleration of the tip of a prong. (4 marks)

AQA, 2007

3 A simple pendulum consists of a 25 g mass tied to the end of a light string
800 mm long. The mass is drawn to one side until it is 20 mm above its rest
position, as shown in **Figure 3**. When released it swings with simple
harmonic motion.

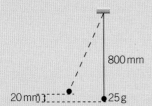

▲ Figure 3

(a) Calculate the period of the pendulum. (2 marks)

(b) Show that the initial amplitude of the oscillations is approximately 0.18 m,
and that the maximum speed of the mass during the first oscillation is
about 0.63 m s⁻¹. (4 marks)

(c) Calculate the magnitude of the tension in the string when the mass passes
through the lowest point of the first swing. (2 marks)

AQA, 2003

4 (a) **Figure 4a** shows a demonstration used in teaching simple harmonic
motion. A sphere rotates in a horizontal plane on a turntable. A lamp
produces a shadow of the sphere. This shadow moves with approximate
simple harmonic motion on the vertical screen.

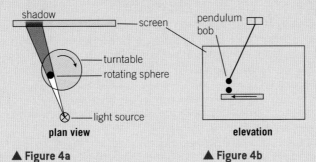

▲ Figure 4a   ▲ Figure 4b

(i) The turntable has a radius of 0.13 m, and the teacher wishes the time taken for one cycle of the motion to be 2.2 s. The mass of the sphere is 0.050 kg.
Calculate the magnitude of the horizontal force acting on the sphere.

(ii) State the direction in which the force acts. *(3 marks)*

(b) **Figure 4b** shows how the demonstration might be extended. A simple pendulum is mounted above the turntable so that the shadows of the sphere and the pendulum bob can be seen to move in a similar way and with the same period.

(i) Calculate the required length of the pendulum.

(ii) Calculate the maximum acceleration of the pendulum bob when its motion has an amplitude of 0.13 m. *(3 marks)*

(c) **Figure 5** is a graph of displacement against time for the pendulum.

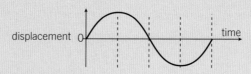

▲ Figure 5

Sketch, for the same interval, graphs of:
(i) acceleration against time for the bob, and
(ii) kinetic energy against time for the bob. *(4 marks)*

AQA, 2005

5 (a) Simple harmonic motion may be represented by the equation

$$a = -\omega^2 x$$

(i) Explain the significance of the minus sign in this equation.

(ii) Copy **Figure 6** and sketch the corresponding $v–t$ graph to show how the phase of velocity v relates to that of the acceleration a. *(2 marks)*

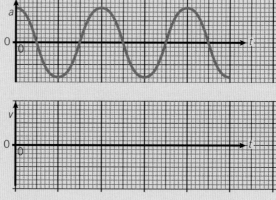

▲ Figure 6

(b) (i) A mass of 24 kg is attached to the end of a spring of spring constant 60 N m^{-1}. The mass is displaced 0.035 m vertically from its equilibrium position and released. Show that the maximum kinetic energy of the mass is about 40 mJ.

(ii) When the mass on the spring is quite heavily damped its amplitude halves by the end of each complete cycle. Sketch a graph to show how the kinetic energy, E_K (mJ), of the mass on the spring varies with time, t (s), over a single period.
Start at time, $t = 0$, with your maximum kinetic energy.
You should include suitable values on each of your scales. *(8 marks)*

AQA, 2004

6   To celebrate the Millennium in the year 2000, a footbridge was constructed across the
    River Thames in London. After the bridge was opened to the public it was discovered
    that the structure could easily be set into oscillation when large numbers of people were
    walking across it.
    (a) What name is given to this kind of physical phenomenon, when caused by
        a periodic driving force?   (*1 mark*)
    (b) Under what condition would this phenomenon become particularly
        hazardous? Explain your answer.   (*4 marks*)
    (c) Suggest **two** measures which engineers might adopt to reduce the size
        of the oscillations of a bridge.   (*2 marks*)
                                                                                    AQA, 2002

7   **Figure 7** shows how the displacement of the bob of a simple pendulum varies
    with time.

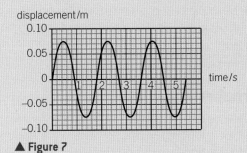

▲ **Figure 7**

    (a) (i)   Calculate the frequency of the oscillation.
        (ii) State the magnitude of the amplitude of the oscillation.
        (iii) State how the frequency and amplitude of a simple pendulum are
              affected by increased damping.   (*5 marks*)
    (b) Draw on a copy of **Figure 7** the displacement–time graph for a pendulum
        that has the same period and amplitude but oscillates 90° ($\frac{\pi}{2}$ radian) out
        of phase with the one shown.   (*2 marks*)
    (c) The pendulum bob has a mass of 8.0×10^{-3} kg. Calculate:
        (i)   the maximum acceleration of the bob during the oscillation,
        (ii) the total energy of the oscillations.   (*5 marks*)
                                                                                    AQA, 2006

8   (a) A spring, which hangs from a fixed support, extends by 40 mm when
        a mass of 0.25 kg is suspended from it.
        (i)   Calculate the spring constant of the spring.
        (ii) An additional mass of 0.44 kg is then placed on the spring and the
              system is set into vertical oscillation. Show that the oscillation
              frequency is 1.5 Hz.   (*4 marks*)

    (b) With both masses still in place, the spring is now suspended from a
        horizontal support rod that can be made to oscillate vertically, as shown
        in **Figure 8**, with amplitude 30 mm at several different frequencies.

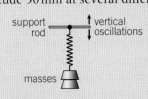

▲ **Figure 8**

        Describe fully, with reference to amplitude, frequency, and phase, the
        motion of the masses suspended from the spring in each of the following
        cases.
        (i)   The support rod oscillates at a frequency of 0.2 Hz.
        (ii) The support rod oscillates at a frequency of 1.5 Hz.
        (iii) The support rod oscillates at a frequency of 10 Hz.   (*6 marks*)
                                                                                    AQA, 2006

Learning objectives:

→ Define internal energy.

→ State the lowest temperature that is possible.

→ Demonstrate the first law of thermodynamics in action.

Specification reference: 3.6.2.1; 3.6.2.2

▲ **Figure 1** *Energy transfer by heating in winter*

When you are outdoors in winter, you need to wrap up well, otherwise energy is transferred by heating from your body to your surroundings. Your body loses energy, and your surroundings gain energy. In summer, if you are in a very hot room, you will heat up because of energy transferred to you from the room.

Energy transfer between two objects takes place if:

• one object exerts a force on the other object and makes it move. In other words, one object does work on the other object.

• one object is hotter than the other object, so energy transfer by heating takes place by means of conduction, convection, or radiation. In other words, energy is transferred by heating because of a temperature difference between two objects.

Internal energy

The brake pads of a moving vehicle become hot if the brakes are applied for a long enough time. The work done by the frictional force between the brake pads and the wheel heats the brake pads, which gain energy from the kinetic energy of the vehicle. The temperature of the brake pads increases as a result, and the internal energy of each brake pad increases.

As explained below, the internal energy of an object is the energy of its molecules due to their individual movements and positions. The internal energy of an object due to its temperature is sometimes called **thermal energy**. However, some of the internal energy of an object might be due to other causes. For example, an iron bar that is magnetised has more internal energy than if it is unmagnetised, because of the magnetic interaction between the iron bar's atoms.

The internal energy of an object is increased because of:

• energy transfer by heating the object, or

• work done on the object, for example work done by electricity.

If the internal energy of an object stays constant, then either:

• there is no energy transfer by heating and no work is done, or

• energy transfer by heating and work done balance each other out.

For example, the internal energy of a lamp filament increases when the lamp is switched on because work is done by the electricity supply pushing electrons through the filament. Because of this, the filament becomes hot. When it reaches its operating temperature, energy is transferred to the surroundings by heating, and the filament radiates light. Work done by the electricity supply pushing electrons through the filament is balanced by the energy transfer and light radiated from the filament.

The first law of thermodynamics

In general, when work is done on or by an object and/or energy is transferred by heating,

**the change of internal energy of the object =
the total energy transfer due to work done and heating**

This general statement is called *the first law of thermodynamics*. When it is applied to an object, the directions of the energy transfers (i.e., *to* or *from* the object) are very important and determine whether the overall internal energy of the object increases or decreases.

About molecules

A molecule is the smallest particle of a pure substance that is characteristic of the substance. For example, a water molecule consists of two hydrogen atoms joined to an oxygen atom.

An atom is the smallest particle of an element that is characteristic of the element. For example, a hydrogen atom consists of a proton and an electron.

- In a solid, the atoms and molecules are held to each other by forces due to the electrical charges of the protons and electrons in the atoms. The molecules in a solid vibrate randomly about fixed positions. The higher the temperature of the solid, the more the molecules vibrate. The energy supplied to raise the temperature of a solid increases the kinetic energy of the molecules. If the temperature is raised enough, the solid melts. This happens because its molecules vibrate so much that they break free from each other and the substance loses its shape. The energy supplied to melt a solid raises the potential energy of the molecules because they break free from each other.

- In a liquid, the molecules move about at random in contact with each other. The forces between the molecules are not strong enough to hold the molecules in fixed positions. The higher the temperature of a liquid, the faster its molecules move. The energy supplied to a liquid to raise its temperature increases the kinetic energy of the liquid molecules. Heating the liquid further causes it to vaporise. The molecules have sufficient kinetic energy to break free and move away from each other.

- In a gas or vapour, the molecules also move about randomly but much further apart on average than in a liquid. Heating a gas or a vapour makes the molecules speed up and so gain kinetic energy.

The internal energy of an object is the sum of the random distribution of the kinetic and potential energies of its molecules.

Increasing the internal energy of a substance increases the kinetic and/or potential energy associated with the random motion and positions of its molecules.

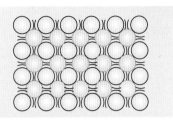

A solid is made up of particles arranged in a 3-dimensional structure. There are strong forces of attraction between the particles. Although the particles can vibrate, they cannot move out of their positions in the structure.

When a solid is heated, the particles gain energy and vibrate more and more vigorously. Eventually they may break away from the solid structure and become free to move around. When this happens, the solid has turned into liquid: it has melted.

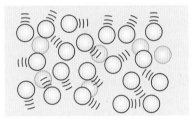

In a liquid the particles are free to move around. A liquid therefore flows easily and has no fixed shape. There are still forces of attraction between the particles. When a liquid is heated, some of the particles gain enough energy to break away from the other particles. The particles which escape from the body of the liquid become a gas.

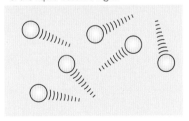

In a gas, the particles are far apart. There are almost no forces of attraction between them. The particles move about at high speed. Because the particles are so far apart, a gas occupies a very much larger volume than the same mass of liquid.

The molecules collide with the container. These collisions are responsible for the pressure which a gas exerts on its container.

▲ **Figure 2** *Particles in a solid, a liquid, and a gas*

Temperature and temperature scales

The temperature of an object is a measure of the degree of hotness of the object. The hotter an object is, the more internal energy it has. Place your hand in cold water, and your hand's internal energy decreases because of energy transferred out by heating the water. Place your hand in warm water, and its internal energy increases because of energy transferred into it by the water heating your hand. If the water is at the same temperature as your hand, no overall energy transfer by heating takes place. Your hand is then in **thermal equilibrium** with the water. For any two objects that are at the same temperature, no overall energy transfer by heating will take place.

A temperature scale is defined in terms of *fixed points*, which are standard degrees of hotness that can be accurately reproduced.

- The **Celsius scale** of temperature, in units of °C, is defined in terms of:

  1 ice point, 0 °C, which is the temperature of pure melting ice,

  2 steam point, 100 °C, which is the temperature of steam at standard atmospheric pressure.

- The **absolute scale** of temperature, in units of kelvins (K), is defined in terms of:

  1 **absolute zero**, 0 K, which is the lowest possible temperature,

  2 the triple point of water, 273.16 K, which is the temperature at which ice, water, and water vapour co-exist in thermodynamic equilibrium.

Because ice point on the absolute scale is 273.15 K and steam point is 100 K higher, then

temperature in °C = absolute temperature in kelvins − 273.15

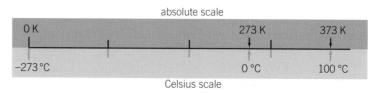

▲ **Figure 3** *Temperature scales*

About absolute zero

The absolute scale of temperature, also called the kelvin scale, is based on absolute zero, which is the lowest possible temperature. No object can have a temperature below absolute zero. *An object at absolute zero has minimum internal energy*, regardless of the substances the object consists of.

The pressure of a fixed mass of an ideal gas in a sealed container of fixed volume decreases as the gas temperature is reduced (see Topic 20.1). If the pressure measured at ice point and at steam point is plotted on a graph as shown in Figure 4, the line between the two points always cuts the temperature axis at −273 °C, regardless of which gas is used or how much gas is used.

Study tip

To change from °C to kelvins, simply add on 273 (.15).

The kelvin scale depends on a fundamental feature of nature, that is, the lowest temperature that is possible in nature. In contrast, the Celsius scale depends on the properties of a substance that was chosen for convenience instead of for any fundamental reason. That substance happens to be water.

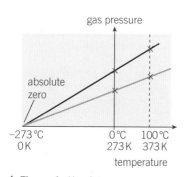

▲ **Figure 4** *Absolute zero*

The coldest places in the world

You don't need to travel to the South Pole to find the coldest places in the world. Simply go to the nearest university physics department that has a low-temperature research laboratory. Substances have very strange properties at very low temperatures. For example, metals cooled to a few degrees within absolute zero become superconductors, which means that they have zero electrical resistance. Superfluids have been discovered that can empty themselves out of containers! Temperatures within a few microkelvins of absolute zero have been reached in these laboratories.

Q: Explain why you can't use a liquid-in-glass thermometer to measure very low temperatures.

▲ **Figure 5** *A low-temperature research laboratory*

Summary questions

1 a Explain why an electric motor becomes warm when it is used.

  b A battery is connected to an electric motor which is used to raise a weight at a steady speed. When in operation, the electric motor is at a constant temperature which is above the temperature of its surroundings. Describe the energy transfers that take place.

2 a Define internal energy.

  b Describe a situation in which the internal energy of an object is constant even though work is done on the object.

3 a State one difference between the motion of the molecules in a solid and the molecules in a liquid.

  b Describe how the motion of the molecules in a solid changes when the solid is heated.

4 a State each of the following temperatures to the nearest degree on the absolute scale:

    i the temperature of pure melting ice,

    ii 20 °C,

    iii −196 °C.

  b Gas thermometers are used to calibrate all thermometers. The pressure of a constant-volume gas thermometer was 100 kPa at a temperature of 273 K.

    i Calculate the temperature, in kelvins, of the gas when its pressure was 120 kPa.

    ii Calculate the pressure of the gas at 100 °C.

A thermometer test

Use a travelling microscope to measure the interval between adjacent graduations on the scale of an accurate liquid-in-glass thermometer. You may be surprised to find that the interval distance is not the same near the middle of the scale as it is near the ends of the scale. This is because the expansion of the liquid is not directly proportional to the change of temperature.

All thermometers are calibrated in terms of the temperature measured by a gas thermometer. This is a thermometer consisting of a dry gas in a sealed container. The pressure of the gas is proportional to the absolute temperature of the gas. By measuring the gas pressure, p_{Tr}, at the triple point of water (273.16 K by definition) and at unknown temperature T/K, the unknown temperature in kelvins can be calculated using

$$\frac{T}{273.16} = \frac{p}{p_{Tr}}$$

where p is the gas pressure at the unknown temperature.

Learning objectives:

→ Explain what is meant by heating up and by cooling down.

→ State which materials heat up and cool down the fastest.

→ Define and measure specific heat capacity.

Specification reference: 3.6.2.1

▼ **Table 1** *Some specific heat capacities*

substance	specific heat capacity / $J\,kg^{-1}\,K^{-1}$
aluminium	900
concrete	850
copper	390
iron	490
lead	130
oil	2100
water	4200

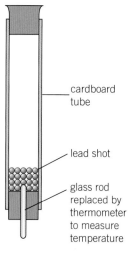

▲ **Figure 1** *The inversion tube experiment*

cardboard tube

lead shot

glass rod replaced by thermometer to measure temperature

Heating and cooling

Sunbathers on the hot sandy beaches of the Mediterranean Sea dive into the sea to cool off. Sand heats up much more readily than water does. Even when the sand is almost too hot to walk barefoot across, the sea water is refreshingly cool. The temperature rise of an object when it is heated depends on:

• the mass of the object

• the amount of energy supplied to it

• the substance or substances from which the object is made.

The specific heat capacity, c, of a substance is the energy needed to raise the temperature of unit mass of the substance by 1 K without change of state. The unit of c is $J\,kg^{-1}\,K^{-1}$.

Specific heat capacities of some common substances are shown in Table 1.

To raise the temperature of mass m of a substance from temperature T_1 to temperature T_2,

$$\text{the energy needed } \Delta Q = mc(T_2 - T_1)$$

For example, to calculate the energy that must be supplied to raise the temperature of 5.0 kg of water from 20 °C to 100 °C, use the above equation, and you will get $\Delta Q = 5.0 \times 4200 \times 80 = 1.7 \times 10^6\,J$.

The inversion tube experiment

In this experiment, the gravitational potential energy of an object falling in a tube is converted into internal energy when it hits the bottom of a tube. Figure 1 shows the idea. The object is a collection of tiny lead spheres.

The tube is inverted each time the spheres hit the bottom of the tube. The temperature of the lead shot is measured initially and after a particular number of inversions.

Let m represent the mass of the lead shot.

For a tube of length L, the loss of gravitational potential energy for each inversion = mgL.

Therefore, for n inversions, the loss of gravitational potential energy = $mgLn$.

The gain of internal energy of the lead shot = $mc\Delta T$, where c is the specific heat capacity of lead and ΔT is the temperature rise of the lead shot.

Assuming that all the gravitational potential energy lost is transferred to internal energy of the lead shot, $mc\Delta T = mgLn$.

Therefore, $c = \dfrac{gLn}{\Delta T}$

So, the experiment can be used to measure the specific heat capacity of lead with no other measurements than the length of the tube, the temperature rise of the lead, and the number of inversions.

Specific heat capacity measurements using electrical methods

Measurement of the specific heat capacity of a metal

A block of the metal of known mass m in an insulated container is used. A 12 V electrical heater is inserted into a hole drilled in the metal and used to heat the metal by supplying a measured amount of electrical energy. A thermometer inserted into a second hole drilled in the metal is used to measure the temperature rise ΔT (= its final temperature − its initial temperature). A small amount of water or oil in the thermometer hole will improve the thermal contact between the thermometer and the metal.

The electrical energy supplied
= heater current I × heater pd V × heating time t

So, assuming no heat loss to the surroundings, $mc\Delta T = IVt$

Therefore, $c = \dfrac{IVt}{m\Delta T}$

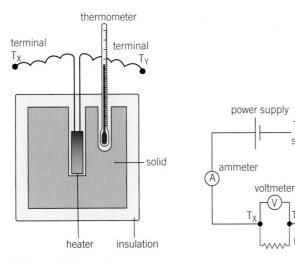

▲ **Figure 2** *Measuring c*

Measurement of the specific heat capacity of a liquid

A known mass of the liquid is used in an insulated calorimeter of known mass and known specific heat capacity. A 12 V electrical heater is placed in the liquid and used to heat it directly. A thermometer inserted into the liquid is used to measure the temperature rise, ΔT.

- The electrical energy supplied = current I × voltage V × heating time t
- The energy needed to heat the liquid = mass of liquid (m_l) × specific heat capacity of liquid (c_l) × temperature rise (ΔT)

Hint

1 Notice that the unit of mass × the unit of c × the unit of temperature change gives the joule. In other words, $kg \times J\,kg^{-1}\,K^{-1} \times K = J$.

2 The **heat capacity, C,** of an object is the heat supplied to raise the temperature of the object by 1 K. Therefore, for an object of mass m made of a single substance of specific heat capacity c, its heat capacity $C = mc$. For example, the heat capacity of 5.0 kg of water is 21 000 J K^{-1} = 5.0 kg × 4200 J kg^{-1} K^{-1}.

Study tip

A temperature change is the same in °C as it is in K. If you are given the initial and final temperatures in °C, just calculate the temperature difference in °C.

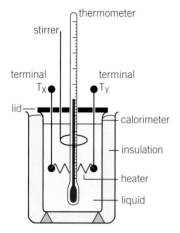

thermometer
stirrer
terminal T_X
terminal T_Y
lid
calorimeter
insulation
heater
liquid

▲ **Figure 3** *Measurement of the specific heat capacity of a liquid*

Study tip

Before you measure a liquid's temperature in a heating experiment, give the liquid a stir.

Study tip

If the volume flow rate is given, you need to know the density of the fluid to calculate the rate of flow of mass $(\frac{m}{t})$. See Topic 11.1, Density.

- The energy needed to heat the calorimeter = mass of calorimeter (m_{cal}) × specific heat capacity of calorimeter (c_{cal}) × temperature rise (ΔT)

  Assuming no heat loss to the surroundings,

  therefore, $IVt = m_l c_l \Delta T + m_{cal} c_{cal} \Delta T$

So, c can be calculated from this equation because all of the other quantities are known.

Continuous flow heating

In an electric shower, water passes steadily through copper coils heated by an electrical heater. The water is hotter at the outlet than at the inlet. This is an example of continuous flow heating. For mass m of liquid passing through the heater in time t at a steady flow rate, and assuming no heat loss to the surroundings:

the electrical energy supplied per second $IV = mc\dfrac{\Delta T}{t}$

where ΔT is the temperature rise of the water and c is its specific heat capacity.

Note that when the outflowing water has attained a steady temperature, the temperature of the copper coils does not change, so no $mc\Delta T$ term is needed for the copper coils in the above equation.

For a solar heating panel, the energy gained per second by heating the liquid that flows through the panel is equal to $mc\dfrac{\Delta T}{t}$.

Summary questions

Use the data in Table 1 for the following calculations.

1 Calculate:

  a the energy needed to heat an aluminium pan of mass 0.30 kg from 15 °C to 100 °C,

  b the energy needed to heat 1.50 kg of water from 15 °C to 100 °C.

2 a Calculate the time taken to heat the water and pan in Q1 from 15 °C to 100 °C using a 2.0 kW electric hot plate, assuming that no energy is transferred to the surroundings by heating.

  b Calculate the energy needed to raise the temperature of 80 kg of water in an insulated copper tank of mass 20 kg from 20 °C to 50 °C.

3 In an inversion tube experiment, 0.50 kg of lead shot at an initial temperature of 18 °C was inverted 50 times in a tube of length 1.30 m. The final temperature of the lead shot was 23 °C. Calculate:

  a the total gravitational potential energy released by the lead,

  b the specific heat capacity of lead. Assume that $g = 9.81\ \text{m s}^{-2}$.

4 An electric shower is capable of heating water from 10 °C to 40 °C when the flow rate is 0.025 kg s^{-1}. Calculate the minimum power of the heater.

19.3 Change of state

When a solid is heated and heated, its temperature increases until it melts. If it is a pure substance, it melts at a well-defined temperature, called its **melting point**. Once all the solid has melted, continued heating causes the temperature of the liquid to increase until the liquid boils. This occurs at a certain temperature, called the **boiling point**. The substance turns to a vapour as it boils away.

The three physical states of a substance, solid, liquid, and vapour, have different physical properties. For example:

- The density of a gas is much less than the density of the same substance in the liquid or the solid state. This is because the molecules of a liquid and of a solid are packed together in contact with each other. In contrast, the molecules of a gas are on average separated from each other by relatively large distances.

- Liquids and gases can flow, but solids can't. This is because the atoms in a solid are locked together by strong force bonds, which the atoms are unable to break free from. In a liquid or a gas, the molecules are not locked together. This is because they have too much kinetic energy, and the force bonds are not strong enough to keep the molecules fixed to each other.

Latent heat

When a solid or a liquid is heated so that its temperature increases, its molecules gain kinetic energy. In a solid, the atoms vibrate more about their mean positions. In a liquid, the molecules move about faster, still keeping in contact with each other, but free to move about.

1 **When a solid is heated at its melting point**, its atoms vibrate so much that they break free from each other. The solid therefore becomes a liquid due to energy being supplied at the melting point. The energy needed to melt a solid at its melting point is called **latent heat of fusion**.

   Latent heat is released when a liquid solidifies. This happens because the liquid molecules slow down as the liquid cools until the temperature decreases to the melting point. At the melting point, the molecules move slowly enough for the force bonds to lock the molecules together. Some of the latent heat released keeps the temperature at the melting point until all the liquid has solidified.

   - *Latent* means *hidden*. Latent heat supplied to melt a solid may be thought of as hidden because no temperature change takes place even though the solid is being heated.

   - Fusion is a word used for the melting of a solid because the solid *fuses* into a liquid as it melts.

2 **When a liquid is heated at its boiling point**, the molecules gain enough kinetic energy to overcome the bonds that hold them close together. The molecules therefore break away from each other to form bubbles of vapour in the liquid. The energy needed to vaporise a liquid is called **latent heat of vaporisation**.

Learning objectives:

→ Define latent heat.

→ Measure latent heat.

→ Explain why the temperature of a substance stays steady when it is changing state.

Specification reference: 3.6.2.1

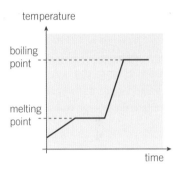

▲ **Figure 1** *Melting and boiling*

Latent heat is released when a vapour condenses. This happens because the vapour molecules slow down as the vapour is cooled. The molecules move slowly enough for the force bonds to pull the molecules together to form a liquid.

Some solids vaporise directly when heated. This process is called **sublimation**.

In general, much more energy is needed to vaporise a substance than to melt it. For example, 2.25 MJ is needed to vaporise 1 kg of water at its boiling point. In comparison, 0.336 MJ is needed to melt 1 kg of ice at its melting point. The energy needed to change the state of unit mass (i.e., 1 kg) of a substance at its melting point (or its boiling point) is called its specific latent heat of fusion (or vaporisation).

The **specific latent heat of fusion**, l_f, of a substance is the energy needed to change the state of unit mass of the substance from solid to liquid without change of temperature.

The **specific latent heat of vaporisation** of a substance is the energy needed to change the state of unit mass of the substance from liquid to vapour without change of temperature.

So, the energy Q needed to change the state of mass m of a substance from solid to liquid (or liquid to vapour) without change of temperature is given by

$$Q = ml$$

where l is the specific latent heat of fusion (or the specific latent heat of vaporisation). The unit of specific latent heat is $J\,kg^{-1}$.

Hint

Where a substance changes its state and changes its temperature, to calculate the energy transferred:

- use $Q = ml$ to calculate the energy transferred when its state changes

- use $Q = mc(T_2 - T_1)$ when its temperature changes.

Synoptic link

If the volume of a substance is given, you need to know its density to find its mass. See Topic 11.1, Density.

Worked example

Calculate the energy needed to melt 5.0 kg of ice at 0 °C and heat the melted ice to 50 °C.

specific latent heat of fusion of ice = $3.36 \times 10^5\,J\,kg^{-1}$

specific heat capacity of water = $4200\,J\,kg^{-1}\,K^{-1}$

Solution

To melt 5.0 kg of ice, energy needed $Q_1 = ml = 5.0 \times 3.36 \times 10^5 = 1.68 \times 10^6\,J$

To heat 5.0 kg of melted ice (i.e., water) from 0 °C to 50 °C, energy needed $Q_2 = mc\,(T_2 - T_1) = 5.0 \times 4200 \times (50 - 0) = 1.05 \times 10^6\,J$

Therefore, the total energy needed $= Q_1 + Q_1 = 2.73 \times 10^6\,J$

Temperature–time graphs

If a pure solid is heated to its melting point and beyond, its temperature–time graph will be as shown in Figure 2.

Assuming that no heat loss occurs during heating, and assuming that energy is transferred to the substance at a constant rate P (i.e., power supplied), then

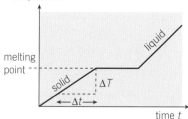

▲ **Figure 2** *Temperature against time for a solid being heated*

- before the solid melts, $P = mc_S\left(\dfrac{\Delta T}{\Delta t}\right)_S$, where $\left(\dfrac{\Delta T}{\Delta t}\right)_S$ is the rise of temperature per second and c_S is the specific heat capacity of the solid.

  So the rise of temperature per second of the solid is $\left(\dfrac{\Delta T}{\Delta t}\right)_S = \dfrac{P}{mc_S}$

- after the solid melts, $P = mc_L\left(\dfrac{\Delta T}{\Delta t}\right)_L$, where $\left(\dfrac{\Delta T}{\Delta t}\right)_L$ is the rise of temperature per second and c_L is the specific heat capacity of the liquid.

  So the rise of temperature per second of the liquid is $\left(\dfrac{\Delta T}{\Delta t}\right)_L = \dfrac{P}{mc_L}$

Therefore, if the solid has a larger specific heat capacity than the liquid, the rate of temperature rise of the solid is less than that of the liquid. In other words, the liquid heats up faster than the solid.

Hint

At the melting point, P = energy supplied per second = $\dfrac{ml}{t}$, where l is the specific latent heat of fusion of the substance and t is the time taken to melt mass m of the substance. Therefore, the time taken to melt the substance is $t = \dfrac{ml}{P}$.

Study tip

For a pure substance, change of state is at constant temperature.

Summary questions

1   **a**   Explain why energy is needed to melt a solid.

   b   Explain why the internal energy of the water in a beaker must be reduced to freeze the water.

2   Calculate the mass of water boiled away in a 3 kW electric kettle in 2 min.

   The specific latent heat of vaporisation of water is 2.25 MJ kg^{-1}.

3   A plastic beaker containing 0.080 kg of water at 15 °C was placed in a refrigerator and cooled to 0 °C in 1200 s.

   a   Calculate how much energy each second was removed from the water in this process. The specific heat capacity of water = 4200 J kg^{-1} K^{-1}.

   b   Calculate how long the refrigerator would take to freeze the water in **a**. The specific latent heat of fusion of ice = 3.36×10^5 J kg^{-1}.

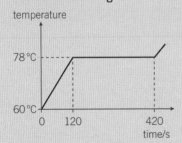

▲ **Figure 3**

4   The temperature–time graph shown in Figure 3 was obtained by heating 0.12 kg of a substance in an insulated container. The specific heat capacity of the substance in the solid state is 1200 J kg^{-1} K^{-1}.

   Calculate:

   a   the energy per second supplied to the substance in the solid state if its temperature increased from 60 °C to its melting point at 78 °C in 120 s,

   b   the energy needed to melt the solid if it took 300 s to melt with energy supplied at the same rate as in **a**.

specific heat capacity of water = 4200 J kg^{-1} K^{-1}
specific heat capacity of ice = 2100 J kg^{-1} K^{-1}
specific latent heat of fusion of ice = 3.4 × 10^5 J kg^{-1}
specific latent heat of vaporisation of water = 2.3 × 10^6 J kg^{-1}

1   A tray containing 0.20 kg of water at 20 °C is placed in a freezer.
   (a) The temperature of the water drops to 0 °C in 10 min.
      Calculate:
      (i)   the energy lost by the water as it cools to 0 °C,
      (ii) the average rate at which the water is losing energy, in J s^{-1}.   *(3 marks)*

   (b) (i)   Estimate the time taken for the water at 0 °C to turn completely
              into ice.
         (ii) State any assumptions you make.   *(3 marks)*

                                                                                         AQA, 2003

2   **(a)** Calculate the energy released when 1.5 kg of water at 18 °C cools to 0 °C
         and then freezes to form ice, also at 0 °C.   *(4 marks)*
   (b) Explain why it is more effective to cool cans of drinks by placing them in
         a bucket full of melting ice rather than in a bucket of water at an initial
         temperature of 0 °C.   *(2 marks)*

                                                                                         AQA, 2006

3   An electrical heater is used to heat a 1.0 kg block of metal, which is well lagged.
   The table shows how the temperature of the block increased with time.

temp / °C	20.1	23.0	26.9	30.0	33.1	36.9
time / s	0	60	120	180	240	300

   (a) Plot a graph of temperature against time.   *(3 marks)*
   (b) Determine the gradient of the graph.   *(2 marks)*
   (c) The heater provides thermal energy at the rate of 48 W. Use your value for
         the gradient of the graph to determine a value for the specific heat capacity
         of the metal in the block.   *(2 marks)*
   (d) The heater in part (c) is placed in some crushed ice that has been placed
         in a funnel as shown in **Figure 1**.

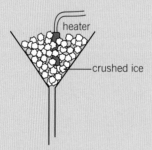

▲ Figure 1

   The heater is switched on for 200 s and 32 g of ice are found to have melted
   during this time.
   Use this information to calculate a value for the specific latent heat of
   fusion for water, stating **one** assumption made.   *(3 marks)*
                                                                                         AQA, 2002

4 In an experiment to measure the temperature of the flame of a Bunsen burner, a lump of copper of mass 0.12 kg is heated in the flame for several minutes. The copper is then transferred quickly to a beaker, of negligible heat capacity, containing 0.45 kg of water, and the temperature rise of the water measured.

specific heat capacity of copper = 390 J kg^{-1} K^{-1}

(a) If the temperature of the water rises from 15 °C to 35 °C, calculate the thermal energy gained by the water. *(2 marks)*

(b) (i) State the thermal energy lost by the copper, assuming no heat is lost during its transfer.

    (ii) Calculate the fall in temperature of the copper.

    (iii) Hence calculate the temperature reached by the copper while in the flame. *(4 marks)*
AQA, 2006

5 A bicycle and its rider have a total mass of 95 kg. The bicycle is travelling along a horizontal road at a constant speed of 8.0 m s^{-1}.

(a) Calculate the kinetic energy of the bicycle and rider. *(2 marks)*

(b) The brakes are applied until the bicycle and rider come to rest. During braking, 60% of the kinetic energy of the bicycle and rider is converted to thermal energy in the brake blocks. The brake blocks have a total mass of 0.12 kg and the material from which they are made has a specific heat capacity of 1200 J kg^{-1} K^{-1}.

    (i) Calculate the maximum rise in temperature of the brake blocks.

    (ii) State an assumption you have made in part (b)(i). *(4 marks)*
AQA, 2004

6 A female runner of mass 60 kg generates thermal energy at a rate of 800 W.

(a) Assuming that she loses no energy to the surroundings and that the average specific heat capacity of her body is 3900 J kg^{-1} K^{-1}, calculate:

    (i) the thermal energy generated in one minute,

    (ii) the temperature rise of her body in one minute. *(3 marks)*

(b) In practice it is desirable for a runner to maintain a constant temperature. This may be achieved partly by the evaporation of sweat. The runner in part (a) loses energy at a rate of 500 W by this process.

Calculate the mass of sweat evaporated in one minute. *(3 marks)*

(c) Explain why, when she stops running, her temperature is likely to fall. *(2 marks)*
AQA, 2005

7 In a geothermal power station, water is pumped through pipes into an underground region of hot rocks. The thermal energy of the rocks heats the water and turns it to steam at high pressure. The steam then drives a turbine at the surface to produce electricity.

(a) Water at 21 °C is pumped into the hot rocks and steam at 100 °C is produced at a rate of 190 kg s^{-1}.

    (i) Show that the energy per second transferred from the hot rocks to the power station in this process is at least 500 MW.

    (ii) The hot rocks are estimated to have a volume of 4.0×10^6 m^3. Estimate the fall of temperature of these rocks in one day if thermal energy is removed from them at the rate calculated in part (i) without any thermal energy gain from deeper underground.

specific heat capacity of the rocks = 850 J kg^{-1} K^{-1}

density of the rocks = 3200 kg m^{-3} *(7 marks)*
AQA, 2006

20 Gases
20.1 The experimental gas laws

Specification reference: 3.6.2.2

Learning objectives:

→ State the experimental gas laws.

→ Calculate the increase of the pressure of a gas when it is heated or compressed.

→ State what is meant by an isothermal change.

→ Calculate the work done in an isobaric process.

When you use a cycle pump to inflate a tyre, you raise the air pressure in the tyre because the pump pushes air through a valve into the tyre. The valve lets the air in but does not allow it out. The tyre is a buffer between the wheel frame and the ground. If the tyre pressure is too low, the wheel frame will rub on the ground when you are cycling.

The **pressure** of a gas is the force per unit area that the gas exerts normally (i.e., at right angles) on a surface. Pressure is measured in pascals (Pa), where $1\,Pa = 1\,N\,m^{-2}$. The pressure of a gas depends on its temperature, the volume of the gas container, and the mass of gas in the container.

Boyle's law

The apparatus shown in Figure 1 can be used to investigate how the pressure of a fixed mass of gas depends on its volume when the temperature stays the same. Measurements using this apparatus show that the gas pressure × its volume is constant for a fixed mass of gas at constant temperature. This is called Boyle's law. Any change at constant temperature is called an *isothermal* change.

Boyle's law states that for a fixed mass of gas at constant temperature,

$$pV = \text{constant}$$

where p = gas pressure and V = gas volume.

The measurements plotted as a graph of pressure against $\dfrac{1}{\text{volume}}$ give a straight line through the origin. This is because Boyle's law can be written as $p = \text{constant} \times \dfrac{1}{V}$, which represents the equation $y = mx$ for a straight-line graph through the origin, if p is plotted on the y-axis and $\dfrac{1}{V}$ is plotted on the x-axis.

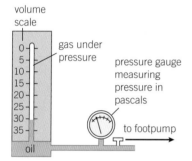

▲ **Figure 1** *Testing Boyle's law*

Hint

1 A graph of pressure against volume is a curve that tends towards each axis, as shown in Figure 2.

2 An **ideal gas** is a gas that obeys Boyle's law.

Synoptic link

A gas at very high pressure does *not* obey Boyle's law. The molecules are so close to each other that the molecules' own volume becomes significant. See Topic 20.3, The kinetic theory of gases.

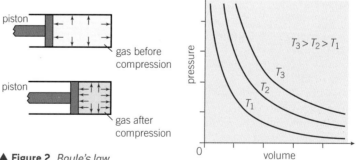

▲ **Figure 2** *Boyle's law*

Charles' law

Using a glass tube open at one end containing dry air trapped by a suitable liquid, you can find out how the volume of a fixed mass of gas at constant pressure varies with temperature. Plotting the measurements of the volume of the gas at 0 °C and 100 °C on a graph shows you the idea behind absolute zero, which you studied in Topic 19.1. No matter how much gas is used, provided the gas is an ideal gas, its volume will be zero at absolute zero, which is −273.15 °C.

Figure 3 shows how the volume of a fixed mass of gas at constant pressure varies with absolute temperature T in kelvins. The graph is a straight line through the origin. The relationship, called **Charles' law**, between the gas volume V and the temperature T in kelvins can therefore be written as

$$\frac{V}{T} = \text{constant}$$

Any change at constant pressure is called an *isobaric* change. When work is done to change the volume of a gas, energy must be transferred by heating to keep the pressure constant, and so the work done by the gas on a piston can be given by the equation

$$\text{Work done} = p\Delta V$$

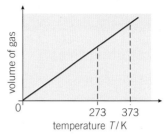

▲ **Figure 3** *Charles' law*

The pressure law

Figure 4 shows how the pressure of a fixed mass of gas at constant volume can be measured at different temperatures. If the measurements are plotted on a graph of pressure against temperature in kelvins, they give a straight line through the origin (as you saw in Topic 19.1). The relationship between pressure p and temperature T, in kelvins, can therefore be written as

$$\frac{p}{T} = \text{constant}$$

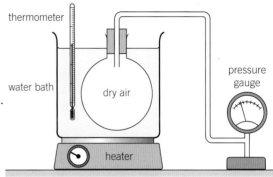

▲ **Figure 4** *The pressure law*

Deep sea diving

The extra pressure on underwater swimmers caused by a few metres of water above them is enough to give them breathing difficulties. They have to use special breathing apparatus to reach much greater depths. Gases in the lungs are compressed by the high pressure on their body, and these gases pass into their blood system. If a diver ascends too fast, dissolved nitrogen is released into their blood system, causing a life-threatening painful condition known as 'the bends'. Divers have to learn how to work out factors such as safe rates of ascent or descent.

Q: The pressure of the atmosphere at the Earth's surface is about 100 kPa, which is about the same as the pressure due to 10 m of water. In 2012, the Australian-built Deep Sea Challenger sea craft successfully descended to a depth of about 11 km. Estimate the extra pressure on it in kPa at this depth.

A: 110 MPa

Summary questions

1   A hand pump of volume $2.0 \times 10^{-4}\,\text{m}^3$ is used to force air through a valve into a container of volume $8.0 \times 10^{-4}\,\text{m}^3$ which contains air at an initial pressure of 101 kPa. Calculate the pressure of the air in the container after one stroke of the pump, assuming the temperature is unchanged.

2   A sealed can of fixed volume contains air at a pressure of 101 kPa at 100 °C. The can is then cooled to a temperature of 20 °C. Calculate the pressure of the air in the can.

3   The volume of a fixed mass of gas at 15 °C was 0.085 m³. The gas was then heated to 55 °C without change of pressure. Calculate the new volume of this gas.

4   A hand pump was used to raise the pressure of the air in a flask of volume $1.20 \times 10^{-4}\,\text{m}^3$, without and then with powder in the flask.

   • Without the powder in the flask, the pressure increased from 110 kPa to 135 kPa.

   • With 0.038 kg of powder in the flask, the pressure increased from 110 kPa to 141 kPa.

   a   Demonstrate that the volume of air in the hand pump initially was $2.7 \times 10^{-5}\,\text{m}^3$.

   b   Calculate the volume and the density of the powder.

20.2 The ideal gas law

Learning objectives:

→ Define an ideal gas.

→ Discuss whether the experimental gas laws can be combined, and if so, how.

→ Distinguish between molar mass and molecular mass.

Specification reference: 3.6.2.2; 3.6.2.3

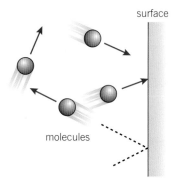

▲ **Figure 1** *Molecules in motion*

Synoptic link

Randomness occurs in radioactive decay as well as in Brownian motion of molecules. You can't predict when a random event will take place. See Topic 26.5, Radioactive decay.

Molecules in a gas

The molecules of a gas move at random with different speeds. When a molecule collides with another molecule or with a solid surface, it bounces off without losing speed. The pressure of a gas on a surface is due to the gas molecules hitting the surface. Each impact causes a tiny force on the surface. Because there are a very large number of impacts each second, the overall result is that the gas exerts a measurable pressure on the surface.

Molecules are too small to see individually. You can see the effect of individual molecules in a gas if you observe smoke particles with a microscope. If a beam of light is directed through the smoke, you will see the smoke particles as tiny specks of light wriggling about unpredictably. This type of motion is called **Brownian motion** after Robert Brown, who first observed it in 1827 with pollen grains in water. The motion of each particle is because it is bombarded unevenly and randomly by individual molecules. The particle therefore experiences forces due to these impacts, which change its magnitude and direction at random. So Brownian motion showed the existence of molecules and atoms.

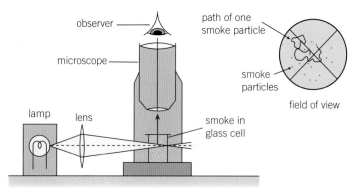

▲ **Figure 2** *Brownian motion*

The Avogadro constant

The density of oxygen gas is 16 times that of hydrogen gas at the same temperature. Therefore, the mass of a certain volume of oxygen is 16 times that of the mass of the same volume of hydrogen at the same temperature. When such measurements were first made in the 19th century, Amadeo Avogadro put forward the hypothesis that equal volumes of gases at the same temperature and pressure contain equal numbers of molecules.

How many molecules are in a particular amount of gas? Avogadro thought of the idea of counting atoms and molecules in terms of the number of atoms in 1 gram of hydrogen. Now we use 12 grams of the carbon isotope $^{12}_{6}C$ as the standard amount because hydrogen gas contains a small proportion of the isotope of hydrogen $^{2}_{1}H$, which cannot easily be removed.

- The **Avogadro constant**, N_A, is defined as the number of atoms in exactly 12 g of the carbon isotope $^{12}_{6}C$. The value of N_A (to four significant figures) is 6.023×10^{23}. So the mass of an atom of $^{12}_{6}C$ is $1.993 \times 10^{-23}\,g\left(= \dfrac{12\,g}{6.023 \times 10^{23}}\right)$.

- One atomic mass unit (u) is $\dfrac{1}{12}$th of the mass of a $^{12}_{6}C$ atom. The mass of a carbon atom is $1.993 \times 10^{-26}\,kg$, so $1\,u = 1.661 \times 10^{-27}\,kg$.

Molar mass

One **mole** of a substance consisting of identical particles is defined as the quantity of substance that contains N_A particles. The number of moles in a given quantity of a substance is its **molarity**. The unit of molarity is the **mol**.

The **molar mass** of a substance is the mass of 1 mol of the substance. The unit of molar mass is $kg\,mol^{-1}$. For example, the molar mass of oxygen gas is $0.032\,kg\,mol^{-1}$. So $0.032\,kg$ of oxygen gas contains N_A oxygen molecules.

Therefore,

1 the number of moles in mass M_S of a substance $= \dfrac{M_S}{M}$, where M is the molar mass of the substance,

2 the number of molecules in mass M_S of a substance $= \dfrac{N_A M_S}{M}$.

For example, because the molar mass of carbon dioxide is $0.044\,kg$ ($= 44\,g$), then n moles of carbon dioxide has a mass of $44\,ng$ and contains nN_A molecules.

The ideal gas equation

As you learnt in Topic 20.1, an **ideal gas** is a gas that obeys Boyle's law. The three experimental gas laws can be combined to give the equation

$$\frac{pV}{T} = \textbf{constant, for a fixed mass of ideal gas}$$

where p is the pressure, V is the volume, and T is the absolute temperature. This equation takes in all situations where the pressure, volume, and temperature of a fixed mass of gas changes.

Equal volumes of ideal gases at the same temperature and pressure contain equal numbers of moles. Further measurements show that one mole of any ideal gas at 273 K and a pressure of 101 kPa has a volume of $0.0224\,m^3$. Therefore, for 1 mol of any ideal gas, the value of $\dfrac{pV}{T}$ for 1 mol is equal to $8.31\,J\,mol^{-1}\,K^{-1}$ $\left(= \dfrac{pV}{T} = \dfrac{101 \times 10^3\,Pa \times 0.0224m^3}{273K}\right)$.

This value is called the **molar gas constant R**. A graph of pV against temperature T for n moles is a straight line through absolute zero and has a gradient equal to nR.

Hint

1 The masses in atomic mass units (to 1 u) of various atoms are: hydrogen H = 1 u, carbon C = 12 u, nitrogen N = 14 u, oxygen O = 16 u, and copper Cu = 64 u.

2 The masses in atomic mass units (to 1 u) of various molecules are: water H_2O = 18 u, carbon monoxide CO = 28 u, carbon dioxide CO_2 = 44 u, and oxygen O_2 = 32 u.

Study tip

Be clear about molar mass and molecular mass.

Hint

The unit of R is the joule per mole per kelvin ($J\,mol^{-1}\,K^{-1}$), which is the same as the unit of $\dfrac{pressure \times volume}{absolute\ temp. \times no.\ of\ moles}$. This is because the unit of pressure (the pascal = $1\,N\,m^{-2}$) × the unit of volume (m^3) is the joule (= $1\,N\,m$).

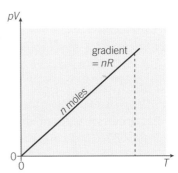

▲ **Figure 3** *A graph of pV against T for an ideal gas*

Summary questions

$N_A = 6.02 \times 10^{23}\,mol^{-1}$
$R = 8.31\,J\,mol^{-1}\,K^{-1}$

1 A gas cylinder has a volume of 0.024 m³ and is fitted with a valve designed to release the gas if the pressure of the gas reaches 125 kPa. Calculate:

   a the maximum number of moles of gas that can be contained by this cylinder at 50 °C,

   b the pressure in the cylinder of this amount of gas at 10 °C.

2 In an electrolysis experiment, $2.2 \times 10^{-5}\,m^3$ of a gas is collected at a pressure of 103 kPa and a temperature of 20 °C. Calculate:

   a the number of moles of gas present,

   b the volume of this gas at 0 °C and 101 kPa.

3 a Sketch a graph to show how the pressure of 2 mol of gas varies with temperature when the gas is heated from 20 °C to 100 °C in a sealed container of volume 0.050 m³.

   b The molar mass of the gas in **a** is 0.032 kg mol⁻¹. Calculate the density of the gas.

4 The molar mass of air is 0.029 kg mol⁻¹.

   a Calculate the density of air at 20 °C and a pressure of 101 kPa.

   b Calculate the number of molecules in 0.001 m³ of air at 20 °C and a pressure of 101 kPa.

So, the combined gas law can be written as

$$pV_m = RT$$

where V_m = volume of 1 mol of ideal gas at pressure p and temperature T.

Therefore, for n moles of ideal gas,

$$pV = nRT$$

where V = volume of the gas at pressure p and temperature T in kelvins.

This equation is called the **ideal gas equation**.

Using the ideal gas equation

- The *mass M_S* of a substance is equal to its molar mass M × the number of moles n. Because $n = \dfrac{pV}{RT}$ for an ideal gas, then $M_S = M \times \left(\dfrac{pV}{RT}\right)$ gives the mass of ideal gas in volume V at pressure p and absolute temperature T.

- The *density of an ideal gas* of molar mass M is $\rho = \dfrac{mass\ M_s}{volume\ V}$
$$= \frac{nM}{V} = \frac{pM}{RT}$$

Therefore, for an ideal gas at constant pressure, its density ρ is inversely proportional to its temperature T.

- In the equation $pV = nRT$, substituting the number of moles $n = \dfrac{N}{N_A}$ gives

$$pV = NkT$$

where the **Boltzmann constant** k is $\dfrac{R}{N_A}$, and N is the number of molecules.

Prove for yourself that $k = 1.38 \times 10^{-23}\,J\,K^{-1}$. We will meet k again in the next topic when we discuss how much kinetic energy a gas molecule has.

Worked example

$R = 8.31\,J\,mol^{-1}\,K^{-1}$

Calculate the number of moles and the mass of air in a balloon when the air pressure in the balloon is 170 kPa, the volume of the balloon is $8.4 \times 10^{-4}\,m^3$, and the temperature of the air in the balloon is 17 °C.

molar mass of air = 0.029 kg mol⁻¹

Solution

$T = 273 + 17 = 290\,K$

Using $pV = nRT$ gives $n = \dfrac{pV}{RT} = \dfrac{170 \times 10^3 \times 8.4 \times 10^{-4}}{8.31 \times 290}$
$$= 5.9 \times 10^{-2}\,mol$$

Mass of air = number of moles × molar mass = $5.9 \times 10^{-2} \times 0.029$
$$= 1.7 \times 10^{-3}\,kg$$

20.3 The kinetic theory of gases

The gas laws are experimental laws, or empirical in nature. This means they were devised by experiments and observations. They can be explained by assuming that a gas consists of point molecules moving about at random, continually colliding with the container walls. Each impact causes a force on the container. The force of many impacts is the cause of the pressure of the gas on the container walls.

Boyle's law can be explained as follows. The pressure of a gas at constant temperature is increased by reducing its volume because the gas molecules travel less distance between impacts at the walls due to the reduced volume. Therefore, there are more impacts per second, and so the pressure is greater.

The **pressure law** can be explained as follows. The pressure of a gas at constant volume is increased by raising its temperature. The average speed of the molecules is increased by raising the gas temperature. Therefore, the impacts of the molecules on the container walls are harder and more frequent. So the pressure is raised as a result.

Molecular speeds

The molecules in an ideal gas have a continuous spread of speeds, as shown in Figure 1. The speed of an individual molecule changes when it collides with another gas molecule. But the distribution stays the same, as long as the temperature does not change.

The **root mean square speed** of the molecules,

$$c_{rms} = \left[\frac{c_1^2 + c_2^2 + \dots + c_N^2}{N}\right]^{\frac{1}{2}},$$

where $c_1, c_2, c_3, \dots c_N$ represent the speeds of the individual molecules, and N is the number of molecules in the gas.

If the temperature of a gas is raised, its molecules move faster, on average. The root mean square speed of the molecules increases. The distribution curve becomes flatter and broader because the greater the temperature, the more molecules there are moving at higher speeds (see Figure 2).

The kinetic theory equation

Kinetic theory was devised by mathematics and theories, instead of by observations and experiments like the gas laws. For an ideal gas consisting of N identical molecules, each of mass m, in a container of volume V, the pressure p of the gas is given by the equation

$$pV = \frac{1}{3}Nm(c_{rms})^2$$

where c_{rms} is the root mean square speed of the gas molecules.

To derive the **kinetic theory equation**, you need to apply the laws of mechanics and statistics to the molecular model of a gas. In doing so, some assumptions must be made about the molecules in a gas:

Learning objectives:

→ Explain the increase of pressure of a gas when it is compressed or heated.

→ Describe the behaviour of a gas.

→ Discuss what the mean kinetic energy of a gas molecule depends on.

Specification reference: 3.6.2.3

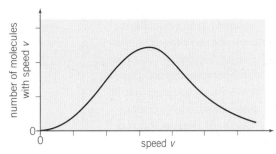

▲ **Figure 1** *Distribution of molecular speeds*

Hint

The root mean square speed of the molecules of a gas is *not* the same as the mean speed. The mean speed is the sum of the speeds divided by the number of molecules.

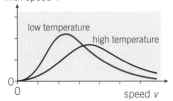

number of molecules with speed v

low temperature

high temperature

speed v

▲ **Figure 2** *The effect of temperature on the distribution of speeds*

53

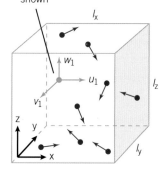

chosen molecule with its velocity components shown

▲ **Figure 3** *Molecules in a box*

 Making assumptions

In deriving the kinetic theory equation, some assumptions are made about the molecules of a gas. The ideal gas equation is then obtained by applying the further assumption that the mean kinetic energy of a gas molecule is directly proportional to the absolute temperature of the gas. Because the ideal gas law is an experimental law, the assumptions made in deriving it from the molecular model of a gas must be valid for any gas that obeys the ideal gas law. Under conditions where a gas does not obey the ideal gas equation (e.g., very high pressure), one or more of the assumptions is no longer valid. For example, a gas at very high pressure does not obey the ideal gas law. Its molecules are so close to each other that the volume of the molecules is significant and they do not behave as 'point molecules'.

Q: When the volume of the gas decreases, there is less space for the molecules to move about in. Explain how this would affect the gas pressure.

1 The molecules are point molecules. The volume of each molecule is negligible compared with the volume of the gas.

2 They do not attract each other. If they did, the effect would be to reduce the force of their impacts on the container surface.

3 They move about in continual random motion.

4 The collisions they undergo with each other and with the container surface are elastic collisions (i.e., there is no overall loss of kinetic energy in a collision).

5 Each collision with the container surface is of much shorter duration than the time between impacts.

Part 1

Consider one molecule of mass m in a rectangular box of dimensions l_x, l_y, and l_z as shown in Figure 3. Let u_1, v_1, and w_1 represent its velocity components in the x, y, and z directions, respectively.

Note that the speed, c_1, of the molecule is given by the following rule for adding perpendicular components (in this case the three velocity components u_1, v_1, and w_1):

$$c_1^2 = u_1^2 + v_1^2 + w_1^2$$

You will need to use this rule in Part 2.

- Each impact of the molecule with the shaded face in Figure 3 reverses the x-component of velocity. So the impact changes the x-component of its momentum from $+mu_1$ to $-mu_1$. Therefore, the change of its momentum due to the impact = final momentum − initial momentum = $(-mu_1) - (mu_1) = -2mu_1$.

- The time, t, between successive impacts on this face is given by the equation

$$t = \frac{\text{the total distance to the opposite face and back}}{x\text{-component of velocity}} = \frac{2l_x}{u_1}$$

Using Newton's second law therefore gives

$$\text{the force on the molecule} = \frac{\text{change of momentum}}{\text{time taken}}$$

$$= \frac{-2mu_1}{\frac{2l_x}{u_1}} = \frac{-mu_1^2}{l_x}$$

Because the force F_1 of the impact on the surface is equal and opposite to the force on the molecule in accordance with Newton's third law, then

$$F_1 = \frac{+mu_1^2}{l_x}$$

- Because pressure = $\frac{\text{force}}{\text{area}}$, the pressure p_1 of the molecule on the surface is given by the equation

$$p_1 = \frac{\text{force}}{\text{area of the shaded face } (l_y l_z)} = \frac{mu_1^2}{l_x l_y l_z} = \frac{mu_1^2}{V}$$

where V = the volume of the box = $l_x l_y l_z$

Part 2

For N molecules in the box moving at different velocities, the total pressure p is the sum of the individual pressures p_1, p_2, p_3, ... p_N, where each subscript refers to each molecule.

So, $p = \dfrac{mu_1^2}{V} + \dfrac{mu_2^2}{V} + \dfrac{mu_3^2}{V} + \dots + \dfrac{mu_N^2}{V}$

$= \dfrac{m}{V}(u_1^2 + u_2^2 + u_3^2 + \dots + u_N^2) = \dfrac{Nm\bar{u}^2}{V}$

where $\bar{u}^2 = \dfrac{u_1^2 + u_2^2 + u_3^2 + \dots + u_N^2}{N}$

Because the motion of the molecules is random, there is no preferred direction of motion. The equation above could equally well have been derived in terms of the y-components of velocity $v_1, v_2, v_3, \dots v_N$, or the z-components of velocity $w_1, w_2, w_3, \dots w_N$.

i.e.,

$p = \dfrac{Nm\bar{v}^2}{V}$ where $\bar{v}^2 = \dfrac{v_1^2 + v_2^2 + v_3^2 + \dots + v_N^2}{N}$

$p = \dfrac{Nm\bar{w}^2}{V}$ where $\bar{w}^2 = \dfrac{w_1^2 + w_2^2 + w_3^2 + \dots + w_N^2}{N}$

Therefore, $p = \dfrac{Nm}{3V}(\bar{u}^2 + \bar{v}^2 + \bar{w}^2)$

The note below shows that, because the motion of the molecules is random, the *root mean square speed* of the gas molecules is given by the equation $c_{rms}^2 = \bar{u}^2 + \bar{v}^2 + \bar{w}^2$. So

$$p = \dfrac{Nm}{3V}c_{rms}^2 \qquad \text{or} \qquad pV = \dfrac{1}{3}Nmc_{rms}^2$$

Note

As explained in Part 1, the speed c of the nth molecule is related to its velocity components according to equations of the form:

$$c_n^2 = u_n^2 + v_n^2 + w_n^2$$

The root mean square speed of the molecules, c_{rms}, is defined by

$(c_{rms})^2 = \dfrac{c_1^2 + c_2^2 + c_3^2 + \dots + c_N^2}{N}$

So, $(c_{rms})^2 = \dfrac{u_1^2 + v_1^2 + w_1^2 + u_2^2 + v_2^2 + w_2^2 + u_3^2 + v_3^2 + w_3^2 + \dots + u_N^2 + v_N^2 + w_N^2}{N}$

$= \bar{u}^2 + \bar{v}^2 + \bar{w}^2$

Molecules and kinetic energy

For an ideal gas, its internal energy is due only to the kinetic energy of the molecules of the gas.

The mean **kinetic energy of a molecule of a gas**

$= \dfrac{\text{total kinetic energy of all the molecules}}{\text{total number of molecules}}$

$= \dfrac{\frac{1}{2}mc_1^2 + \frac{1}{2}mc_2^2 + \frac{1}{2}mc_3^2 + \dots + \frac{1}{2}mc_N^2}{N}$

$= \dfrac{\frac{1}{2}m(c_1^2 + c_2^2 + c_3^2 + \dots + c_N^2)}{N} = \frac{1}{2}mc_{rms}^2$

The higher the temperature of a gas, the greater the mean kinetic energy of a molecule of the gas.

Classical physics

Scientists regularly check and use each other's work. Theories are devised to understand experimental observations. Experiments then test the predictions from the theories.

The kinetic theory equation is the result of theories mainly made in the mid-19th century. Scottish physicist James Clerk Maxwell derived an equation for the distribution curve of the speeds of the molecules of a gas shown in Figure 2. Avogadro theorised that equal volumes of gas at the same temperature and pressure contain equal numbers of molecules, but couldn't tell *how many* molecules there were. In 1865, Austrian physicist Josef Loschmidt estimated the number of molecules in 1 cm^3 of gas at 0 °C and at atmospheric pressure — now known as Avogadro's number. It was redefined twice before today's definition, including by Einstein, who used the idea of Brownian motion.

Maxwell established what we now call classical physics, which explains all the properties of matter and radiation by using Newton's laws, the laws of thermodynamics, and Maxwell's equations. By the late 19th century, many physicists thought that the laws of physics had all been discovered. But further discoveries were made that couldn't be explained by classical physics. Max Planck hit on a revolutionary new theory that energy is quantised, but didn't like the idea very much. A few years later, Einstein used the idea to explain another failure of classical theory, the photoelectric effect, which you learnt in Topic 3.1. Even Planck was convinced. Quantum theory took off!

Worked example

$k = 1.38 \times 10^{-23}\,\text{J mol}^{-1}\,\text{K}^{-1}$,
$N_A = 6.02 \times 10^{23}\,\text{mol}^{-1}$

Calculate the root mean square speed of oxygen molecules at $0\,^{\circ}\text{C}$. The molar mass of oxygen = $0.032\,\text{kg mol}^{-1}$

Solution

$T = 273\,\text{K}$

The mass of an oxygen molecule, $m = \dfrac{0.032}{6.02 \times 10^{23}}$

$= 5.3 \times 10^{-26}\,\text{kg}$

Rearranging $\dfrac{1}{2}m(c_{\text{rms}})^2 = \dfrac{3}{2}kT$

gives $(c_{\text{rms}})^2 = \dfrac{3kT}{m}$

$= \dfrac{3 \times 1.38 \times 10^{-23} \times 273}{5.3 \times 10^{-26}}$

$= 2.13 \times 10^5\,\text{m}^2\,\text{s}^{-2}$

Therefore, the root mean square speed $c_{\text{rms}} = \sqrt{2.13 \times 10^5}$

$= 460\,\text{m s}^{-1}$ (2 s.f.)

For an ideal gas, by assuming that the mean kinetic energy of a molecule is $\dfrac{1}{2}m(c_{\text{rms}})^2 = \dfrac{3}{2}kT$ where $k = \dfrac{R}{N_A}$, then $3kT = m(c_{\text{rms}})^2$.

Substituting $3kT$ for $m(c_{\text{rms}})^2$ in the kinetic theory equation

$pV = \dfrac{1}{3}Nm(c_{\text{rms}})^2$ therefore gives $pV = \dfrac{1}{3}N \times 3kT = NkT$

Because $Nk = \dfrac{NR}{N_A} = nR$, you then get the ideal gas equation $pV = nRT$.

You have derived the ideal gas equation (which is an experimental law) from the kinetic theory equation by assuming that the mean kinetic energy of an ideal gas molecule is $\dfrac{3}{2}kT$.

So, you can say that for an ideal gas at absolute temperature T,

the **mean kinetic energy of a molecule of an ideal gas** = $\dfrac{3}{2}kT$,

where $k = \dfrac{R}{N_A}$. Recall that the constant k is called the Boltzmann constant. Its value $\left(= \dfrac{R}{N_A}\right)$ is $1.38 \times 10^{-23}\,\text{J K}^{-1}$.

Notes:

Using the above equation for an ideal gas,

- the total kinetic energy of one mole = $N_A \times \dfrac{3}{2}kT = \dfrac{3}{2}RT$ (as $k = \dfrac{N_A}{R}$),

- the total kinetic energy of n moles of an ideal gas = $n \times \dfrac{3}{2}RT = \dfrac{3}{2}nRT$.

The total kinetic energy of n moles of an ideal gas = $\dfrac{3}{2}nRT$

Therefore, for n moles of an ideal gas at temperature T (in kelvins),

$$\textbf{internal energy} = \dfrac{3}{2}nRT$$

Summary questions

$N_A = 6.02 \times 10^{23}\,\text{mol}^{-1}$, $R = 8.31\,\text{J mol}^{-1}\,\text{K}^{-1}$,
$k = 1.38 \times 10^{-23}\,\text{J K}^{-1}$

1 a Explain in molecular terms why the pressure of a gas in a sealed container increases when its temperature is raised.

 b The molar mass of oxygen is $0.032\,\text{kg mol}^{-1}$. A cylinder of volume $0.025\,\text{m}^3$ contains oxygen gas at a pressure of $120\,\text{kPa}$ and a temperature of $373\,\text{K}$. Calculate:

 i the number of moles of oxygen in the cylinder,

 ii the total kinetic energy of all the gas molecules in the container.

2 For a hydrogen molecule (molar mass = $0.002\,\text{kg mol}^{-1}$) at $0\,^{\circ}\text{C}$, calculate:

 a its mean kinetic energy

 b its root mean square speed.

3 Air consists mostly of nitrogen and oxygen in proportions $1:4$ by mass.

 a Explain why the mean kinetic energy of a nitrogen molecule in air is the same as that of an oxygen molecule in the same sample of air.

 b Demonstrate that the root mean square speed of a nitrogen molecule in air is $1.07 \times$ that of an oxygen molecule in the same sample of air.

 nitrogen = $0.028\,\text{kg mol}^{-1}$
 oxygen = $0.032\,\text{kg mol}^{-1}$

4 An ideal gas of molar mass $0.028\,\text{kg mol}^{-1}$ is in a container of volume $0.037\,\text{m}^3$ at a pressure of $100\,\text{kPa}$ and a temperature of $300\,\text{K}$. Calculate:

 a the number of moles,

 b the mass of gas present,

 c the root mean square speed of the molecules.

1 The escape velocity v_{esc} of an object from a planet or moon is the minimum velocity the object must have to escape from the planet. For a planet or moon of radius R, it can be shown that $v_{esc} = (2gR)^{\frac{1}{2}}$, where g is the gravitational field strength at the surface of the planet or moon.

   (a) The radius of the Earth's moon is 1740 km, and its surface gravitational field strength is $1.62\,N\,kg^{-1}$. Calculate the escape velocity from the Earth's moon. *(2 marks)*

   (b) The average temperature of the lunar surface during the lunar day is about 400 K.

   (i) Calculate the mean kinetic energy of a molecule of an ideal gas at 400 K.

   (ii) Demonstrate that the root mean square speed of a molecule of oxygen gas at this temperature is $560\,m\,s^{-1}$.

   (iii) Explain why gas molecules released on the lunar surface escape into space. *(8 marks)*

   (c) Astronomers have discovered the existence of water vapour in a giant gas planet orbiting a star 64 light years from Earth. The astronomers observed the spectrum of infrared light from the star and discovered absorption lines due to water vapour which are present only when the planet passes across the face of the star.

   (i) Why did astronomers conclude that the absorption lines were due to the planet rather than the star?

   (ii) Give **one** reason why it might not have been possible to detect such absorption lines if the planet's surface had been at the same temperature but the planet had been much smaller in diameter and in mass. *(6 marks)*

2 (a) (i) Sketch a graph of pressure against volume for a fixed mass of ideal gas at constant temperature. Label this graph O.

   On the *same axes* sketch **two** additional curves A and B, if the following changes are made:

   (ii) The same mass of gas at a lower constant temperature (label this A).

   (iii) A greater mass of gas at the original constant temperature (label this B). *(3 marks)*

   (b) A cylinder of volume $0.20\,m^3$ contains an ideal gas at a pressure of 130 kPa and a temperature of 290 K. Calculate:

   (i) the amount of gas, in moles, in the cylinder,

   (ii) the total kinetic energy of a molecule of gas in the cylinder,

   (iii) the total kinetic energy of the molecules in the cylinder. *(5 marks)*

   AQA, 2005

3 (a) State the equation of state for an ideal gas. *(1 mark)*

   (b) A fixed mass of an ideal gas is heated whilst its volume is kept constant. Sketch a graph to show how the pressure, p, of the gas varies with the absolute temperature, T, of the gas. *(2 marks)*

   (c) Explain in terms of molecular motion, why the pressure of the gas in part (b) varies with the absolute temperature. *(4 marks)*

   (d) Calculate the average kinetic energy of the gas molecules at a temperature of 300 K. *(2 marks)*

   AQA, 2004

4 (a) The molecular theory model of an ideal gas leads to the derivation of the equation

$$pV = \frac{1}{3}Nmc_{rms}^2$$

   Explain what each symbol in the equation represents. *(4 marks)*

(b) One assumption used in the derivation of the equation stated in part (a) is that molecules are in a state of random motion.
  (i) Explain what is meant by random motion.
  (ii) State **two** more assumptions used in this derivation. *(4 marks)*
(c) Describe how the motion of gas molecules can be used to explain the pressure exerted by a gas on the walls of its container. *(4 marks)*

AQA, 2002

5 The number of molecules in one cubic metre of air decreases as altitude increases. The table shows how the pressure and temperature of air compare at sea level and at an altitude of 10 000 m.

altitude	pressure / Pa	temperature / K
sea level	1.0×10^5	300
10 000 m	2.2×10^4	270

(a) Calculate the number of moles of air in a cubic metre of air at:
  (i) sea level,
  (ii) 10 000 m. *(3 marks)*
(b) In air, 23% of the molecules are oxygen molecules. Calculate the number of extra oxygen molecules there are per cubic metre at sea level compared with a cubic metre of air at an altitude of 10 000 m. *(2 marks)*

AQA, 2006

6 **(a)** **Figure 1** shows a helium atom of mass 6.8×10^{-27} kg about to strike the wall of a container. It rebounds with the same speed.

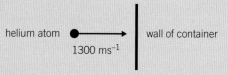

helium atom
1300 ms^{-1}
wall of container

▲ **Figure 1**

  (i) Calculate the momentum change of the helium atom.
  (ii) Calculate the number of collisions per second on each cm^2 of the container wall that will produce a pressure of 1.5×10^5 Pa. *(5 marks)*
(b) The molar mass of gaseous nitrogen is 0.028 kg mol^{-1}. The average kinetic energy for nitrogen molecules in a sample is 8.6×10^{-21} J.
  (i) Calculate the temperature of the sample.
  (ii) Calculate the mean square speed of the nitrogen molecules. *(5 marks)*

AQA, 2006 and 2007

7 The graph in **Figure 2** shows the best fit line for the results of an experiment in which the volume of a fixed mass of gas was measured over a temperature range from 20 °C to 100 °C. The pressure of the gas remained constant throughout the experiment.

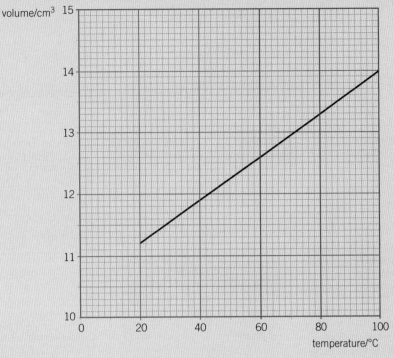

▲ **Figure 2**

(a) Use the graph in **Figure 2** to calculate a value for the absolute zero of temperature in °C.
Demonstrate clearly your method of working. *(4 marks)*

(b) Use data from the graph to calculate the mass of gas used in the experiment.
You may assume that the gas behaved like an ideal gas throughout the experiment.
gas pressure throughout the experiment = 1.0×10^5 Pa
molar mass of the gas used = 0.044 kg mol^{-1} *(5 marks)*

(c) Use the kinetic theory of gases to explain why the pressure of an ideal gas decreases
  (i) when it is expanded at constant temperature,
  (ii) when its temperature is lowered at constant volume. *(5 marks)*
AQA, 2005 and 2006

8 (a) A cylinder of fixed volume contains 15 mol of an ideal gas at a pressure of 500 kPa and a temperature of 290 K.
  (i) Demonstrate that the volume of the cylinder is 7.2×10^{-2} m^3.
  (ii) Calculate the average kinetic energy of a gas molecule in the cylinder. *(4 marks)*

(b) A quantity of gas is removed from the cylinder, and the pressure of the remaining gas falls to 420 kPa. If the temperature of the gas is unchanged, calculate the amount, in mol, of gas remaining in the cylinder. *(2 marks)*

(c) Explain in terms of the kinetic theory why the pressure of the gas in the cylinder falls when gas is removed from the cylinder. *(4 marks)*
AQA, 2003

Section 6 Summary

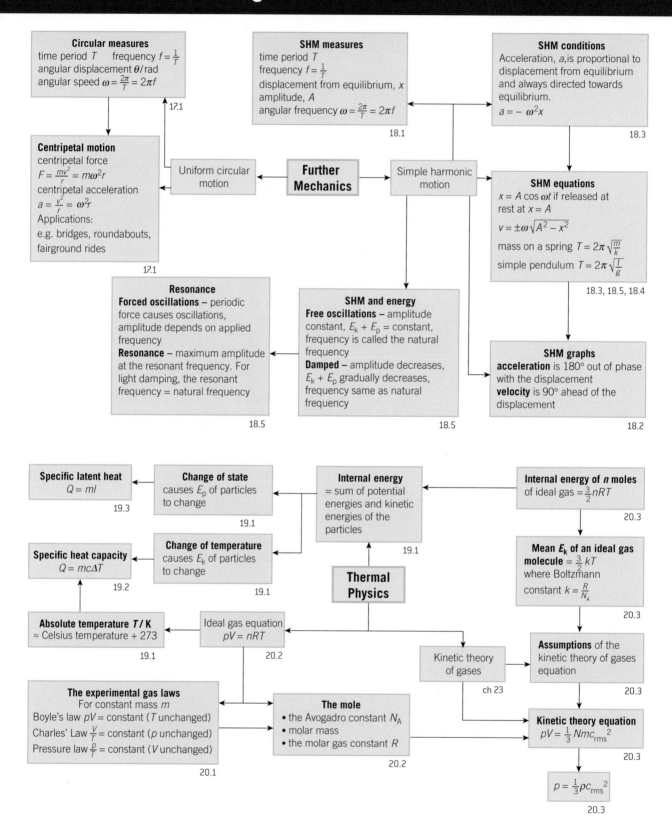

Circular measures
time period T frequency $f = \frac{1}{T}$
angular displacement θ/rad
angular speed $\omega = \frac{2\pi}{T} = 2\pi f$

17.1

SHM measures
time period T
frequency $f = \frac{1}{T}$
displacement from equilibrium, x
amplitude, A
angular frequency $\omega = \frac{2\pi}{T} = 2\pi f$

18.1

SHM conditions
Acceleration, a, is proportional to displacement from equilibrium and always directed towards equilibrium.
$a = -\omega^2 x$

18.3

Centripetal motion
centripetal force
$F = \frac{mv^2}{r} = m\omega^2 r$
centripetal acceleration
$a = \frac{v^2}{r} = \omega^2 r$
Applications:
e.g. bridges, roundabouts, fairground rides

17.1

Uniform circular motion

Further Mechanics

Simple harmonic motion

SHM equations
$x = A \cos \omega t$ if released at rest at $x = A$
$v = \pm\omega\sqrt{A^2 - x^2}$
mass on a spring $T = 2\pi\sqrt{\frac{m}{k}}$
simple pendulum $T = 2\pi\sqrt{\frac{l}{g}}$

18.3, 18.5, 18.4

Resonance
Forced oscillations – periodic force causes oscillations, amplitude depends on applied frequency
Resonance – maximum amplitude at the resonant frequency. For light damping, the resonant frequency = natural frequency

18.5

SHM and energy
Free oscillations – amplitude constant, $E_k + E_p$ = constant, frequency is called the natural frequency
Damped – amplitude decreases, $E_k + E_p$ gradually decreases, frequency same as natural frequency

18.5

SHM graphs
acceleration is 180° out of phase with the displacement
velocity is 90° ahead of the displacement

18.2

Specific latent heat
$Q = ml$

19.3

Change of state
causes E_p of particles to change

19.1

Internal energy
= sum of potential energies and kinetic energies of the particles

19.1

Internal energy of n moles
of ideal gas $= \frac{3}{2}nRT$

20.3

Specific heat capacity
$Q = mc\Delta T$

19.2

Change of temperature
causes E_k of particles to change

19.1

Thermal Physics

Mean E_k of an ideal gas molecule $= \frac{3}{2}kT$
where Boltzmann constant $k = \frac{R}{N_A}$

20.3

Absolute temperature T / K
$\approx$ Celsius temperature + 273

19.1

Ideal gas equation
$pV = nRT$

20.2

Kinetic theory of gases

ch 23

Assumptions of the kinetic theory of gases equation

20.3

The experimental gas laws
For constant mass m
Boyle's law $pV = $ constant (T unchanged)
Charles' Law $\frac{V}{T} = $ constant (p unchanged)
Pressure law $\frac{p}{T} = $ constant (V unchanged)

20.1

The mole
• the Avogadro constant N_A
• molar mass
• the molar gas constant R

20.2

Kinetic theory equation
$pV = \frac{1}{3}Nmc_{rms}^2$

20.3

$p = \frac{1}{3}\rho c_{rms}^2$

20.3

Practical skills

In this section you have met the following skills:

- use of a stopwatch or stopclock to measure the time period of an object in periodic motion (e.g. simple harmonic motion)
- use of a thermometer or a temperature sensor connected to a data logger safely and accurately in experiments involving heating or cooling
- use of electrical equipment safely and correctly in electrical heating experiments (e.g. measurement of specific heat capacity)
- use of a manometer and a pressure gauge to measure the pressure of a gas (e.g. variation of gas pressure with temperature at constant volume)
- use of a capillary tube to measure the expansion of a gas at different temperatures.

Maths skills

In this section you have met the following skills:

- use of the radian and its conversion to and from degrees (e.g. angular displacement calculations)
- calculation of a mean from a set of repeat readings (e.g. time period of a pendulum)
- interpret graphs in terms of relevant physics variables (e.g. simple harmonic graphs relating acceleration, velocity and displacement to each other)
- use of sines and cosines in simple harmonic motion problems
- use a calculator or a spreadsheet to carry out calculations using data and appropriate formulae, expressing answers to an appropriate number of significant figures and with appropriate units
- plot a straight line graph using experimental data given an equation of the form $y = k\,x^n$, where n is known, and determine the gradient and intercept and what each of these represents (e.g. finding g using a simple pendulum)
- plot a log-log graph using experimental data given an equation of the form $y = k\,x^n$ to find n (e.g. time period of the oscillations of a mass-spring system)
- use appropriate formulae and experimental data from heating experiments to determine thermal properties (e.g. specific latent heat)
- carry out calculations using the ideal gas equation and the kinetic theory of gases equation.

Extension task

Resonance is discussed in this section and in Topic 4.5, Stationary and progressive waves. Research further into resonance using other books and the internet to produce a presentation to show your class. Things to talk about include

- another example of resonance in physics not already covered in this book
- how resonance is made use of in your example
- the general features of a resonant system

1   In a football match, a player kicks a stationary football of mass 0.44 kg and gives it a speed of 32 m s^{-1}.

  (a) (i)   Calculate the change of momentum of the football.
       (ii) The contact time between the football and the footballer's boot was
             9.2 ms. Calculate the average force of impact on the football.   (3 marks)
  (b) A video recording showed that the toe of the boot was moving on a circular arc of
       radius 0.62 m centred on the knee joint when the football was struck. The force of
       the impact slowed the boot down from a speed of 24 m s^{-1} to a speed of 15 m s^{-1}.
       (i)   Calculate the deceleration of the boot along the line of the impact force when it
             struck the football.
       (ii) Calculate the centripetal acceleration of the boot just before impact.
       (iii) Discuss briefly the radial force on the knee joint before impact and
             during the impact.   (4 marks)

       AQA, 2005

2   When the wheels of a car rotate at 6.5 revolutions per second the external rear view
    mirror vibrates violently. This is because the centre of mass of one of the wheels is not at
    the centre of the wheel. To correct this, a mass of 0.015 kg is attached to the rim of the
    wheel 0.25 m from the centre of the wheel as shown in **Figure 1**.

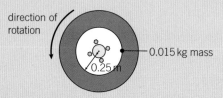

direction of rotation

0.015 kg mass

0.25 m

▲ Figure 1

  (a) (i)   Calculate the force exerted by the wheel on the 0.015 kg mass due to the
             rotation of the wheel.
       (ii) Copy **Figure 1** and draw an arrow to show the direction the 0.015 kg mass
             would move if it became detached when in the position shown in **Figure 1**
             whilst the wheel is rotating.   (4 marks)
  (b) Sketch a graph to show how the vertical component of the force acting on the
       0.015 kg mass due to the rotation of the wheel varies with time, t, during one
       complete rotation of the wheel. At $t = 0$ the 0.015 kg mass is in the position
       shown in **Figure 1**. Show upwards force as positive and downwards force as
       negative, and include a suitable time scale.   (2 marks)
  (c) Without the mass in place, the rotation of the wheel makes the external rear-view
       mirror of the car undergo forced vibrations.
       Explain what is meant by *forced vibrations* and state and explain how these vibrations
       will vary as the car increases in speed from rest.   (7 marks)

       AQA, 2004

3   The rotor of a model helicopter has four blades. Each of the blades is 0.55 m long with
    a uniform cross-sectional area of 3.5×10^{-4} m^2 and negligible mass. An end-cap of mass
    1.5 kg is attached to the outer end of each blade.
  (a) (i)   Demonstrate that there is a force of about 7 kN acting on each end-cap when the
             blades rotate at 15 revolutions per second.
       (ii) State the direction in which the force acts on the end-cap.
       (iii) Demonstrate that this force leads to a longitudinal stress in the blade of about
             20 MPa.
       (iv) Calculate the change in length of the blade as a result of its rotation.
             Young modulus of the blade material = 6.0×10^{10} Pa
       (v)   Calculate the total strain energy stored in **one** of the blades due to
             its extension.   (10 marks)

(b) The model helicopter can be made to hover above a point on the ground by directing the air from the rotors vertically downwards at speed v.

    (i)   Demonstrate that the change in momentum of the air each second is $A\rho v^2$, where A is the area swept out by the blades in one revolution and ρ is the density of air.

    (ii) The model helicopter has a weight of 900 N. Calculate the speed of the air downwards when the helicopter has no vertical motion.

         Density of air = $1.3 \, \text{kg m}^{-3}$

(5 marks)
AQA, 2003

4 When the pump of a central heating system reaches a certain speed after being switched on, a straight section of a pipe vibrates strongly.

  (a) Explain why the pipe vibrates strongly at a certain pump speed.

  (b) State and explain **one** way to reduce these vibrations of the pipe.

(5 marks)
AQA, 2006

5 The International Space Station (ISS) moves in a circular orbit around the Earth at a speed of $7.68 \, \text{km s}^{-1}$ and at a height of 380 km above the Earth's surface.

  (a) Calculate the centripetal acceleration of the ISS.

(3 marks)

  (b) Explain why a scientist working on board the ISS experiences 'apparent weightlessness'.

(2 marks)

      This state of apparent weightlessness makes the space station an ideal laboratory for experiments in 'zero gravity', conditions such as the study of lattice vibrations in solids.

  (c) **Figure 2** shows a mass–spring system which, in zero gravity, provides a good model of forces acting on an atom in a solid lattice.

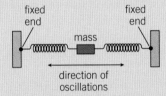

▲ **Figure 2**

When the mass is displaced and released, it oscillates as shown. The motion is very similar to the motion in one dimension of an atom in a crystalline solid. The springs behave like the bonds between adjacent atoms.

    (i)   The mass in the model system is 2.0 kg, and it oscillates with a period of 1.2 s. Demonstrate that the stiffness of the spring system is about $55 \, \text{N m}^{-1}$.

    (ii) The bonds between the atoms in a particular solid have the same stiffness as the model system, and the mass of the oscillating atom is $4.7 \times 10^{-26} \, \text{kg}$. Calculate the frequency of oscillation of the atom.

(4 marks)
AQA, 2005

6   Whilst investigating the oscillations of a helical spring, a student carried out
    measurements when various masses were suspended from the spring. For each
    mass, the length l of the spring was measured and 50 vertical oscillations were
    timed. The results are shown in Table 1.   (2 marks)

▼ Table 1

length l / mm	time for 50 oscillations / s	time period T / s	T^2 / s^2
316	12.5	0.25	0.063
333	17.5		
349	22.0		
364	25.5		
381	28.5		
397	31.0		

(a)   Copy and complete the table.
(b)   (i)   Assuming that the spring obeys Hooke's law, demonstrate that $T^2 = 4\pi^2 \dfrac{(l - l_0)}{g}$

          where l_0 is the length of the unloaded spring and T is the time period of vertical
          oscillations.
      (ii) Plot a graph of T^2 against l.
      (iii) Use the graph to determine values for g and l_0.   (9 marks)
(c)   Estimate the value of l that would give a time period of 1.00 s. State and explain
      one reason why the behaviour of the spring may cause your estimated value to be
      incorrect.   (4 marks)
                                                                               AQA, 2005

7   A beaker contains 1.3×10^{-4} m^3 of water at a temperature of 18 °C. The beaker is placed in
    a freezer. The water cools to 0 °C and freezes in a total time of 1700 s.
    (a)   (i)   Calculate the mass of water in the beaker.
                density of water = 1000 kg m^{-3}
          (ii) Calculate the average energy per second transferred from the water. Assume the
                beaker has negligible heat capacity.
                specific heat capacity of water = 4200 J kg^{-1} K^{-1}
                specific latent heat of fusion of ice = 3.3×10^5 J kg^{-1} (4 marks)
    (b)   The freezer uses electrical energy from the mains at a rate of 25 J s^{-1}.
          Calculate the total energy transferred to the surroundings from the time
          the beaker of water is placed in the freezer to when it freezes completely.   (2 marks)
                                                                               AQA, 2005

8   A long vertical tube contains several small particles of lead, which rest on its base. When
    the tube is inverted, the particles of lead fall freely through a vertical height equal to the
    length of the tube, which is 1.2 m.
    (a)   Describe the energy changes that take place in the lead particles during one
          inversion of the tube.   (3 marks)
    (b)   The tube is made from an insulating material and is used in a experiment to
          determine the specific heat capacity of lead. The following results are obtained.
          mass of lead: 0.025 kg   number of inversions: 50
          length of tube: 1.2 m   change in temperature of the lead: 4.5 K
          Calculate:
          (i)   the change in potential energy of the lead as it falls after one inversion down
                the tube,
          (ii) the total change in potential energy after 50 inversions,
          (iii) the specific heat capacity of the lead.   (4 marks)
                                                                               AQA, 2005

9 An electric shower heats the water flowing through it from $10\,^{\circ}$C to $42\,^{\circ}$C when the volume flow rate is $5.2 \times 10^{-5}\,\text{m}^3\,\text{s}^{-1}$.
   (a) (i) Calculate the mass of water flowing through the shower each second.
           density of water $= 1000\,\text{kg}\,\text{m}^{-3}$
       (ii) Calculate the power supplied to the shower, assuming all the electrical energy supplied to it is gained by the water as thermal energy.
           specific heat capacity of water $= 4200\,\text{J}\,\text{kg}^{-1}\,\text{K}^{-1}$. (*4 marks*)
   (b) A jet of water emerges horizontally at a speed of $2.5\,\text{m}\,\text{s}^{-1}$ from a hole in the shower head. The hole is $2.0\,\text{m}$ above the floor of the shower. Calculate the horizontal distance travelled by this jet. Assume air resistance is negligible. (*3 marks*)
       AQA, 2004

10 (a) The air in a room of volume $27.0\,\text{m}^3$ is at a temperature of $22\,^{\circ}$C and a pressure of $105\,\text{kPa}$.
       Calculate:
       (i) the temperature, in K, of the air,
       (ii) the number of moles of air in the room,
       (iii) the number of gas molecules in the room. (*5 marks*)
   (b) The temperature of an ideal gas in a sealed container falls. State, with a reason, what happens to the
       (i) mean square speed of the gas molecules,
       (ii) pressure of the gas. (*4 marks*)
       AQA, 2004

11 The table gives the average kinetic energy E_k of gas molecules at certain temperatures T.

$E_k/10^{-21}\,\text{J}$	6.21	6.62	7.04	7.45	7.87	8.28
T/K	300	320	340	360	380	400

   (a) (i) Plot a graph of E_k against T.
       (ii) Determine the gradient of your graph and hence calculate a value for the Boltzmann constant. Demonstrate all your working. (*8 marks*)
   (b) One of the assumptions of the kinetic theory is that collisions of gas molecules are elastic.
       (i) State what is meant by an elastic collision.
       (ii) State another assumption of the kinetic theory.
       (iii) Explain how the data in the table leads to the concept of absolute zero. (*4 marks*)
       AQA, 2002

12 A sample of air has a density of $1.24\,\text{kg}\,\text{m}^{-3}$ at a pressure of $1.01 \times 10^5\,\text{Pa}$ and a temperature of $300\,\text{K}$.
   (a) Calculate the mean kinetic energy of an air molecule under these conditions. (*2 marks*)
   (b) Calculate the mean square speed for the air molecules. (*3 marks*)
   (c) Explain why, when the temperature of the air is increased to $320\,\text{K}$, some of the molecules will have speeds much less than that suggested by the value you calculated in part (b). (*2 marks*)
       AQA, 2007

Section 7
Fields

Introduction

This section extends your A Level Physics Year 1 studies on mechanics, electricity, and energy. You will deepen your understanding of the gravitational force, the electrostatic force between charged objects, and the forces that magnetic fields exert on current-carrying conductors and moving charges. In all these situations, objects exert forces on each other without being in direct contact, so these forces 'act at a distance'. Action at a distance is described in terms of a field surrounding an object due to its mass or charge or due to an electric current passing through it and exerting a force on any similar object in the field. In your studies of exchange particles in Section 1 of A Level Year 1, you will have gained some appreciation of the actions of these forces. In this section, you will deepen your understanding of the differences and similarities between gravitational, electric, and magnetic fields, and the effects they have on objects. Both Newton's law of gravitation and its application to satellite motion and Coulomb's law of force between charged objects are considered in depth. The study of electric and magnetic fields applies knowledge and understanding to a range of applications and devices. These include the capacitor, which is commonly used in radio circuits and timing circuits, and alternating current generators and transformers, which are used to generate and supply electricity to our homes.

Working scientifically

In this part of the course, you will apply your understanding of important concepts in physics, such as force fields and field strength. You will use your knowledge from previous parts of the course in the context of vectors and electricity and apply this knowledge to important applications and devices in electrostatics and electromagnetism, such as capacitors and transformers. You will develop and broaden your practical skills by, for example, using oscilloscopes and signal generators in studying capacitor charging and discharging, and using data loggers in studying electromagnetic induction. You will also develop further skills in analysing and evaluating practical measurements.

The maths skills you will develop in this section include arithmetical, graphical, and algebraic skills. All of these skills involve inverse, inverse-square, and exponential functions, as well as logarithms. You will also use vectors and field lines to represent fields, and develop your awareness beyond dynamics by studying rates of change in capacitor discharge and electromagnetic induction. These maths skills will improve your confidence and prepare you for the final parts of the A Level Physics course, such as when you meet exponential change again.

What you already know

From your A Level Physics Year 1 studies on mechanics, you should know that:

- ☐ A vector quantity has magnitude and direction, whilst a scalar quantity has magnitude only.
- ☐ Displacement, velocity, acceleration, and force are vector quantities.
- ☐ A vector can be resolved parallel and perpendicular to a line at angle θ to the direction of the force into a parallel component $F\cos\theta$ and a perpendicular component $F\sin\theta$.
- ☐ The resultant force on an object = its mass × its acceleration, and its weight = its mass × g.
- ☐ The work done W on an object is given by $W = Fs\cos\theta$, where θ is the angle between the direction of the force and the displacement of the object.
- ☐ Power = rate of transfer of energy.
- ☐ The area under a graph of force against distance represents work done.

From your A Level Physics Year 1 studies on electricity, you should know that:

- ☐ Charge = current × time for a constant current.
- ☐ An electric current in a metallic conductor is a flow of free electrons due to a potential difference (pd) across the ends of the conductor.
- ☐ The pd between two points = work done per unit charge to move a charged object from one point to the other.
- ☐ Electrical power = current × pd.
- ☐ Resistance = pd/current.
- ☐ The symbols Q, I, t, V, P, and R, respectively, represent charge, current, time, pd, power, and resistance.
- ☐ For components in *series*, the current is the same in each component, and the sum of the pds across the components is equal to the total pd.
- ☐ For components in *parallel*, the pd is the same across each component, and the sum of the currents through the components is equal to the total current.
- ☐ The electromotive force (emf) of a source of pd is the electrical energy per unit charge produced by the source.

Learning objectives:

→ Illustrate a gravitational field.

→ Explain what is meant by the strength of a gravitational field.

→ Define a radial field and a uniform field.

Specification reference: 3.7.2.2

▲ **Figure 1** *Earth in space*

Study tip

Remember that *g* is a vector, even though it is called strength.

What goes up must come down – or must it? Throw a ball into the air and it returns to you because of the Earth's gravity. The force of gravity on the ball pulls it back to Earth. The force of attraction between the ball and the Earth is an example of gravitational attraction, which exists between any two masses. It isn't obvious that there is a force of attraction between you and any object near you, but it is true. Any two masses exert a gravitational pull on each other. But the force is usually too weak to be noticed unless at least one of the masses is very large.

The mass of an object creates a force field around itself. Any other mass placed in the field is attracted towards the object. The second mass also has a force field around itself and this pulls on the first object with an equal force in the opposite direction. The force field round a mass is called a **gravitational field**.

If a small test mass is placed close to a massive body, the small mass and the body attract each other with equal and opposite forces. However, this force is too small to move the massive body noticeably. The small mass, assuming it is free to move, is pulled by the force towards the massive body. The path which the smaller mass would follow is called a **field line** or sometimes a **line of force**. Figure 2 shows the field lines near a planet. The lines are directed to the centre of the planet because a small object released near the planet would fall towards its centre.

The strength of a gravitational field, *g*, is the force per unit mass on a small test mass placed in the field.

The test mass needs to be small, otherwise it might pull so much on the other object that it changes its position and alters the field. In general, the force on a small mass in a gravitational field varies from one position to another. If a small test mass, *m*, is at a particular position in a gravitational field where it is acted on by a gravitational force *F*, the gravitational field strength at that position is given by

$$g = \frac{F}{m}$$

The unit of gravitational field strength is the newton per kilogram (N kg^{-1}). For example, the gravitational field strength of the Earth at the surface of the Earth is 9.8 N kg^{-1}.

Free fall in a gravitational field

The weight of an object is the force of gravity on it. If an object of mass *m* is in a gravitational field, the gravitational force on the object is $F = mg$, where *g* is the gravitational field strength at the object's position. If the object is not acted on by any other force, it accelerates with

$$\text{acceleration } a = \frac{\text{force}}{\text{mass}} = \frac{mg}{m} = g$$

The object therefore falls freely with acceleration *g*. So, *g* may also be described as the acceleration of a freely falling object.

An object that falls freely is unsupported. Although the object in this situation is commonly described as being weightless, it is better to describe it as unsupported because it is acted on by the force of gravity alone.

The unit of gravitational field strength is $N\,kg^{-1}$, and the unit of the acceleration of free fall is $m\,s^{-2}$.

Field patterns

1   A **radial field** is where the field lines are like the spokes of a wheel, always directed to the centre. Figure 2 shows an example of a radial field. The force of gravity on a small mass near a much larger spherical mass is always directed to the centre of the larger mass. For example, the force on a small object near a spherical planet always acts towards the centre of the planet, regardless of the position of the object. The magnitude of g in a radial field decreases with increasing distance from the massive body.

2   A **uniform field** is where the gravitational field strength is the same in magnitude and direction throughout the field. The field lines are therefore parallel to one another and equally spaced.

Synoptic link

See Topic 7.4, Free fall, for more about the measurement of g.

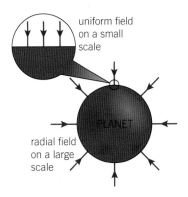

▲ **Figure 2**   *Field patterns*

   Is the Earth's gravitational field uniform or radial?

The force of gravity due to the Earth on a small mass decreases with distance from the Earth. So the gravitational field strength of the Earth falls with increasing distance from the Earth. The field is therefore radial. However, over small distances much less than the Earth's radius, the change of gravitational field strength is insignificant. For example, the measured value of g has the same magnitude $(= 9.8\,N\,kg^{-1})$ and direction (downwards) 100 m above the Earth as it is on the surface. In theory, g is smaller higher up, but the difference is too small to be noticeable – provided we don't go too high! Only over distances that are small compared with the Earth's radius can the Earth's field be considered uniform.

Summary questions

1   **a**   Explain what is meant by a field line or a line of force of a gravitational field.

   b   With the aid of a diagram in each case, explain what is meant by

   i   a radial field,   **ii**   a uniform field.

2   **a**   Calculate the gravitational force on:

   i   an object of mass 3.5 kg in a gravitational field at a position where $g = 9.5\,N\,kg^{-1}$,

   ii   an object of mass 100 kg in a gravitational field at a position where $g = 1.6\,N\,kg^{-1}$.

   b   Calculate the gravitational field strength at a position in a gravitational field where

   i   an object of mass 2.5 kg experiences a force of 40 N,

   ii   an object of mass 18 kg experiences a force of 72 N.

3   Demonstrate that the acceleration of an object falling freely in a gravitational field is equal to g, where g is the gravitational field strength at that position.

4   Figure 3 represents a small part of the Earth's surface. Sketch the lines of force near this part of the Earth's surface:

   a   if the density of the Earth in this part is uniform,

   b   if there is a large mass of dense matter under this part of the surface.

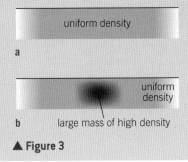

▲ **Figure 3**

Learning objectives:

→ Define gravitational potential.

→ Calculate the gravitational potential difference between two points.

→ Explain where an object would have to be placed for its gravitational potential energy to be zero.

Specification reference: 3.7.2.3

▲ **Figure 1** *Into space*

Synoptic link

Gravitational fields act on objects due to their mass. Gravitational potential is defined in terms of work done per unit mass. Electric fields act on charged objects. Electric potential is defined in terms of work done per unit positive charge. See Topic 22.3, Electric potential.

Note

In general, if a small object of mass m is moved from gravitational potential V_1 to gravitational potential V_2, its change of gpe $\Delta E_P = m(V_2 - V_1) = m\Delta V$, where $\Delta V = (V_2 - V_1)$.

Imagine you are in a space rocket, about to blast off from the surface of a planet. The planet's gravitational field extends far into space although it becomes weaker with increased distance from the planet. To escape from the planet's pull due to gravity, the rocket must do work against the force of gravity on it due to the planet. If the rocket fuel doesn't provide enough energy to escape, you are doomed to return! The planet's gravitational field is like a trap which the rocket must climb out from to escape. As the rocket rises, its gravitational potential energy (gpe) increases. If it falls back, its gpe decreases. If the rocket carried a gpe meter which was set to zero when it was far away from the planet, the meter reading would be negative when the rocket is on the surface of the planet and would increase towards zero as the rocket moves away from the planet.

Gravitational potential energy is the energy of an object due to its position in a gravitational field. The position for zero gpe is at infinity – in other words, the object would be so far away that the gravitational force on it is negligible. Your rocket climbing out of the planet's field needs to increase its gpe to zero to escape completely. At the surface, its gpe is negative, so it needs to do work to escape from the field completely.

The **gravitational potential** at a point in a gravitational field is the gpe per unit mass of a small test mass. This is equal to the work done per unit mass to move a small object from infinity to that point as the gravitational potential at infinity is zero. So, you can define gravitational potential at a point as *the work done per unit mass to move a small object from infinity to that point*.

The gravitational potential, V, at a point is the work done per unit mass to move a small object from infinity to that point.

Therefore, for a small object of mass m at a position where the gravitational potential is V, the work W that must be done to enable it to escape completely is given by $W = mV$. Rearranging this gives

$$V = \frac{W}{m}$$

The unit of gravitational potential is $J\,kg^{-1}$.

Suppose our rocket has a payload mass of 1000 kg and the gravitational potential at the surface of the planet is $-100\,MJ\,kg^{-1}$. Assume that the fuel is used quickly to boost the rocket to high speed. For the rocket to escape completely, the gpe of the 1000 kg payload must increase from $-100 \times 1000\,MJ$ to zero. So the work done on the payload must be at least 100 000 MJ to escape. If the rocket payload is only given 40 000 MJ of kinetic energy from the fuel, then it can only increase its gpe by 40 000 MJ. So it can only reach a position in the field where the gravitational potential is $-60\,MJ\,kg^{-1}$.

Because the work done ΔW to move it from V_1 to V_2 is equal to the change of its gravitational potential energy, then $\Delta W = m\Delta V$.

Potential gradients

Equipotentials are surfaces of constant potential. Because of this, no work needs to be done to move along an equipotential surface. Hillwalkers know all about equipotentials because the contour lines on a map are lines of constant potential. A contour line joins points of equal height above sea level, so a hillwalker following a contour line has constant potential energy. Sensible hillwalkers take great care where the contour lines are very close to each other. One slip and their gravitational potential energy might fall dramatically!

The equipotentials near the Earth are circles as shown in Figure 2. At increasing distance from the surface, the gravitational field becomes weaker. So the gain of gravitational potential energy per metre of height gain becomes less. In other words, away from the Earth's surface, the equipotentials for equal increases of potential are spaced further apart.

However, near the surface over a small region, the equipotentials are horizontal (i.e., parallel to the ground) as shown in Figure 3. This is because the gravitational field over a small region is uniform. A 1 kg mass raised from the surface by 1 m gains 9.8 J of gravitational potential energy. If it is raised another 1 m, it gains another 9.8 J. So its gpe rises by 9.8 J for every metre of height it gains above the surface.

The **potential gradient** at a point in a gravitational field is the change of potential per metre at that point.

Near the Earth's surface, the potential changes by $9.8\,\mathrm{J\,kg^{-1}}$ for every metre of height gained. So the potential gradient near the surface of the Earth is constant and equal to $9.8\,\mathrm{J\,kg^{-1}\,m^{-1}}$. However, further from the Earth's surface, the potential gradient becomes less and less.

In general, for a change of potential ΔV over a small distance Δr, the potential gradient $= \dfrac{\Delta V}{\Delta r}$.

Consider a test mass m being moved away from a planet as shown in Figure 3. To move m a small distance Δr in the opposite direction to the gravitational force F_{grav} on it, its gravitational potential energy must be increased

- by an equal and opposite force F acting through the distance Δr,
- by an amount of energy equal to the work done by F, that is, $\Delta W = F\Delta r$.

For the test mass, the change of potential ΔV of the test mass $= \dfrac{\Delta W}{m}$, then $\Delta V = \dfrac{F\Delta r}{m}$, so $F = \dfrac{m\Delta V}{\Delta r}$. So, $F_{\mathrm{grav}} = -\dfrac{m\Delta V}{\Delta r}$ as $F_{\mathrm{grav}} = -F$.

Gravitational field strength $g = \dfrac{F_{\mathrm{grav}}}{m} = -\dfrac{\Delta V}{\Delta r}$

Gravitational field strength g is the negative of the potential gradient.

$$g = -\frac{\Delta V}{\Delta r}$$

where the minus sign here shows you that g acts in the opposite direction to the potential gradient.

Study tip

$\Delta E_{\mathrm{P}} = mg\Delta h$ can only be applied for values of Δh that are very small compared with the Earth's radius. $\Delta E_{\mathrm{P}} = m\Delta V$ can *always* be applied. Remember that V is a scalar. You can use the equation $\Delta E_{\mathrm{P}} = mg\Delta h$ to estimate how much potential energy is gained by a mountaineer who climbs to the top of any mountain on Earth. Even the height of Mount Everest, which is the highest mountain on Earth and about 10 km high, is much less than the Earth's radius, which is about 6360 km.

Q: A 70 kg mountaineer climbs a vertical distance 10 km. Estimate how many 100 g chocolate bars she would need to eat to release the same amount of energy as her gain of potential energy. Assume the energy released per gram of chocolate is 30 kJ.

A: 2

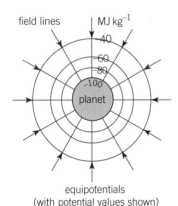

▲ **Figure 2** *Equipotentials near a planet*

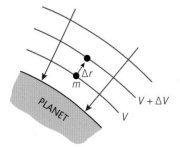

▲ **Figure 3** *Potential gradients*

Potential gradients are like contour gradients on a map. The closer the contours are on a map, the steeper the hill. Likewise, the closer the equipotentials are, the greater the potential gradient is and the stronger the field is. Where the equipotentials show equal changes of potential for equal changes of spacing, the potential gradient is constant. So the gravitational field strength is constant, and the field is uniform. Notice also that the gradient is always at right angles to the equipotentials, so the direction of the lines of force of the gravitational field are always perpendicular to the equipotentials.

Summary questions

$g = 9.8\,\text{N}\,\text{kg}^{-1}$

1 a Calculate the gain of gravitational potential energy of an object of mass 12 kg when its centre of mass is raised through a height of 2.0 m.

   b Demonstrate that the gravitational potential difference between the Earth's surface and a point 2.0 m above the surface is $19.6\,\text{J}\,\text{kg}^{-1}$.

2 A rocket of mass 35 kg launched from the Earth's surface gains 70 MJ of gravitational potential energy when it reaches its maximum height.

   a Calculate the gravitational potential difference between the Earth's surface and the highest point reached by the rocket.

   b The gravitational potential of the Earth's gravitational field at the surface of the Earth is $-63\,\text{MJ}\,\text{kg}^{-1}$. Calculate:

      i the gravitational potential at the highest point reached by the rocket,

      ii the work that would need to have been done by the rocket to escape from the Earth's gravitational field.

3 Figure 4 shows the equipotentials near a non-spherical object.

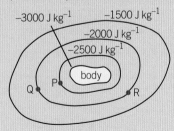

▲ Figure 4

a Calculate the gravitational potential energy of a 0.1 kg object at:

   i P,

   ii Q,

   iii R.

b Calculate how much work must be done on the object to move it from:

   i P to Q,

   ii Q to R.

4 Figure 5 shows equipotentials at a spacing of 1.0 km near a planet. The point labelled X is on the $-500\,\text{kJ}\,\text{kg}^{-1}$ equipotential.

   a Demonstrate that the potential gradient at X is $5.0\,\text{J}\,\text{kg}^{-1}\,\text{m}^{-1}$,

   b Hence calculate the gravitational field strength at X.

   c Calculate the work that would need to be done to remove an object of mass 50 kg from X to infinity.

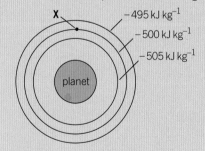

▲ Figure 5

We owe our understanding of gravitation to Isaac Newton. 'The notion of gravity was occasioned by the fall of an apple!', said Newton when asked what made him develop the idea of gravity. Newton's theory of gravitation was an enormous leap forward because it explains events from the down-to-earth falling apple to the motion of the planets. Like any good theory, it can be used to make predictions. For example, the return of a comet and its exact path can be calculated using Newton's theory of gravitation. So successful was his theory that it became known as Newton's law of gravitation.

How Newton established his theory of gravitation

Newton realised that gravity is universal. Any two masses exert a force of attraction on each other. He knew about the careful measurements of planetary motion made by astronomers like Johannes Kepler. Forty or more years before Newton established the theory of gravitation, Kepler had shown that the motion of the planets was governed by a set of laws. Kepler used his own and other astronomers' measurements of motion of each planet to show that each planet orbits the Sun. The measurements that he made for each planet were its time period T (i.e., the time for one complete orbit of the Sun) and the average radius r of its orbit. He used his measurements to show that the value of $\frac{r^3}{T^2}$ was the same for all the planets. This is called **Kepler's third law**.

Learning objectives:

→ Describe how gravitational attraction varies with distance.

→ Explain what is meant by an inverse-square law.

→ Discuss whether spherical objects, for example planets, can be treated as point masses.

Specification reference: 3.7.2.1

▼ **Table 1** *Kepler's third law*

	Mercury	Venus	Earth	Mars	Jupiter	Saturn
Average radius r of orbit / 10^{10} m	6	11	15	23	78	143
Time T for one orbit / 10^7 s	0.8	1.95	3.2	5.9	37.4	93.0
$\frac{r^3}{T^2}$ / 10^{16} m^3 s^{-2}	337	350	330	349	340	338

To explain Kepler's third law, Newton started by assuming that the planets and the Sun were point masses. A scale model of the Solar System with the Sun represented by a 1p piece would put the Earth about a metre away, represented by a grain of sand! Newton realised that there is a force of gravitational attraction between any planet and the Sun, and that it caused the planet to orbit the Sun. He assumed that the force of gravitation between a planet and the Sun varies inversely with the square of their distance apart. In other words, if the force is F at distance d apart, then

- at distance $2d$ apart, the force is $\frac{F}{4}$,

- at distance $3d$ apart, the force is $\frac{F}{9}$,

- at distance $4d$ apart, the force is $\frac{F}{16}$.

Notes

1 Work out for yourself the gravitational force between two point masses, each of mass 10 kg apart at 0.1 m apart. The values of m_1, m_2, and r must be put into the equation in units of kilograms and metres. The force of gravitational attraction works out at 6.7×10^{-7} N, which is far too small to notice except with extremely sensitive equipment. Only if one of the masses is very large does the force become noticeable, unless special techniques are used, as you will see later.

2 The equation for Newton's law of gravitation can be applied to any two objects provided r is the distance between their centres of mass.

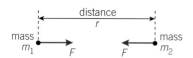

▲ **Figure 1** *Newton's law of gravitation*

Study tip

Think carefully before writing about the 'separation' of two objects. Do you mean distance between centres, or distance between surfaces?

Using this inverse-square law of force, Newton was able to prove that $\frac{r^3}{T^2}$ was the same for all of the planets. Newton then went on to use the inverse-square law of force to explain and make predictions for many other events involving gravity.

Newton's law of gravitation assumes that the gravitational force between any two **point** objects is

• always an attractive force,
• proportional to the mass of each object,
• proportional to $\frac{1}{r^2}$, where r is their distance apart.

These last two requirements can be summarised as

$$\text{gravitational force } F = \frac{Gm_1m_2}{r^2}$$

where m_1 and m_2 are masses of the two objects (see Figure 1).

The constant of proportionality, G, in the above equation, is called the **universal constant of gravitation**. The unit of G can be worked out from the equation above.

Rearranged, the equation gives $G = \frac{Fr^2}{m_1 \, m_2}$.

So G can be given units of $\text{N m}^2\text{kg}^{-2}$. The value of G is $6.67 \times 10^{-11}\,\text{N m}^2\text{kg}^{-2}$.

Worked example

$G = 6.67 \times 10^{-11}\,\text{N m}^2\text{kg}^{-2}$.

The distance from the centre of the Sun to the centre of the Earth is 1.5×10^{11} m. The mass of the Sun is 2.0×10^{30} kg, and the mass of the Earth is 6.0×10^{24} kg.

a The Earth has a diameter of 1.3×10^7 m. The Sun has a diameter of about 1.4×10^9 m. Explain why it is reasonable to consider the Sun and the Earth at a distance of 1.5×10^{11} m apart as point masses on this distance scale.

b Calculate the force of gravitational attraction between the Sun and the Earth.

Solution

a On a scale model where the centre of the Sun was 1 m away from the centre of the Earth, the Sun would be a sphere of diameter about 1 cm, and the Earth would be a sphere of diameter about 0.1 mm, no larger than a dot. The distance from the Earth to any part of the Sun is therefore the same to within 1%. Therefore this shows that they can be treated as point objects.

b $F = \dfrac{6.67 \times 10^{-11} \times 2.0 \times 10^{30} \times 6.0 \times 10^{24}}{(1.5 \times 10^{11})^2} = 3.6 \times 10^{22}\,\text{N}$

Cavendish's measurement of G

The first accurate measurement of G was made by Henry Cavendish in 1798. He devised a torsion balance made of two small lead balls at either end of a rod. The rod was suspended horizontally by a torsion wire, as in Figure 2. The wire was calibrated by measuring the couple required to twist it per degree. Then, with the rod at rest in equilibrium, two massive lead balls were brought near the torsion balance to make the wire twist. By measuring the angle it twisted through, the force of attraction between each massive lead ball and the small ball nearest to it was calculated. The distance between the centres of the small and large masses was also measured. Then G was calculated using the equation for the law of gravitation.

Q: Explain why the forces on the small balls make the torsion balance turn.

Worked example

Estimate the mass of a lead sphere of radius 5 cm. The density of lead is $11\,300\,\mathrm{kg\,m^{-3}}$.

Solution

Volume ≈ (diameter)^3 = $(2 \times 0.05\,\mathrm{m})^3 = 10^{-3}\,\mathrm{m^3}$

Mass = volume × density = $10^{-3}\,\mathrm{m^3} \times 11\,300\,\mathrm{kg\,m^{-3}} \approx 10\,\mathrm{kg}$.

▲ **Figure 2** *Cavendish's measurement of* G

Summary questions

$G = 6.67 \times 10^{-11}\,\mathrm{N\,m^2\,kg^{-2}}$

1. a Calculate the force of gravitational attraction between two 'point objects' of masses 60 kg and 80 kg at a distance of 0.5 m apart.

   b Calculate the distance between two identical point objects, each of mass 0.20 kg, which exert a force of $9.0 \times 10^{-8}\,\mathrm{N}$ on each other.

2. a Calculate the force of gravitational attraction between the Earth and an object of mass 80 kg on the surface of the Earth where $g = 9.8\,\mathrm{N\,kg^{-1}}$.

   b Use the result of your calculation in **a** to estimate the mass of the Earth. Assume that the mass of the Earth is concentrated at its centre.

   The radius of the Earth = $6.4 \times 10^6\,\mathrm{m}$.

3. The Sun exerted a force of 6.0 N on a 1000 kg comet when it was at a distance of $1.5 \times 10^{11}\,\mathrm{m}$ from the Sun. Calculate the force due to the Sun on the comet when it was at a distance of:

   a $0.5 \times 10^{11}\,\mathrm{m}$ from the Sun,

   b $7.5 \times 10^{11}\,\mathrm{m}$ from the Sun.

4. A space rocket of mass 1500 kg travelled from the Earth to the Moon, a distance of $3.8 \times 10^8\,\mathrm{m}$.

   a When the space rocket was mid-way between the Earth and the Moon, calculate the force of gravitational attraction on it:

      i due to the Earth,

      ii due to the Moon.

   b Calculate the magnitude and direction of the force of gravity of the Earth and the Moon on the space rocket when it was mid-way between the Earth and the Moon.

   The mass of the Earth = $6.0 \times 10^{24}\,\mathrm{kg}$. The mass of the Moon = $7.4 \times 10^{22}\,\mathrm{kg}$.

Synoptic link

Newton's law of gravitation and Coulomb's law of force between point charges are both inverse-square laws. For example, doubling the separation of two point masses or two point charges causes the force to reduce to a quarter. See Topic 22.4.

Learning objectives:

→ Describe the shape of a graph of g against r for points outside the surface of a planet.

→ Compare this graph with the graph of V against r.

→ Explain the significance of the gradient of the V–r graph.

Specification reference: 3.7.2.2;
3.7.2.4

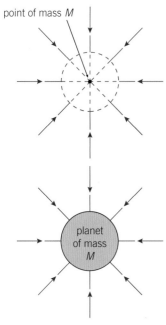

point of mass M

planet of mass M

▲ **Figure 1** *Comparing fields*

Gravitational field strength

The law of gravitation can be used to determine the **gravitational field strength** at any point in the field of a planet or any other spherical mass. Newton showed that the field of a spherical mass is the same as if the mass were concentrated at its centre. The field lines of a spherical mass are always directed towards the centre, so the field pattern is just the same as for a point mass, as shown in Figure 1.

- **For a point mass M**, the magnitude of the force of attraction on a test mass m (where $m << M$) at distance r from M is given by Newton's law of gravitation $F = \dfrac{GMm}{r^2}$.

  Therefore, the magnitude of the gravitational field strength at distance r is given by

$$g = \frac{F}{m} = \frac{GM}{r^2}$$

- **For a spherical mass M of radius R**, the force of attraction on a test mass m at distance r **from the centre of M** is the same as if mass M was concentrated at its centre. Therefore, the force of attraction between m and M is

$$F = \frac{GMm}{r^2}$$

  So, the magnitude of the gravitational field strength at distance r is given by $g = \dfrac{F}{m} = \dfrac{GM}{r^2}$, only if distance r is greater than or equal to the radius R of the sphere.

The **magnitude of the gravitational field strength is $g = \dfrac{GM}{r^2}$** at distance r from a point object or from the centre of a sphere of mass M.

Note:

The gravitational field strength at the surface of a sphere of radius R and mass M is given by $g_s = \dfrac{GM}{r^2}$ because the distance from the centre, r, is equal to R at the surface.

Worked example

$G = 6.67 \times 10^{-11}\,\mathrm{N\,m^2\,kg^{-2}}$

The gravitational field strength at the surface of the Earth is $9.8\,\mathrm{N\,kg^{-1}}$. Calculate:

a the mass of the Earth,

b the gravitational field strength of the Earth at a height of 1000 km above the surface.

The radius of the Earth = 6400 km.

Solution

a Rearranging $g_s = \dfrac{GM}{R^2}$ gives $M = \dfrac{g_s R^2}{G}$

$$= \frac{9.8 \times (6400 \times 10^3)^2}{6.67 \times 10^{-11}}$$

$$= 6.0 \times 10^{24}\,\text{kg}$$

b At height $h = 1000\,\text{km}$, $r = R + h = 7400\,\text{km}$

$$\therefore g = \frac{GM}{r^2} = \frac{6.67 \times 10^{-11} \times 6.0 \times 10^{24}}{(7400 \times 10^3)^2} = 7.3\,\text{N kg}^{-1}$$

The variation of *g* with distance from the centre of a spherical planet (or star)

Consider the gravitational field beyond the surface of a spherical planet of mass M and radius R. As you saw on the previous page, for any position at or beyond the surface, the magnitude of the gravitational field strength is given by $g = \dfrac{GM}{r^2}$, where r is the distance from the position to the centre of the sphere.

Because the surface gravitational field strength $g_s = \dfrac{GM}{R^2}$, then $GM = g_s R^2$

Therefore, $g = \dfrac{g_s R^2}{r^2}$

The equation shows how g changes with increase of distance r from the centre of the planet.

*   At distance $r = 2R$, $g = \dfrac{g_s R^2}{(2R)^2} = \dfrac{g_s}{4}$

*   At distance $r = 3R$, $g = \dfrac{g_s R^2}{(3R)^2} = \dfrac{g_s}{9}$

*   At distance $r = 4R$, $g = \dfrac{g_s R^2}{(4R)^2} = \dfrac{g_s}{16}$

Figure 2 shows how g varies with distance r. The shape of the curve beyond $r = R$ is an *inverse-square law* curve because g decreases in inverse proportion to r^2.

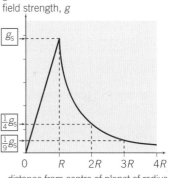

gravitational field strength, g

distance from centre of planet of radius R

▲ **Figure 2** *Gravitational field strength*

Inside a planet

From the equation $g = \dfrac{GM}{r^2}$, you might think that the magnitude of g inside the Earth becomes ever larger and larger as r becomes smaller and smaller. However, inside the planet, only the mass in the sphere of radius r contributes to g. The remainder of the mass outside r up to the surface gives no resultant force. So as r becomes smaller, the mass M that contributes to g becomes smaller too. At the centre, the mass that contributes to g is zero. So g is zero at the centre. Figure 2 also shows how g inside the planet varies with distance from the centre. For a spherical planet of density ρ, demonstrate that $g = \dfrac{4\pi G \rho r}{3}$ inside the planet.

Gravitational potential near a spherical planet

At or beyond the surface of a spherical planet, the gravitational potential V at distance r from the centre of the planet of mass M is given by

$$V = -\frac{GM}{r}$$

Applying this equation to the surface of the Earth with $M = 6.0 \times 10^{24}\,\text{kg}$ and $r = 6.4 \times 10^6\,\text{m}$ gives a value of $-63\,\text{MJ kg}^{-1}$. This means that $63\,\text{MJ}$ of work needs to be done to remove a $1\,\text{kg}$ mass from the surface of the Earth to infinity.

The **escape velocity** from a planet is the minimum velocity an object must be given to escape from the planet when projected vertically from the surface.

- At the surface of a planet of radius R, the potential $V = -\frac{GM}{R}$. At infinity, the potential is, by definition, zero. Therefore, to move an object of mass m from the surface to infinity, the work that must be done is $\Delta W = m\Delta V = \frac{GMm}{R}$.

- If the object is projected at speed v, then for it to be able to escape from the planet, its initial kinetic energy $\frac{1}{2}mv^2 \geq \Delta W$.

So, $\frac{1}{2}mv^2 \geq \Delta W$, which gives $\frac{1}{2}mv^2 \geq \frac{GMm}{R}$

Therefore, $v^2 \geq \frac{2GM}{R}$

So, the escape velocity, $v_{\text{esc}} = \sqrt{\frac{2GM}{R}}$

Because $g = \frac{GM}{R^2}$, substituting gR^2 for GM therefore gives

$$v_{\text{esc}} = \sqrt{2gR}$$

Figure 3 shows how the force of gravity on a $1\,\text{kg}$ mass (i.e., g) varies with distance from the surface of a planet. As you have just learnt, the equation for this curve is $g = \frac{g_s R^2}{r^2}$.

The area under the curve also represents the work done to move the $1\,\text{kg}$ mass from infinity to the surface. Therefore the area under the curve in Figure 3 gives the magnitude of the gravitational potential at the surface.

- Each grid square in Figure 3 represents a $1\,\text{N}$ force acting for a distance of $2.5 \times 10^6\,\text{m}$. So each grid square represents $2.5\,\text{MJ}$ ($= 1\,\text{N} \times 2.5 \times 10^6\,\text{m}$) of work done.

- Therefore, the work done to move the $1\,\text{kg}$ mass from infinity to the surface can be estimated by counting the number of grid squares under the curve and multiplying this number by $2.5\,\text{MJ}$.

By counting part-filled squares that are half-filled or more as wholly-filled squares, and neglecting part-filled squares that are less than half-filled, show for yourself that this method gives an estimate for the work done that is very close to the value of $63\,\text{MJ}$ determined above.

In effect, the area method used above is an application of *work done = force × distance moved*, with a variable force $F = \frac{GMm}{r^2}$ and $m = 1\,\text{kg}$.

Consider one small step in moving the $1\,\text{kg}$ mass from infinity to the surface. Suppose the distance from the centre of the Earth decreases from r_1 to r_2 in making this small step of distance Δr.

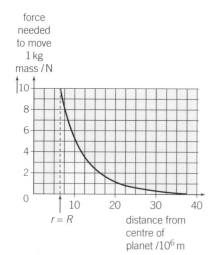

force needed to move $1\,\text{kg}$ mass / N

$r = R$ distance from centre of planet / $10^6\,\text{m}$

▲ **Figure 3** *Work done*

Hint

Figure 3 shows how g changes with distance from the centre of a planet. The area under any section of the curve gives the corresponding change of potential for this section. This is because the change of potential between any two points on the curve is equal to the work done to move a $1\,\text{kg}$ mass from one point to the other.

The work done ΔW to make this small step is given by

$$\Delta W = F\Delta r = \frac{GMm}{r^2}\Delta r$$

In Figure 3, the work done ΔW in making each small step $\Delta r = 2.5 \times 10^6$ m is represented by each column of grid squares under the curve. So the total work done to move from infinity to the surface is given by the total area under the curve.

Where does the equation for gravitational potential V = −GM/r come from?

You *don't* need to know the proof below. It is given here just to give you some further insight into the use of maths in physics.

The change of potential ΔV for the small step (from $r_1 = r$ to $r_2 = r - \Delta r$) $= V_2 - V_1$, where V_2 is the potential at r_2 and V_1 is the potential at r_1

$\Delta V = \frac{\Delta W}{m} = F\frac{\Delta r}{m} = -\frac{GM}{r^2}\Delta r$, where the minus sign indicates a decrease in r.

Because $\dfrac{1}{r_1} - \dfrac{1}{r_2} = \dfrac{r_2 - r_1}{r_1 r_2} = \dfrac{(r - \Delta r) - r}{r(r - \Delta r)} = \dfrac{-\Delta r}{r^2}$, only if $\Delta r << r$,

then $\Delta V = V_2 - V_1 = GM\left(\dfrac{1}{r_1} - \dfrac{1}{r_2}\right)$.

So, $V_2 = -\dfrac{GM}{r_2}$ and $V_1 = -\dfrac{GM}{r_1}$

Therefore, in general, the potential V at distance r from the centre of a planet is given by

$$V = -\frac{GM}{r}$$

Potential gradients near a spherical planet

The gravitational potential V near a spherical planet is inversely proportional to the distance r from the centre of the planet because $V = -\frac{GM}{r}$. Figure 4 shows how the gravitational potential of the Earth varies with distance. Note the potential curve is a $\frac{1}{r}$ curve, not an inverse-square (i.e., $\frac{1}{r^2}$) curve like the field strength curve in Figure 2. So the potential:

- at distance $2R$ from the centre is $0.50 \times$ the potential at distance R from the centre,

- at distance $3R$ from the centre is $0.33 \times$ the potential at distance R from the centre,

- at distance $4R$ from the centre is $0.25 \times$ the potential at distance R from the centre, and so on.

Synoptic link

The gravitational potential at distance r from the centre of a spherical mass is proportional to $\frac{1}{r}$; the electric potential at distance r from a point charge is also proportional to $\frac{1}{r}$. See Topic 22.3.

Note

The gradient of the potential curve at any point is equal to $-g$, where g is the gravitational field strength at the point. This is because $g = -$ the potential gradient, as you learnt in Topic 21.2. The potential curve becomes less steep as the distance from the centre decreases. At any point on the curve, the gradient of the curve can be found by drawing a tangent to the curve at that point and measuring the gradient of the tangent.

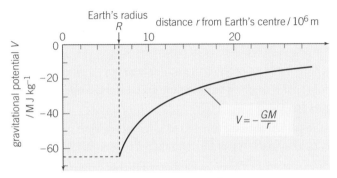

▲ **Figure 4** *Gravitational potential near the Earth*

▲ **Figure 5** *A rocket during takeoff*

Multistage rockets

The Russian physicist Konstantin Tsiolovski predicted in 1895 that space rockets would need to consist of parts in many stages, that is, be multistage. The reason is that rocket fuel releases no more than about $30\,MJ\,kg^{-1}$ when it burns, and this is not enough to enable a single-stage rocket to climb into space from the Earth's surface, as the surface potential is about $-63\,MJ\,kg^{-1}$. A multistage rocket jettisons the lowest stage when the fuel contained in that part has all been burnt, and only the top stage (the payload) escapes from the Earth's gravitational pull.

Study tip

Know how to draw $\left(\frac{1}{r^2}\right)$ and $-\left(\frac{1}{r}\right)$ graphs, and note how g changes much more sharply with distance than V does.

Summary questions

$G = 6.67 \times 10^{-11}\,N\,m^2\,kg^{-2}$

1   The Moon has a radius of 1740 km, and its surface gravitational field strength is $1.62\,N\,kg^{-1}$ to three significant figures.

   a   Calculate the mass of the Moon.

   b   The Moon's gravitational pull on the Earth causes the ocean tides. Demonstrate that the gravitational pull of the Moon on the Earth's oceans is approximately three millionths of the gravitational pull of the Earth on its oceans. Assume that the distance from the Earth to the Moon is 380 000 km.

2   The Sun has a mass of 2.0×10^{30} kg and a mean diameter of 1.4×10^9 m. Calculate:

   a   its gravitational field strength at

      i   its surface,

      ii   the Earth's orbit, which is at a distance of 1.5×10^{11} m from the Sun.

   b   The Earth has a mass of 6.0×10^{24} kg. Show that the gravitational field strength of the Earth is equal and opposite to the gravitational field strength of the Sun at a distance of 260 000 km from the centre of the Earth.

3   The tip of the tallest mountain on the Earth, Mount Everest, is 9 km above sea level. The mean radius of the Earth to the nearest kilometre is 6378 km.

   a   Calculate the difference between the gravitational field strength of the Earth at sea level and the top of Mount Everest.

   b   Discuss whether it is reasonable to assume that the Earth's gravitational field is uniform between the surface and a height of 10 km above the surface.

   c   Calculate the gain of potential energy of a mountaineer of mass 80 kg who travels to the top of the mountain from sea level.

4   Use the data in question **1** to calculate the gravitational potential at the surface of the Moon and hence calculate the work done to launch a 500 kg rocket from the surface so that it escapes from the Moon's gravitational field.

21.5 Satellite motion

On any clear night you should be able to see satellites passing overhead in the night sky. Although they are pinpoints of light, they are noticeable because they move steadily through the constellations. You can find information on the Internet to help you identify some of them from their directions. However, satellite motion is not confined to artificial satellites orbiting the Earth. Any small mass that orbits a larger mass is a satellite. The Moon is the Earth's only natural satellite. Mars has two moons, Phobos and Deimos. Jupiter has at least 14 satellites including the four satellites Io, Callisto, Ganymede, and Europa, first observed by Galileo four centuries ago.

Newton knew that the time period T of a planet orbiting the Sun depends on the mean radius r of the orbit in accordance with Kepler's third law, $T^2 \propto r^3$.

Newton realised that the force of gravitational attraction between each planet and the Sun is the centripetal force that keeps the planet on its orbit. By assuming that the gravitational force is given by $\frac{GMm}{r^2}$, the gravitational field strength $\frac{GM}{r^2}$ is therefore equal to the centripetal acceleration $\frac{v^2}{r}$, where M is the mass of the Sun, m is the mass of the planet, r is the radius of the orbit, and v is the speed of the planet.

Therefore, the speed of the planet is given by $v^2 = \frac{GM}{r}$.

Because speed $v = \dfrac{\text{circumference of the orbit}}{\text{time period}} = \dfrac{2\pi r}{T}$,

then $\dfrac{(2\pi r)^2}{T^2} = \dfrac{GM}{r}$.

Rearranging this equation gives $\dfrac{r^3}{T^2} = \dfrac{GM}{4\pi^2}$.

Because $\frac{GM}{4\pi^2}$ is the same for all of the planets, then $\frac{r^3}{T^2}$ is the same for all of the planets.

So, by assuming that the force of attraction F varies with distance according to the inverse-square law (i.e., $F \propto \frac{1}{r^2}$), Newton was able to prove Kepler's third law.

The mass of the Sun

Newton's theory not only explains Kepler's laws, but it also allows the mass M to be calculated if the value of G is known. The Earth orbits the Sun once per year on a circular orbit of radius 1.5×10^{11} m. So you can prove for yourself that the value of $\frac{r^3}{T^2}$ for any planet is 3.4×10^{18} m^3 s^{-2}.

Q: Given $G = 6.67 \times 10^{-11}$ N m^2 kg^{-2}, demonstrate that the mass of the Sun is 2.0×10^{30} kg.

Learning objectives:

→ State the condition needed for a satellite to be in a stable orbit.

→ Describe what happens to the speed of a satellite if it moves closer to the Earth.

→ Discuss why a geostationary satellite must be in an orbit above the Equator.

Specification reference: 3.7.2.4

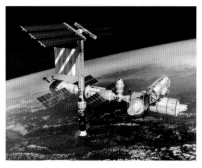

▲ **Figure 1** *Space station in orbit*

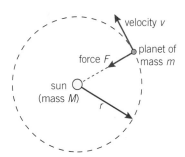

▲ **Figure 2** *Explanation of planetary motion*

Vehicle tracking

Vehicles on the road in the UK are tracked by a network of thousands of cameras that automatically read and record vehicle number plates. Millions of number plates are recorded on a central database every day. GPS satellites pinpoint the location of each camera as well as the time and date of each record.

The benefits of nationwide vehicle tracking include the following.

- it ensures that vehicle registration and MOT test certificates and motor insurance policies are up to date
- it eliminates the need for paper tax discs on windscreens
- it identifies stolen vehicles and vehicles suspected to have been used in criminal activities.

Concerns expressed by MPs and the public include the following.

- the movements of millions of people may be stored on the database for many years
- The data might be misused, so that innocent people suffer further direct surveillance.

The benefits are clear, but the concerns raise issues about the way we use science. The concerns need to be addressed to ensure people have confidence in the system of vehicle tracking.

Synoptic link

By assuming that the satellite is in uniform circular motion, you can equate $\dfrac{mv^2}{r}$ to the gravitational force on the satellite. See Topic 17.2, Centripetal acceleration.

Geostationary satellites

A geostationary satellite orbits the Earth directly above the equator and has a time period of exactly 24 h. It therefore remains in a fixed position above the equator because it has exactly the same time period as the Earth's rotation. The radius of orbit of a geostationary satellite can be calculated using the equation $\dfrac{r^3}{T^2} = \dfrac{GM}{4\pi^2}$.

So, $T = 24\,\text{h} = 24 \times 3600\,\text{s} = 86\,400\,\text{s}$

$r^3 = \dfrac{GMT^2}{4\pi^2} = \dfrac{6.67 \times 10^{-11} \times 6.0 \times 10^{24} \times (86\,400)^2}{4\pi^2} = 7.6 \times 10^{22}\,\text{m}^3$

$r = 4.2 \times 10^7\,\text{m} = 42\,000\,\text{km}.$

The radius of the Earth is 6400 km. Therefore, the height of a geostationary satellite above the Earth is 36 000 km (= 42 000 − 6400 km).

Note

Note that a geosynchronous orbit is a 24 hour orbit inclined to the equator.

SATNAV at work

Vehicle satellite navigation (SATNAV) units receive time signals from a system of global positioning satellites (GPS) that orbit the Earth at a height of just over 22 000 km. The signals are synchronised, and the receiver in each unit is programmed to measure the time differences between the signals and use the differences to pinpoint its

▲ **Figure 3** *SATNAV at work*

position. The software in the unit is designed to display the local road map showing the vehicle's position and route ahead to the destination supplied by the driver. But beware – SATNAVs have been known to send heavy goods vehicles down unsuitable narrow roads!

Q: Estimate how far a vehicle moving at $30\,\text{m s}^{-1}$ could travel in the time it takes for a radio signal to travel 44 000 km.

A: About 4 m.

The energy of an orbiting satellite

Consider a satellite of mass m in a circular orbit of radius r, about a spherical planet or star of mass M. The speed of the satellite v is given by $v^2 = \dfrac{GM}{r}$, as explained earlier in this topic.

So the kinetic energy E_K of the satellite $= \dfrac{1}{2}mv^2 = \dfrac{1}{2}m \times \dfrac{GM}{r} = \dfrac{GMm}{2r}$.

The satellite is at distance r from the centre of the planet. At this distance, the gravitational potential $V = -\dfrac{GM}{r}$. So the satellite's gravitational potential energy, $E_P = mV = -\dfrac{GMm}{r}$

Therefore the total energy of the satellite $E = E_P + E_K$
$$= -\frac{GMm}{r} + \frac{GMm}{2r} = -\frac{GMm}{2r}$$

For a satellite in a circular orbit of radius r, its total energy

$$E = -\frac{GMm}{2r}$$

Figure 4 shows how E, E_P and E_K vary with r. Notice that all three curves are $\dfrac{1}{r}$ curves and that the total energy is always negative.

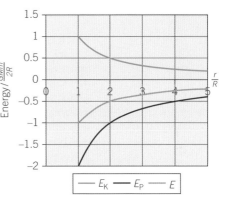

▲ **Figure 4** *Satellite energy*

Summary questions

$G = 6.67 \times 10^{-11}\,\text{N}\,\text{m}^2\,\text{kg}^{-2}$

The radius of the Earth $= 6400\,\text{km}$, and $g = 9.8\,\text{N}\,\text{kg}^{-1}$ at the Earth's surface.

1 **a** Two satellites X and Y are seen from the ground crossing the night sky at the same time. Satellite X crosses the sky faster than Y. State with a reason which satellite is higher.

   b Explain why satellite TV dishes must be aligned carefully so they always point to the same position above the equator.

2 A space probe moving at a speed of $3.2\,\text{km}\,\text{s}^{-1}$ is in a circular orbit about a spherical planet. The time period of the satellite is $110\,\text{min}$. Calculate:

   a the radius of the orbit,

   b the centripetal acceleration of the satellite,

   c the mass of the planet.

3 **a** A satellite moves at speed v in a circular orbit of radius r.

   i Write down an expression for the centripetal acceleration of the satellite.

   ii Demonstrate that the speed of the satellite is given by the equation $v^2 = gr$, where g is the gravitational field strength at the orbit.

   b A satellite orbits the Earth in a circular orbit at a height of $100\,\text{km}$. Calculate:

   i the gravitational field strength of the Earth at this distance,

   ii the speed of the satellite,

   iii the time period of the satellite.

4 **a** Demonstrate that the speed v of a satellite in a circular orbit of radius r about a planet of mass M is given by the equation $v^2 = \dfrac{GM}{r}$.

   b A weather satellite is in a polar orbit above the Earth at a height of $1600\,\text{km}$.

   i Demonstrate that its speed is $7.1\,\text{km}\,\text{s}^{-1}$.

   ii Calculate its time period.

   iii Explain why such a satellite can survey global weather patterns every day.

Black holes

A black hole is an astronomical body that is so massive that nothing can escape from it. Anything that falls into a black hole would never be seen again. The radius of a black hole is defined by its *event horizon*, an imaginary spherical surface surrounding the body. Nothing inside the event horizon can escape, not even light. Radiation can be detected from matter drawn into a black hole before it is trapped inside the event horizon. Astronomers reckon there might be a black hole at the centre of our own galaxy, the Milky Way.

▲ **Figure 1** *The Milky Way*

The idea of a black hole was first thought up by John Michell in 1783, although the term itself was first used much later by the American physicist John Wheeler. Michell's idea was not tested until after Einstein published his *General Theory of Relativity* in 1916 in which he predicted mathematically that a strong gravitational field distorts space and time and bends light. He calculated that light grazing the Sun from a star would be deflected by 0.0005° due to the Sun's gravity. The prediction was confirmed by Sir Arthur Eddington in 1919, who led a scientific expedition to South America to test the prediction by photographing stars close to the Sun during a total solar eclipse. The eclipse photographs revealed that the star images were displaced relative to each other, just as Einstein had predicted.

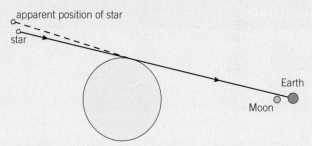

▲ **Figure 2** *Bending starlight*

The modern theory of black holes was started by Karl Schwarzschild, who used Einstein's theory to prove light could not escape from an object with a sufficiently strong gravitational field. He showed that such an object is surrounded by an event horizon, and that the radius, R, of the event horizon is given by $R = \dfrac{2GM}{c^2}$, where M is the mass of the black hole and c is the speed of light in free space.

Evidence for black holes has been obtained by astronomers. The central region of the M87 galaxy is rotating so fast that there must be a massive black hole at its centre. The X-ray source Cygnus X1 is a binary system consisting of a supergiant star accompanied by a very dense invisible star thought to be a black hole pulling matter from its companion.

$c = 3.0 \times 10^8 \, \mathrm{m\,s^{-1}}$, $G = 6.7 \times 10^{-11} \, \mathrm{N\,m^2\,kg^{-2}}$

1 (a) (i)   The mass of the Earth is 6.0×10^{24} kg. Show that the radius of the event horizon of a black hole that has the same mass as the Earth would be 8.9 mm.
       (ii) Calculate the density of an object with the same mass as the Earth and a radius of 9.0 mm. The mass of the Earth = 6.0×10^{24} kg.   (*5 marks*)
   (b) (i)   Use the scientific examples described above to explain what is meant by a scientific hypothesis.
       (ii) What key discovery was made by Eddington in 1919 and what was the significance of this discovery?   (*5 marks*)

(c) The Hubble Space Telescope has led to many new astronomical discoveries. Suppose your Government commits itself to establishing a manned space observatory on the Moon by 2025. Some people take the view this project could lead to many new discoveries. Other people take the view that the money for the project would be better spent helping to improve living conditions in poorer countries. Discuss **one** argument in support of each of these views and use your arguments to decide whether or not you would welcome such a project. *(5 marks)*

2 (a) (i) Explain what is meant by the *gravitational field strength* at a point in a gravitational field.
   (ii) State the SI unit of gravitational field strength. *(2 marks)*

(b) Planet P has mass M and radius R. Planet Q has a radius $3R$. The values of the gravitational field strengths at the surfaces of P and Q are the same.
   (i) Determine the mass of Q in terms of M.
   (ii) **Figure 3** shows how the gravitational field strength above the surface of planet P varies with distance from its centre.

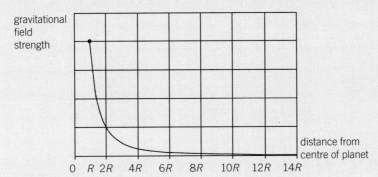

▲ Figure 3

Copy the diagram and draw the variation of the gravitational field strength above the surface of Q over the range shown. *(6 marks)*

AQA, 2006

3 (a) Artificial satellites are used to monitor weather conditions on Earth, for surveillance and for communications. Such satellites may be placed in a *geo-stationary* orbit or in a low polar orbit.
Describe the properties of the geo-stationary orbit and the advantages it offers when a satellite is used for communications. *(3 marks)*

(b) A satellite of mass m travels at angular speed ω in a circular orbit at a height h above the surface of a planet of mass M and radius R.
   (i) Using these symbols, give an equation that relates the gravitational force on the satellite to the centripetal force.
   (ii) Use your equation from part (b)(i) to show that the orbital period, T, of the satellite is given by
   $$T^2 = \frac{4\pi^2(R+h)^3}{GM}$$
   (iii) Explain why the period of a satellite in orbit around the Earth cannot be less than 85 minutes. Your answer should include a calculation to justify this value. *(6 marks)*

(c) Describe and explain what happens to the speed of a satellite when it moves to an orbit that is closer to the Earth. *(2 marks)*

AQA, 2006

4   (a)   (i)   Show that the gravitational field strength of the Earth at height
              h above the surface is given by

$$g = g_s \left(\frac{R}{R+h} \right)^2$$

              where g_s is the gravitational field strength at the surface and R is the
              radius of the Earth.

        (ii) Calculate the gravitational field strength of the Earth at a height
              of 200 km above its surface.   (*5 marks*)

    (b)   An astronaut floats in a spacecraft which is in a circular orbit around the Earth.
          Discuss whether or not the astronaut is weightless in this situation.   (*3 marks*)

                                                                          AQA, 2007

5   NASA wishes to recover a satellite, at present stranded on the Moon's
    surface, and to place it in orbit around the Moon.

    (a)   (i)   **Figure 4** shows a graph of how gravitational field strength due to
              the Moon varies with distance from the centre of the Moon. Mark
              on the figure the area that corresponds to the energy needed to move
              1 kg from the surface of the Moon to a vertical height of 4000 km
              above the surface.
              radius of the Moon = 1700 km

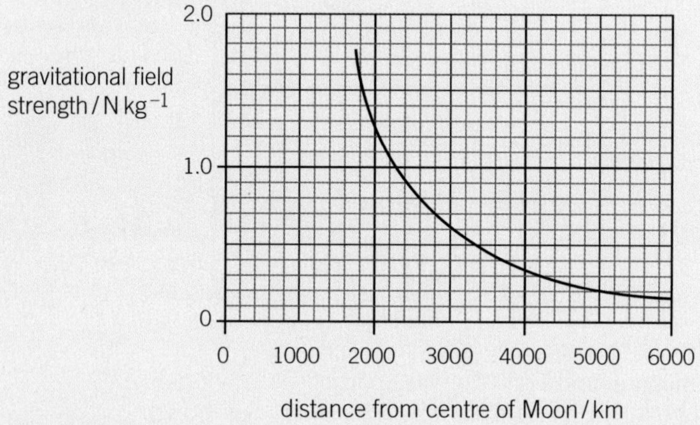

▲ Figure 4

        (ii) The satellite has a mass of 450 kg. Estimate the change in gravitational
              potential energy of the satellite when it is moved from the surface of
              the Moon to a vertical height of 4000 km above the surface.   (*6 marks*)

    (b)   NASA now decides to bring the satellite back to Earth. Explain why
          the amount of fuel required to return the satellite to Earth will be
          much less than the amount required to send it to the Moon originally.   (*5 marks*)

                                                                          AQA, 2004

6   (a)   Explain what is meant by the *gravitational potential* at a point in a
          gravitational field.   (*2 marks*)

    (b)   Use the following data to calculate the gravitational potential at
          the surface of the Moon.
          mass of Earth = 81 × mass of Moon
          radius of Earth = 3.7 × radius of Moon
          gravitational potential at surface of the Earth = −63 MJ kg⁻¹   (*3 marks*)

    (c)   Sketch a graph using axes as in Figure 5 to indicate how the gravitational
          potential due to the Moon varies with distance along a line outwards from
          the surface of the Earth to the surface of the Moon.   (*3 marks*)

                                                                          AQA, 2005

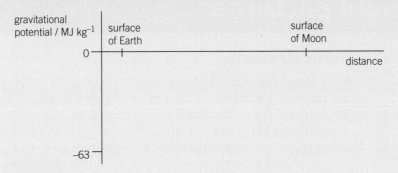

▲ **Figure 5**

7 (a) State the law that governs the magnitude of the force
        between two point masses.   (*2 marks*)
    (b) Table 1 shows how the gravitational potential varies for three points
         above the centre of the Sun.

▼ **Table 1**

distance from centre of Sun / 10^8 m	gravitational potential / 10^{10} J kg^{-1}
7.0 (surface of Sun)	−19
16	−8.3
35	−3.8

    (i)   Show that the data suggest that the potential is inversely proportional
          to the distance from the centre of the Sun.   (*2 marks*)
    (ii) Use the data to determine the gravitational field strength near
          the surface of the Sun.   (*3 marks*)
    (iii) Calculate the change in gravitational potential energy needed for
          the Earth to escape from the gravitational attraction of the Sun.
          mass of the Earth = 6.0×10^{24} kg
          distance of Earth from centre of Sun = 1.5×10^{11} m   (*3 marks*)
    (iv) Calculate the kinetic energy of the Earth due to its orbital speed
          around the Sun and hence find the minimum energy that would be
          needed for the Earth to escape from its orbit. Assume that the Earth
          moves in a circular orbit.   (*3 marks*)
                                                                          AQA, 2005

8   For an object, such as a space rocket, to escape from the gravitational attraction
    of the Earth it must be given an amount of energy equal to the gravitational
    potential energy that it has on the Earth's surface. The minimum initial vertical
    velocity at the surface of the Earth that it requires to achieve this is known as
    the escape velocity.
    (a) (i)   Write down the equation for the gravitational potential energy of a
               rocket when it is on the Earth's surface. Take the mass of the Earth to be
               M, that of the rocket to be m and the radius of the Earth to be R.
         (ii) Show that the escape velocity, v, of the rocket is given by the equation

$$v = \sqrt{\frac{2GM}{R}}$$ (*3 marks*)

    (b) The nominal escape velocity from the Earth is 11.2 km s^{-1}. Calculate a
         value for the escape velocity from a planet of mass four times that of the
         Earth and radius twice that of the Earth.   (*2 marks*)
    (c) Explain why the actual escape velocity from the Earth would be greater
         than the nominal value calculated from the equation given in part (a)(ii). (*2 marks*)
                                                                          AQA, 2004

Learning objectives:

→ Explain how to charge a metal object.

→ Describe what the direction of an electric field line shows concerning a test charge.

→ Illustrate the strength of an electric field by using field lines.

Specification reference: 3.7.3.2

Static electricity

Most plastic materials can be charged quite easily by rubbing with a dry cloth. When charged, they can usually pick up small bits of paper. The bits of paper are attracted to the charged piece of plastic. Do charged pieces of plastic material attract each other? Figure 1 shows an arrangement to test for attraction. A charged perspex ruler will attract a charged polythene comb, but two charged rods of the same material always repel one another.

Like charges repel; unlike charges attract.

Electrons are responsible for charging in most situations. An uncharged atom contains equal numbers of protons and electrons. Add one or more electrons to an uncharged atom and it becomes negatively charged. Remove one or more electrons from an uncharged atom and it becomes positively charged. An uncharged solid contains equal numbers of electrons and protons. To make it negatively charged, electrons must be added to it. To make it positively charged, electrons must be removed from it. When an uncharged perspex rod is rubbed with an uncharged dry cloth, electrons transfer from the rod to the cloth. So the rod becomes positively charged, and the cloth becomes negatively charged.

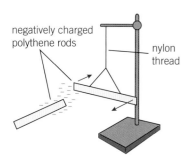

negatively charged polythene rods

nylon thread

▲ **Figure 1** *Electrostatic forces*

• **Electrical conductors** such as metals contain lots of **free electrons**. These are electrons which move about inside the metal and are not attached to any one atom. To charge a metal, it must be first isolated from the Earth. Otherwise, any charge given to it is neutralised by electrons transferring between the conductor and the Earth. Then the isolated conductor can be charged by direct contact with any charged object. If an isolated conductor is charged positively and then earthed, electrons transfer from the Earth to the conductor to neutralise or discharge it (see Figure 2).

• **Electrically insulating materials** do not contain free electrons. All the electrons in an insulator are attached to individual atoms. Some insulators, such as perspex or polythene, are easy to charge because their surface atoms easily gain or lose electrons.

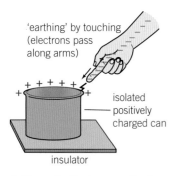

'earthing' by touching (electrons pass along arms)

isolated positively charged can

insulator

▲ **Figure 2** *Discharge to Earth*

The shuttling ball experiment shows that an electric current is a flow of charge. A conducting ball is suspended by an insulating thread between two vertical plates, as in Figure 3. When a high voltage is applied across the two plates, the ball bounces back and forth between the two plates. Each time it touches the negative plate, the ball gains some electrons and becomes negatively charged. It is then repelled by the negative plate and pulled across to the positive plate. When contact is made, electrons on the ball transfer to the positive plate so the ball now becomes positively charged and is repelled back to the negative plate to repeat the cycle. Therefore, the electrons from the high-voltage supply pass along the wire to the negative plate. There, they are ferried across to the other plate by the ball. Then they pass along the wire back to the supply. A microammeter in series with the plates shows that the shuttling ball causes a current round the circuit.

If the plates are brought closer together, the ball shuttles back and forth even more rapidly. As a result, the microammeter reading increases because charge is ferried across at a faster rate.

Suppose the ball shuttles back and forth at frequency f. The time taken for each cycle is therefore $\frac{1}{f}$. The amount and type of charge on the ball depends on the voltage of the plate it last made contact with. Therefore, if the charge ferried across the gap each cycle is Q, the average current round the circuit is

$$I = Qf \left(= \frac{\textbf{charge } Q}{\textbf{time for 1 cycle}} \right)$$

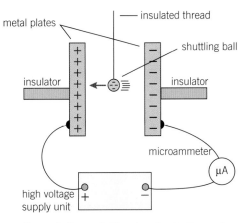

▲ **Figure 3** *The shuttling ball experiment*

Worked example

$e = 1.6 \times 10^{-19}$ C

In a shuttling ball experiment, the microammeter reading was 20 nA when the frequency of the shuttling ball was 4.0 Hz. Calculate:

a the charge carried by the ball,
b the number of electrons needed for the charge calculated in **a**.

Solution

a $\quad Q = \dfrac{I}{f} = \dfrac{20 \times 10^{-9}}{4.0} = 5.0 \times 10^{-9}$ C

b $\quad$ number of electrons $= \dfrac{Q}{e} = \dfrac{5.0 \times 10^{-9}}{1.6 \times 10^{-19}} = 3.1 \times 10^{10}$

The **gold leaf electroscope** is used to detect charge. If a charged object is in contact with the metal cap of the electroscope, some of the charge on the object transfers to the electroscope. As a result, the gold leaf and the metal stem which is attached to the cap gain the same type of charge and the leaf rises because it is repelled by the stem.

If another object with the same type of charge is brought near the electroscope, the leaf rises further because the object forces some charge on the cap to transfer to the leaf and stem.

▲ **Figure 4** *The charged gold leaf electroscope*

Chips and charge

People handling microchips wear antistatic clothing and they work in rooms fitted with antistatic floors. This is because a tiny amount of charge on the pins of an electronic chip can be enough to destroy circuits inside the chip. This can happen if the pins are touched in the presence of a charged body. The pins are earthed when they are touched. As a result, electrons transfer between the pins and Earth. If the connection to Earth is removed, the pins remain charged when the charged body is removed. Microchips are stored in antistatic packets and handled with special tools. Antistatic materials allow charge to flow across the surface.

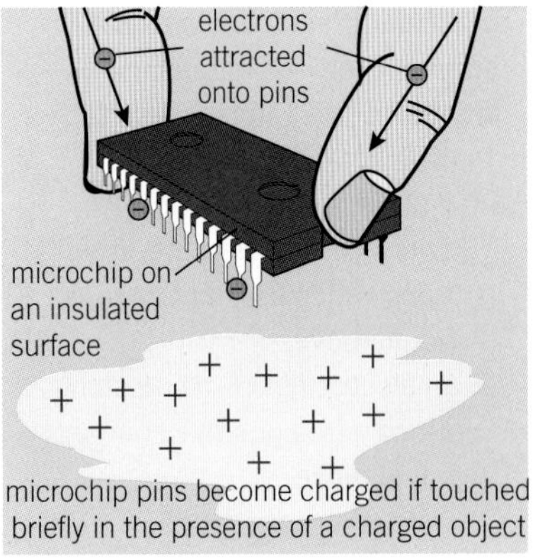

▲ **Figure 5** *Microchip damage*

Study tip

Note that an earthed conductor will become charged if a charged object is placed near it.

Synoptic link

The words *field* or *force field* are usually used where objects exert forces on each other without being in direct contact with each other.

- Objects exert gravitational forces of attraction on each other because they interact due to their mass. See Topic 21.1, Gravitational field strength.
- Charged objects exert an electric force on each other because they interact due to their charge. The force between two charged objects is attractive if the objects are oppositely charged. The force is repulsive if both objects have the same type of charge (both positive or both negative).
- Charged particles in motion exert magnetic forces on each other depending on their direction of motion and their relative speed. You will learn much more about magnetic fields in Chapter 24, Magnetic fields.

You can represent a field by lines of force and/or by vectors that represent the magnitude and direction of the force on a suitable small object in the field.

A line of force (or field line) of:
- a gravitational field is a line along which a free point mass would move,
- an electric field is a line along which a free positive charge would move.

You will see a more detailed comparison of electric fields and gravitational fields in Topic 22.6, Comparing electric fields and gravitational fields.

Field lines and patterns

Any two charged objects exert equal and opposite forces on each other without being directly in contact. An electric field is said to surround each charge. Suppose a small positive charge is placed as a test charge near a body with a much bigger charge which is also positive. If the test charge is free to move, it will follow a path away from the body with the bigger charge. The path a free positive test charge follows is called a **line of force** or a **field line**.

The field lines of an electric field are the lines which positive test charges follow. The direction of an electric field line is the direction a positive test charge would move along. Figure 6 shows the patterns of fields around different charged objects. Each pattern is produced by semolina grains sprinkled on oil. An electric field is set up across the surface of the oil by connecting two metal conductors in the oil to the output terminals of a high-voltage supply unit. The grains line up along the field lines, like plotting compasses in a magnetic field.

- Oppositely charged objects create a field as shown in Figure 6a. The field lines become concentrated at the points. A positive test charge released from an off-centre position would follow a curved path to the negative point charge.

- A point object near an oppositely charged flat plate produces a field as shown in Figure 6b. The field lines are concentrated at the point object, but they are at right angles to the plate where they meet. The field is strongest where the lines are most concentrated.

- Two oppositely charged plates create a field as shown in Figure 6c. The field lines run parallel from one plate to the other, meeting the plates at right angles, except near the edges of the field. The field is *uniform* between the plates because the field lines are parallel to each other.

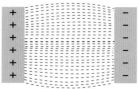

a *oppositely charged points*

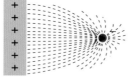

b *a point near a plate*

c *oppositely charged plates*

▲ **Figure 6** *Electric field patterns*

Summary questions

$e = 1.6 \times 10^{-19}\,\text{C}$

1 Explain each of the following observations in terms of transfer of electrons:

   a An insulated metal can is given a positive charge by touching it with a positively charged rod.

   b A negatively charged metal sphere suspended on a thread is discharged by connecting it to the ground using a wire.

2 a In the shuttling ball experiment, explain why the ball shuttles faster if:

      i the potential difference between the plates is increased,

      ii the plates are brought closer together.

   b A ball shuttles between two oppositely charged metal plates at a frequency of 2.5 Hz. The ball carries a charge of 30 nC each time it shuttles from one plate to the other. Calculate:

      i the average current in the circuit,

      ii the number of electrons transferred each time the ball makes contact with a metal plate.

3 An insulated metal conductor is earthed before a negatively charged object is brought near to it.

   a Explain why the free electrons in the conductor move as far away from the charged object as they can.

   b The conductor is then briefly earthed. The charged object is then removed from the vicinity of the conductor. Explain why the conductor is left with an overall positive charge.

4 a A positively charged point object is placed near an earthed metal plate, as shown in Figure 7.

      i Explain why electrons gather at the surface of the metal plate near the object.

      ii Explain why there is a force of attraction between the object and the metal plate.

   b Sketch the pattern of the field lines of the electric field:

      i between two oppositely charged parallel plates,

      ii between a positively charged point object and an earthed metal plate.

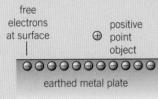

▲ Figure 7

Learning objectives:

→ Describe how to measure, in principle, the strength of an electric field.

→ Discuss whether electric field strength E is a scalar or a vector, and describe how this affects the sign of a test charge that you should use.

→ Explain why E should be described as the force per unit charge instead of the force that acts on one coulomb of charge.

Specification reference: 3.7.3.2

A charged object in an electric field experiences a force due to the field. Provided the object's size and charge are both sufficiently small, the object may be used as a 'test' charge to measure the strength of the field at any position in the field.

The electric field strength, E, at a point in the field is defined as the force per unit charge on a positive test charge placed at that point.

The unit of E is the newton per coulomb (N C^{-1}).

If a positive test charge Q at a certain point in an electric field is acted on by force F due to the electric field, the electric field strength E at that point is given by the equation

$$E = \frac{F}{Q}$$

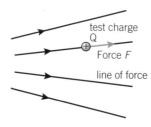

electric field strength, $E = \frac{F}{Q}$ (at Q)

▲ **Figure 1** *Electric field strength*

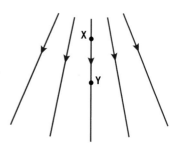

▲ **Figure 2** *A non-uniform field*

Hint

1 Rearranging this equation gives $F = QE$ for the force F on a test charge Q at a point in the electric field, where the electric field strength is E.

2 Electric field strength is a **vector** in the same direction as the force on a positive test charge. In other words, the direction of a field line at any point is the direction of the electric field strength at that point. The force on a small charge in an electric field is:

   • in the same direction as the electric field if the charge is positive,

   • in the opposite direction to the electric field if the charge is negative.

3 The charge of a test charge needs to be very much less than one coulomb because this amount of charge would affect the charges that cause the field, and so it would alter the electric field strength.

Worked example

$e = 1.6 \times 10^{-19}\,\text{C}$

Figure 2 shows the field lines of an electric field.

a   Calculate the magnitude of electric field strength at X if a $+3.5\,\mu\text{C}$ test charge at X experiences a force of $70\,\text{mN}$.

b   At a second position Y in the field, the electric field strength is $15\,000\,\text{N C}^{-1}$ in a direction downwards. Calculate the force on an electron at Y and state the direction of the force.

Solution

a $F = 70 \times 10^{-3}\,\text{N}$, $Q = 3.5 \times 10^{-6}\,\text{C}$,

   $E = \dfrac{F}{Q} = \dfrac{70 \times 10^{-3}}{3.5 \times 10^{-6}} = 2.0 \times 10^{4}\,\text{N C}^{-1}$

b $F = QE = 1.6 \times 10^{-19} \times 15\,000 = 2.4 \times 10^{-15}\,\text{N}$

   The direction of the force on an electron at Y is directly upwards because the field line at Y is directly downwards and the charge on an electron is negative.

The lightning conductor

Air is an insulator provided it is not subjected to an electric field that is too strong. Such a field ionises the air molecules by pulling electrons out of the molecules. In a thunderstorm, a lightning strike to the ground occurs when a cloud becomes more and more charged and the electric field in the air becomes stronger and stronger. The insulating property of air suddenly breaks down and a massive discharge of electric charge occurs between the cloud and the ground. A lightning conductor is a metal rod at the top of a tall building. The rod is connected to the ground by means of a thick metal conductor. When a charged cloud is overhead, it creates a very strong electric field near the tip of the conductor. Air molecules near the tip are ionised by this very strong field. The ions discharge the thundercloud making a lightning strike less likely.

> **Q:** A lightning strike transfers 30 000 C through a pd of 50 000 V. Estimate how much energy is released.

A: 1500 MJ

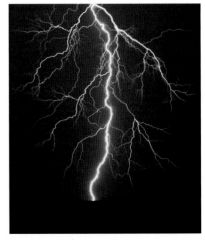

▲ **Figure 3** *Lightning*

The electric field between two parallel plates

Figure 6c in Topic 22.1 shows that the field lines between two oppositely charged flat conductors are parallel to each other and at right angles to the plates. The field pattern for two oppositely charged flat plates is similar, as shown in Figure 4. The field lines are:

- parallel to each other,
- at right angles to the plates,
- from the positive plate to the negative plate.

The field between the plates is uniform. This is because the electric field strength has the same magnitude and direction everywhere between the plates. The electric field strength E can be calculated from the potential difference V between the plates and their separation d using the equation

$$E = \frac{V}{d}$$

To prove this equation, consider a small charge Q between the plates, as in Figure 4.

1. The force F on a small charge Q in the field is given by $F = QE$, where E is the electric field strength between the plates.
2. If the charge is moved from the positive to the negative plate, the work done W by the field on Q is given by $W =$ force $F \times$ distance moved $d = QEd$
3. By definition, the potential difference between the plates, V, is the work done per unit charge when a small charge is moved through potential difference V.

Therefore, $V = \dfrac{W}{Q} = \dfrac{QEd}{Q} = Ed$

Rearranging $V = Ed$ gives $E = \dfrac{V}{d}$

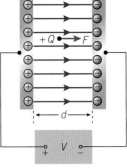

$$F = QE = \frac{QV}{d}$$

▲ **Figure 4** *The electric field strength between two parallel plates*

> ### Hint
>
> The unit of E can be written either as the newton per coulomb ($N\,C^{-1}$) or as the volt per metre ($V\,m^{-1}$).
>
> The link between the two can be seen by combining $F = QE$ and
>
> $E = \dfrac{V}{d}$ to give $F = \dfrac{QV}{d}$.
>
> Rearranging this equation gives
>
> $$\frac{F}{Q} = \frac{V}{d} \, (= E)$$
>
> Therefore, the newton per coulomb and the volt per metre are both acceptable as the unit of E.

Worked example

A pair of parallel plates at a separation of 80 mm are connected to a high-voltage supply unit which maintains a constant pd of 6000 V between the plates. Calculate:

a the electric field strength between the plates,

b the magnitude and direction of the electrostatic force on an ion of charge 4.8×10^{-19} C when it is between the plates.

Solution

a $E = \dfrac{V}{d} = \dfrac{6000}{80 \times 10^{-3}} = 7.5 \times 10^4 \, \text{V m}^{-1}$

b $F = QE = 4.8 \times 10^{-19} \times 7.5 \times 10^4 = 3.6 \times 10^{-14} \, \text{N}$

The force on the ion is directly towards the negative plate.

Synoptic link

E is the same everywhere in a uniform electric field, and g is the same everywhere in a uniform gravitational field. See Topic 21.2, Gravitational potential.

Field factors

An electric field exists near any charged body. The greater the charge on the body, the stronger the electric field is. For a charged metal conductor, the charge on it is spread across its surface. The more concentrated the charge is on the surface, the greater the strength of the electric field is above the surface.

- Figure 5 shows the electric field pattern between a V-shaped conductor opposite a flat plate when a constant pd is applied between the plate and the conductor. The field lines are concentrated at the tip of the V because that is where charge on the V-shaped conductor is most concentrated.

- The electric field between two oppositely charged parallel plates depends on the concentration of charge on the surface of the plates. The charge on each plate is spread evenly across the surface of the plate facing the other plate. Measurements show that the electric field strength between the plates is proportional to the charge per unit area on the facing surfaces.

Therefore, for charge Q on a plate of surface area A, the electric field strength E between the plates is proportional to $\dfrac{Q}{A}$.

Introducing a constant of proportionality, ε_0 (called epsilon nought), into this equation gives $\dfrac{Q}{A} = \varepsilon_0 E$. The value of ε_0 is 8.85×10^{-12} farads per metre (F m^{-1}), where the farad (the unit of capacitance – see Topic 23.1) is 1 coulomb per volt. It is called the permittivity of free space. It represents that charge per unit area on a surface in a vacuum that produces an electric field strength of one volt per metre between the plates. You *don't* need to know the equation $\dfrac{Q}{A} = \varepsilon_0 E$, but you *do* need to know about ε_0, which you will meet in Topic 22.4.

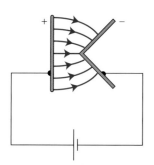

▲ **Figure 5** *The electric field near a metal tip*

Study tip

Field lines should have arrows (+ to –) and, when not straight, they are smooth curves.

Summary questions

$e = 1.6 \times 10^{-19}\,\text{C}$

1 A $+40\,\text{nC}$ point charge Q_1 is placed in an electric field.

   a Calculate the magnitude of the force on Q_1 if the electric field strength where Q_1 is placed is $3.5 \times 10^4\,\text{V}\,\text{m}^{-1}$.

   b Q_1 is moved to a different position in the electric field. The force on Q_1 at this position is $1.6 \times 10^{-3}\,\text{N}$. Calculate the magnitude of the electric field strength at this position.

2 Figure 6 shows the path of a charged dust particle in an electric field.

   a The electric field strength at X is $65\,\text{kV}\,\text{m}^{-1}$. The force due to the field on the particle when it is at X is $8.2 \times 10^{-3}\,\text{N}$ towards the metal surface.

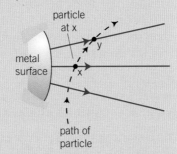

particle at x

metal surface

y

x

path of particle

   i Describe the type of charge carried by X.

   ii Calculate the charge carried by the particle.

▲ Figure 6

   b i Calculate the magnitude of the force on the particle when it is at Y where the electric field strength is $58\,\text{kV}\,\text{m}^{-1}$.

   ii State the direction of the force on the particle when it is at Y.

3 A high-voltage supply unit is connected across a pair of parallel plates which are at a separation of $50\,\text{mm}$.

   a The voltage is adjusted to $4.5\,\text{kV}$. Calculate:

   i the electric field strength between the plates,

   ii the electrostatic force on a droplet in the field that carries a charge of $8.0 \times 10^{-19}\,\text{C}$.

   b The separation between the plates is altered without changing the pd between the plates. The droplet in **a** is now acted on by a force of $4.5 \times 10^{-14}\,\text{N}$. Calculate the new separation between the plates.

4 A certain gas in a tube is subjected to an electric field of increasing strength. The gas becomes conducting when the electric field reaches a strength of $35\,\text{kV}\,\text{m}^{-1}$.

   a The electrodes in the tube are at a spacing of $84\,\text{mm}$. Assuming the field between the electrodes is uniform before the gas conducts, calculate the potential difference between the electrodes that is necessary to produce an electric field of strength $35\,\text{kV}\,\text{m}^{-1}$ in the tube.

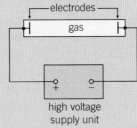

electrodes

gas

+   −

high voltage supply unit

▲ Figure 7

   b i Calculate the force on an electron in the tube when the electric field strength is $35\,\text{kV}\,\text{m}^{-1}$.

   ii Explain why the gas becomes conducting only when the electric field strength in the tube reaches a certain value.

Learning objectives:

→ Explain why potential is defined in terms of the work done per unit + charge.

→ Calculate the electric potential difference between two points.

→ Describe how to find the change in electric potential energy from pd.

→ Explain why potential (and pd) is measured in V.

Specification reference: 3.7.3.3

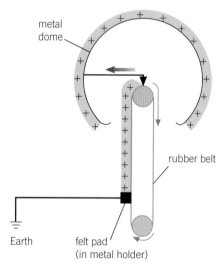

metal dome

rubber belt

Earth

felt pad (in metal holder)

▲ **Figure 1** *The Van de Graaff generator*

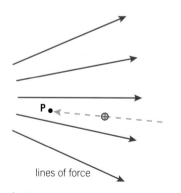

P •

lines of force

▲ **Figure 2**

The Van de Graaff generator

A Van de Graaff generator can easily produce sparks in air several centimetres in length. Figure 1 shows how a Van de Graaff generator works. Charge created when the rubber belt rubs against a pad is carried by the belt up to the metal dome of the generator. As charge gathers on the dome, the potential difference between the dome and Earth increases until sparking occurs.

A spark suddenly transfers energy from the dome. Work must be done to charge the dome because a force is needed to move the charge on the belt up to the dome. So the electric potential energy of the dome increases as it charges up. Some or all of this energy is transferred from the dome when a spark is created.

In general, work must be done to move a charged object X towards another object Y that has the same type of charge. Their electric potential energy increases as X moves towards Y.

The electric potential energy of X increases from zero if it is moved from infinity towards Y. The electric field of Y causes a force of repulsion to act on X. This force must be overcome to move X closer to Y.

The electric potential at a certain position in any electric field is defined as the work done per unit positive charge on a positive test charge (i.e., a small positively charged object) when it is moved from infinity to that position. By definition, the position of zero potential energy is infinity. Thus the electric potential at a certain position is the potential energy per unit positive charge of a positive test charge at that position.

The unit of electric potential is the volt (V), equal to $1\,\text{J}\,\text{C}^{-1}$.

For a positive test charge Q placed at a position in an electric field where its electric potential energy is E_P, the electric potential V at this position is given by

$$V = \frac{E_P}{Q}$$

Note that rearranging this equation gives $E_P = QV$.

Suppose a $+1\,\mu\text{C}$ test charge moves into an electric field from infinity to reach a certain position P, where the electric potential is $-1000\,\text{V}$. The electric potential energy of the test charge at P is therefore $-1.0 \times 10^{-3}\,\text{J}$ ($= QV = 1.0 \times 10^{-6}\,\text{C} \times -1000\,\text{V}$). In other words, $1.0 \times 10^{-3}\,\text{J}$ of work is done by the test charge when it moves from infinity to P.

Note:
If a test charge $+Q$ is moved in an electric field from one position where the electric potential is V_1 to another where the electric potential is V_2, then the work done ΔW on it is given by $\Delta W = Q(V_2 - V_1)$.

Potential gradients

Equipotentials are surfaces of constant potential. A test charge moving along an equipotential has constant potential energy. No work is done by the electric field on the test charge because the force due to the field is at right angles to the equipotential. In other words, the lines of force of the electric field cross the equipotential lines at right angles.

The equipotentials for an electric field are like equipotentials for a gravitational field. Both are lines of constant potential energy for the appropriate test object, in one case a test charge, and in the other case a test mass.

Figure 3 shows the equipotentials of the electric field due to two positively charged objects.

Suppose a $+2.0\,\mu C$ test charge is moved from X to Y.

The potential at X, V_X, is $+1000\,V$, so the test charge at X has potential energy equal to $+2.0 \times 10^{-3}\,J$ ($= QV_X = +2.0 \times 10^{-6}\,C \times +1000\,V$).

The potential at Y, V_Y, is $+400\,V$, so the test charge at Y has potential energy equal to $+8.0 \times 10^{-4}\,J$ ($= QV_Y = +2.0 \times 10^{-6}\,C \times +400\,V$).

Therefore, moving the test charge from X to Y lowers its potential energy by $1.2 \times 10^{-3}\,J$.

Note that if the test charge is moved from Y to Z along the $+400\,V$ equipotential, its potential energy stays constant at $+8.0 \times 10^{-4}\,J$.

The **potential gradient** at any position in an electric field is the change of potential per unit change of distance in a given direction.

1   If the field is non-uniform as in Figure 3, the potential gradient varies according to position and direction. The closer the equipotentials are, the greater the potential gradient is at right angles to the equipotentials.

2   If the field is uniform, such as the field between the two oppositely charged parallel plates shown in Figure 4, the equipotentials *between the plates* are equally spaced lines parallel to the plates. Figure 4 also shows how the potential relative to the negative plate changes with perpendicular distance x from the negative plate.

The graph shows that the potential relative to the negative plate is proportional to distance x. In other words, the potential gradient is:

*   constant,
*   such that the potential increases in the opposite direction to the electric field,
*   equal to $\dfrac{V}{d}$.

As you learnt in Topic 22.2, the electric field strength E between the plates is equal to $\dfrac{V}{d}$ and is directed from the $+$ to the $-$ plate. In other words,

the electric field strength is equal to the negative of the potential gradient.

Synoptic link

Equipotentials in an electric field are like map contours that are lines of constant height and are therefore also lines of gravitational equipotential. See Topic 21.2, Gravitational potential.

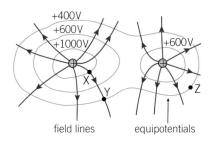

▲ **Figure 3** *Equipotentials*

Study tip

Remember that V is a scalar. Equipotentials always meet field lines at right angles.

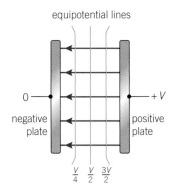

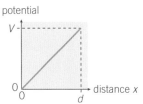

▲ **Figure 4** *A uniform potential gradient*

If the field is non-uniform, this statement still holds. However, the potential gradient at any position in an electric field is given by $\frac{\Delta V}{\Delta x}$ (instead of $\frac{V}{d}$), where ΔV is the change of potential between two closely spaced points at distance apart Δx.

Because the potential gradient is in the opposite direction to the lines of force of the electric field, then in general,

$$\textbf{electric field strength } E = -\frac{\Delta V}{\Delta x}$$

Summary questions

$e = 1.6 \times 10^{-19}$ C

1 An electron in a beam is accelerated from a potential of -50 V to a potential of $+450$ V. Calculate:

 a the potential energy of the electron at

  i -50 V,

  ii $+450$ V,

 b the change of potential energy of the electron.

2 In Figure 3, a test charge q is moved from X to Z. Calculate the change of potential energy of the test charge

 a if $q = +3.0\,\mu$C,

 b if $q = -2.0\,\mu$C.

3 An oil droplet carrying a charge of $+2e$ is in air between two parallel metal plates separated by a distance of 20 mm. The pd between the plates is 5.0 V.

 a Calculate:

  i the potential gradient between the two plates,

  ii the force on the droplet.

 b Calculate the change of electric potential energy of the oil droplet if it moves from the midpoint of the plates to the negative plate.

4 a Define electric potential and state its unit.

 b Two parallel horizontal metal plates are placed one above the other at a separation of 20 mm. A potential difference of $+60$ V is applied between the plates with the top plate positive.

  i Calculate the electric field strength between the plates.

  ii Sketch a graph to show how the electric potential V between the plates varies with height h above the lower plate.

22.4 Coulomb's law

Coulomb's experiment

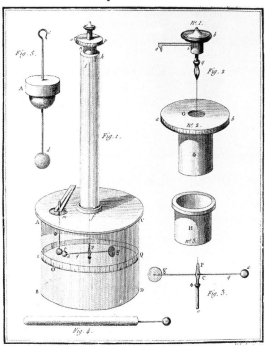

▲ **Figure 1** *Coulomb's torsion balance*

Learning objectives:
→ Describe how the force between two point charges depends on distance.

→ Calculate the force between two charged objects.

→ Explain what the sign of the force (+ or −) indicates.

Specification reference: 3.7.3.1

Like charges repel, and unlike charges attract. The force between two charged objects depends on how close they are to each other. The exact link was first established by Charles Coulomb in France in 1784. He devised a very sensitive torsion balance to measure the force between charged pith balls. Figure 1 shows the arrangement. A needle with a ball made of pith (a substance obtained from plants) at one end and a counterweight at the other end was suspended horizontally by a vertical wire. Another pith ball on the end of a thin vertical rod could be placed in contact with the first ball.

The pith balls were small enough to be considered as point objects. The ball on the rod was charged and then placed in contact with the other ball on the needle. The contact between them charged the second ball, which was then repelled by the ball fixed on the rod. This caused the wire to twist until the electrical repulsion was balanced by the twist built up in the wire. By turning the torsion head at the top of the wire, the distance between the two balls could be set at any required value. The amount of turning needed to achieve that distance gave the force. Some of Coulomb's many measurements are in Table 1.

The measurements fit the link that the force F is proportional to $\frac{1}{r^2}$. All of the other measurements made by Coulomb fitted the same link. Because the force is also proportional to the charge on each ball, Coulomb deduced the following equation, called **Coulomb's law**, for the force F between two point charges Q_1 and Q_2

$$F = \frac{kQ_1Q_2}{r^2}$$

where r is their distance apart.

Note

▼ **Table 1** *Some of Coulomb's results*

distance r	36	18	$8\frac{1}{2}$
force F	36	144	567

Measurements for both variables were actually made in degrees, so the above values are in relative units. Can you make out a pattern for these measurements? Halving the distance from 36 to 18 makes the force increase by a factor of 4.

Halving the distance from 18 to $8\frac{1}{2}$ (near enough 9) increases the force again by a factor of about 4.

a *unlike charges attract*

b *like charges repel*

$$F = \frac{1}{4\pi\varepsilon_0} \frac{Q_1 Q_2}{r^2}$$

▲ **Figure 2** *Coulomb's law*

Synoptic link

Coulomb's law of force between point charges and Newton's law of gravitation are both inverse-square laws. For example, doubling the separation of two point masses or two point charges causes the force to reduce to a quarter. See Topic 21.3, Newton's law of gravitation.

Synoptic link

You can use Newton's law of gravitation to compare the electrostatic force and the gravitational force between a proton (mass = 1.67×10^{-27} kg) and an electron (mass = 9.11×10^{-31} kg) at the same distance. If you need to, see Topic 21.4, Planetary fields. You should find that the electrostatic force is almost 2.3×10^{39} times stronger than the gravitational force.

The constant of proportionality, k, can be shown to be equal to $\frac{1}{4\pi\varepsilon_0}$, where ε_0 is the permittivity of free space. You first encountered this in Topic 22.2.

Coulomb's law is therefore written as

$$F = \frac{1}{4\pi\varepsilon_0} \frac{Q_1 Q_2}{r^2}$$

where r = distance between the two point charges Q_1 and Q_2.

As you saw in Topic 22.2, $\varepsilon_0 = 8.85 \times 10^{-12}$ F m^{-1}

so $\frac{1}{4\pi\varepsilon_0} = 9.0 \times 10^9$ m F^{-1}

Why do salt crystals dissolve in water?

A salt crystal is an ionic crystal. The sodium ions and the chlorine ions in a salt crystal are oppositely charged and the electrostatic forces between them hold them together. Salt crystals dissolve in water because the water weakens the electrostatic forces between the ions at the surface of the crystal. So the ions break free from the surface, and the crystal gradually dissolves. In fact, the force between two ions in water is about 80 times weaker than the force would be in air for the same distance apart. Note that Coulomb's law as stated above applies strictly to a vacuum. In terms of the force between two point charges in air, the force is effectively the same as it would be in a vacuum.

Worked example

$e = 1.60 \times 10^{-19}$ C, $\varepsilon_0 = 8.85 \times 10^{-12}$ F m^{-1}

Calculate the magnitude of the force between a proton and an electron at a separation of 3.00×10^{-10} m.

Solution

$$F = \frac{1}{4\pi\varepsilon_0} \frac{Q_1 Q_2}{r^2} = \frac{1.60 \times 10^{-19} \times 1.60 \times 10^{-19}}{4\pi \times 8.85 \times 10^{-12} \times (3.00 \times 10^{-10})^2} = 2.56 \times 10^{-9}$ N$$

More on $k = \frac{1}{4\pi\varepsilon_0}$

In Topic 22.2, ε_0 was introduced as the constant of proportionality in the equation $\frac{Q}{A} = \varepsilon_0 E$ linking the electric field strength E at the surface of a flat conductor where the charge Q is evenly distributed over a surface area A.

As you will see in Topic 22.5, if Coulomb's law is applied to the force on a test charge q at distance r from a point charge Q, the force on the test charge $F = \frac{kQq}{r^2}$, so the electric field strength at distance r is given by

$$E = \frac{F}{Q} = \frac{kQ}{r^2}$$

By introducing $\dfrac{1}{4\pi\varepsilon_0}$ as k, the equation $E = \dfrac{kQ}{r^2}$ may be written as

$$\frac{Q}{4\pi r^2} = \varepsilon_0 E \text{ or } \frac{Q}{A} = \varepsilon_0 E$$

where $A = 4\pi r^2 =$ the surface area of a sphere of radius r.

So, the general equation $\dfrac{Q}{A} = \varepsilon_0 E$ gives the surface charge density $\dfrac{Q}{A}$ needed on the surface of a conducting sphere in air to produce an electric field of strength E at the surface.

Q: Estimate the charge per unit area $\left(\dfrac{Q}{A}\right)$ on a surface at which the electric field strength is $1\,\text{V m}^{-1}$.

A: $8.85 \times 10^{-12}\,\text{C m}^{-2}$

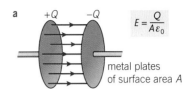

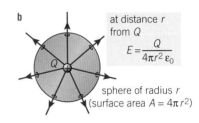

▲ **Figure 3** *Comparison of surface electric field strengths*

Summary questions

$\varepsilon_0 = 8.85 \times 10^{-12}\,\text{F m}^{-1}$,
$e = 1.6 \times 10^{-19}\,\text{C}$

1 Calculate the force between an electron and

   a a proton at a distance of $2.5 \times 10^{-9}\,\text{m}$,

   b a nucleus of a nitrogen atom (charge $+7e$) at a distance of $2.5 \times 10^{-9}\,\text{m}$

2 **a** Two point charges $Q_1 = +6.3\,\text{nC}$ and $Q_2 = -2.7\,\text{nC}$ exert a force of $3.2 \times 10^{-5}\,\text{N}$ on each other when they are at a certain distance, d, apart. Calculate:

     i the distance d between the two charges,

     ii the force between the two charges if they are moved to distance $3d$ apart.

   b A charge of $+4.0\,\text{nC}$ is added to each charge in part **a**. Calculate the force between Q_1 and Q_2 when they are at separation d.

3 A $+30\,\text{nC}$ point charge is at a fixed distance of $6.2\,\text{mm}$ from a point charge Q. The charges attract each other with a force of $4.3 \times 10^{-2}\,\text{N}$.

   a Calculate the magnitude of charge Q and state whether Q is a positive or a negative charge.

   b The two charges are moved $2.5\,\text{mm}$ further part. Calculate the force between them in this new position.

4 Two point objects, X and Y, carry equal and opposite amounts of charge at a fixed separation of $3.6 \times 10^{-2}\,\text{m}$. The two objects exert a force on each other of $5.1 \times 10^{-5}\,\text{N}$.

   a Calculate the magnitude, Q, of each charge, and state whether the charges attract or repel each other.

   b The charge of each object is increased by adding a positive charge of $+2Q$ to each object. Calculate the separation at which the two objects would exert a force of $5.1 \times 10^{-5}\,\text{N}$ on each other, and state whether the objects attract or repel each other.

Study tip

Remember to square r when calculating F.

When substituting charges, get the powers of 10 correct: $\mu = 10^{-6}$, $n = 10^{-9}$, $p = 10^{-12}$.

You can quickly check the solution given in the example opposite by approximating the calculation

$$\frac{1.60 \times 10^{-19} \times 1.60 \times 10^{-19}}{4\pi \times 8.85 \times 10^{-12} \times (3.0 \times 10^{-10})^2}$$

to $\dfrac{2 \times 10^{-38}}{100 \times 10^{-31}}$ to give $2 \times 10^{-9}\,\text{N}$.

This shows that there are no errors to the power of 10 in the calculation.

22.5 Point charges

Learning objectives:

→ State the equation that gives the electric field strength near a point charge.

→ State the equation that gives the potential associated with a point charge.

→ Explain why E is equal to zero inside a charged sphere.

Specification reference: 3.7.3.1; 3.7.3.3

A point charge is a convenient expression for a charged object in a situation where distances under consideration are much greater than the size of the object. The same idea applies to a distant star which is considered as a point object because its diameter is much smaller than the distance to it from the Earth. A test charge in an electric field is a point charge that does not alter the electric field in which it is placed. Such an alteration would happen if an object with a sufficiently large charge is placed in an electric field and it causes a change in the distribution of charge that creates the field.

Consider the electric field due to a point charge $+Q$, as shown in Figure 1. The field lines radiate from Q because a test charge $+q$ in the field would experience a force directly away from Q wherever the test charge was placed. Coulomb's law gives the force F on the test charge q at distance r from Q as

$$F = \frac{1}{4\pi\varepsilon_0} \frac{Qq}{r^2}$$

Therefore, because, by definition electric field strength $E = \frac{F}{q}$, the electric field strength at distance r from Q is given by

$$E = \frac{Q}{4\pi\varepsilon_0 r^2}$$

Note that if Q is negative, the above equation gives a negative value of E corresponding to the field lines pointing inwards towards Q.

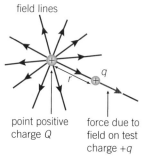

field lines

point positive charge Q

force due to field on test charge $+q$

▲ **Figure 1** *Force near a point charge Q*

Worked example

$\varepsilon_0 = 8.85 \times 10^{-12}\,\mathrm{F\,m^{-1}}$, $e = 1.6 \times 10^{-19}\,\mathrm{C}$

Calculate the electric field strength due to a nucleus of charge $+82e$ at a distance of $0.35\,\mathrm{nm}$.

Solution

$$E = \frac{Q}{4\pi\varepsilon_0 r^2} = \frac{+82 \times 1.6 \times 10^{-19}}{4\pi \times 8.85 \times 10^{-12} \times (0.35 \times 10^{-9})^2} = 9.6 \times 10^{11}\,\mathrm{V\,m^{-1}}$$

Electric field strength as a vector

If a test charge is in an electric field due to several point charges, each charge exerts a force on the test charge. The resultant force per unit charge $\frac{F}{q}$ on the test charge gives the resultant electric field strength at the position of the test charge. Consider the following situations:

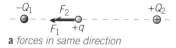

a *forces in same direction*

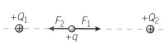

b *forces in opposite direction*

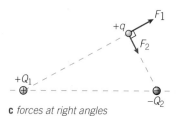

c *forces at right angles*

▲ **Figure 2** *Combined electric fields*

- **Forces in the same direction:** Figure 2a shows a test charge $+q$ on the line between a negative point charge Q_1 and a positive point charge Q_2. The test charge experiences a force $F_1 = qE_1$ where E_1 is the electric field strength due to Q_1, and a force $F_2 = qE_2$, where E_2 is the electric field strength due to Q_2. The two forces act in the same direction because Q_1 attracts q and Q_2 repels q. So the resultant force $F = F_1 + F_2 = qE_1 + qE_2$.

Therefore, the resultant electric field strength

$$E = \frac{F}{q} = \frac{qE_1 + qE_2}{q} = E_1 + E_2$$

- **Forces in opposite directions:** Figure 2b shows a test charge $+q$ on the line between two positive point charges Q_1 and Q_2. The forces on the test charge are the same in magnitude as in Figure 2a but opposite in direction because Q_1 repels q and Q_2 repels q. So the resultant force $F = F_1 - F_2 = qE_1 - qE_2$.

Therefore, the resultant electric field strength

$$E = \frac{F}{q} = \frac{qE_1 - qE_2}{q} = E_1 - E_2$$

- **Forces at right angles to each other:** Figure 2c shows a test charge $+q$ on perpendicular lines from two positive point charges Q_1 and Q_2. The forces on the test charge are smaller in magnitude than in Figures 2a and 2b (because the distances to Q_1 and Q_2 are larger) and are perpendicular to each other. The magnitude of the resultant force F is given by Pythagoras' equation $F^2 = F_1{}^2 + F_2{}^2$.

As the resultant electric field strength $E = \dfrac{F}{q}$, then $E^2 = E_1{}^2 + E_2{}^2$

can be used to calculate the resultant electric field strength.

In general, the resultant electric field strength is the vector sum of the individual electric field strengths.

More about radial fields

The electric field lines of force surrounding a point charge Q are radial. The equipotentials are therefore concentric circles centred on Q. For a charged sphere, you can say that the charge is at the centre of the sphere.

At distance r from Q, the electric field strength $E = \dfrac{Q}{4\pi\varepsilon_0 r^2}$

The equation was derived at the start of this topic. It shows that the electric field strength E is inversely proportional to the square of the distance r. Figure 3 shows how E varies with distance r from Q from a position that is at distance r_0 from Q. The curve is an *inverse-square law* curve because E is proportional to $\dfrac{1}{r^2}$.

Notice that the equation is of the same form as the gravitational field strength equation $g = \dfrac{GM}{r^2}$ for the gravitational field strength at distance r from the centre of a spherical planet. See Topic 21.2 if you need to. The field strength equations are both inverse-square relationships because both the force between two point charges (Coulomb's law) and the force between two point masses (Newton's law of gravitation) vary with distance according to the inverse-square law.

The electric potential $V = \dfrac{Q}{4\pi\varepsilon_0 r}$ at distance r from Q.

Because both Coulomb's law $F = \dfrac{1}{4\pi\varepsilon_0} \dfrac{Q_1 Q_2}{r^2}$ and Newton's law $F = \dfrac{Gm_1 m_2}{r^2}$ are inverse-square relationships, the forces vary with distance in the same way. Therefore, the equation for electric potential near a point charge $V = \dfrac{Q}{4\pi\varepsilon_0 r}$ is of the same form as the gravitational potential near a point mass (or spherical mass). That is, $V = -\dfrac{GM}{r}$, which you derived in Topic 21.4. The equation shows that the electric potential V is inversely proportional to the distance r.

Synoptic link

Remember how to calculate the resultant force of two perpendicular forces. See Topic 6.1, Vectors and scalars.

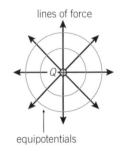

lines of force

equipotentials

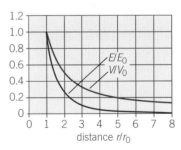

▲ **Figure 3** *The electric field and potential near a point charge*

Study tip

A negative E indicates a field that acts towards a negative charge. But a negative V indicates a value less than zero: E is a vector, whilst V is a scalar. Note that E varies with distance more sharply than V does.

Sparks and shocks

When a spherical metal conductor insulated from the ground is charged, the charge spreads out across the surface with the greatest concentration where the surface is most curved. This is why charge gathers at the tip of a lightning conductor when a charged cloud is overhead. As you saw in Topic 22.2, the electric field at the tip then becomes so strong that air molecules near the tip become ionised and the air conducts. This effect also explains why a fatal accident can happen if a conducting rod is held near an overhead high-voltage cable. An angler walking along a footpath near a railway line was electrocuted because the end of his carbon fibre fishing rod was inadvertently too close to the overhead cable along the track.

Figure 3 also shows how V varies with distance r from Q. The curve is *not* an inverse-square law curve, because V is proportional to $\frac{1}{r}$.

However, the gravitational potential in a gravitational field is always negative, because the force is always attractive. But the electric potential in the electric field near a point charge Q can be positive or negative according to whether Q is a positive or a negative charge.

The relationship between electric field strength and electric potential

1   Electric field strength = − the gradient of a potential against distance graph.

In Topic 22.3, you saw that at any position in an electric field, the electric field strength $E = -\frac{\Delta V}{\Delta x}$, where $\frac{\Delta V}{\Delta x}$ is the potential gradient at that position.

2   Change of potential = area under an electric field strength against distance graph.

Because electric field strength is the force per unit charge on a small positive test charge, a graph of electric field strength against distance shows how the force per unit charge on a positive test charge varies with distance. So the area under any section of the graph gives the work done per unit charge (i.e., change of potential) when a positive test charge is moved through the distance represented by that section.

Summary questions

$\varepsilon_0 = 8.85 \times 10^{-12}\,\text{F}\,\text{m}^{-1}$

1   a   Calculate the electric field strength at a distance of 3.2 mm from a +6.0 nC point charge.

   b   Calculate the distance from the point charge in **a** at which the electric field strength is $5.4 \times 10^5\,\text{V}\,\text{m}^{-1}$.

2   A +25 µC point charge Q_1 is at a distance of 60 mm from a +100 µC charge Q_2.

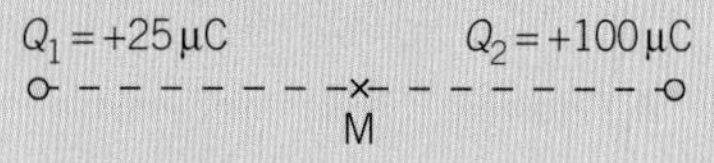

$Q_1 = +25\,\mu\text{C}$ $Q_2 = +100\,\mu\text{C}$
o – – – – – – – –×– – – – – – – –o
               M

▲ Figure 4

   a   A +15 pC charge q is placed at M, 25 mm from Q_1 and 35 mm from Q_2. Calculate:

      i   the resultant electric field strength at M,

      ii   the magnitude and direction of the force on q.

   b   Show that the electric field strength due to Q_1 and Q_2 is zero at the point which is 20 mm from Q_1 and 40 mm from Q_2.

3   A +15 µC point charge Q_1 is at a distance of 20 mm from a +10 µC charge Q_2.

   a   Calculate the resultant electric field strength:

      i   at M, the midpoint between the two charges,

      ii   at the point P along the line between Q_1 and Q_2 which is 25 mm from Q_1 and 45 mm from Q_2.

   b   i   Explain why there is a point along the line between the two charges at which the electric field strength is zero.

      ii   Calculate the distance from this point to Q_1 and to Q_2.

4   A +15 µC point charge Q_1 is at a distance of 30 mm from a −30 µC charge Q_2.

   a   Calculate electric potential at the midpoint of the two charges.

   b   i   Show that the electric potential is zero at a point between the two charges which is 10 mm from Q_1 and 20 mm from Q_2.

      ii   Calculate the electric field strength at this position and state its direction.

22.6 Comparing electric fields and gravitational fields

The similarities and differences between the two types of fields are listed in Table 1. In the mid-19th century, James Clerk Maxwell showed that electric and magnetic forces are different manifestations of the electromagnetic force. About two decades ago, physicists proved that the electromagnetic force and the nuclear force responsible for radioactive decay are different manifestations of a more fundamental force, the electroweak force. As yet, the gravitational force has still not been incorporated into this theoretical framework, despite repeated attempts by physicists to establish a unified theory. The fundamental nature of the gravitational force remains mysterious, even though we use it in everyday situations more than any other force.

Table 1 summarises the conceptual links between electric and gravitational fields.

Learning objectives:

→ State which electrical quantity is analogous to mass.

→ State the main similarities between electric and gravitational fields.

→ State the principal differences between electric and gravitational fields.

Specification reference: 3.7.1

▼ **Table 1** *Similarities and differences between gravitational and electric fields*

	Gravitational fields	Electrostatic fields
Similarities		
line of force or a field line	path of a free test mass in the field	path of a free positive test charge in the field
inverse-square law of force	Newton's law of gravitation $F = \dfrac{Gm_1 m_2}{r^2}$	Coulomb's law of force $F = \dfrac{Q_1 Q_2}{4\pi\varepsilon_0 r^2}$
field strength	force per unit mass, $g = \dfrac{F}{m}$	force per unit + charge, $E = \dfrac{F}{q}$
unit of field strength	$N\,kg^{-1}$ or $m\,s^{-2}$	$N\,C^{-1}$ or $V\,m^{-1}$
uniform fields	g is the same everywhere, field lines parallel and equally spaced	E is the same everywhere, field lines are parallel and equally spaced
potential	gravitational potential energy per unit mass	electric potential energy per unit + charge
unit of potential	$J\,kg^{-1}$	$V\ (= J\,C^{-1})$
potential energy of two point mass or charges	$E_P = \dfrac{-Gm_1 m_2}{r}$	$E_P = \dfrac{Q_1 Q_2}{4\pi\varepsilon_0 r}$
radial fields	due to a point mass or a uniform spherical mass M, $g = \dfrac{GM}{r^2}$ $V = \dfrac{-GM}{r}$	due to a point charge Q, $E = \dfrac{Q}{4\pi\varepsilon_0 r^2}$ $V = \dfrac{Q}{4\pi\varepsilon_0 r}$
Differences		
action at a distance	between any two masses	between any two charged objects
force	attracts only	unlike charges attract; like charges repel
constant of proportionality in force law	G	$\dfrac{1}{4\pi\varepsilon_0}$

1   **(a)**   (i)   Define the electric field strength, E, at a point in an electric field.
        (ii)   State whether E is a scalar or a vector quantity.   *(3 marks)*
   (b)   Point charges of +4.0 nC and −8.0 nC are placed 80 mm apart, as shown in **Figure 1**.

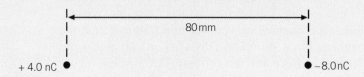

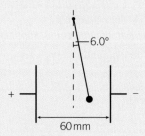

▲ **Figure 1**

      (i)   Calculate the magnitude of the force exerted on the +4.0 nC charge by the −8.0 nC charge.
     (ii)   Determine the distance from the +4.0 nC charge to the point, along the straight line between the charges, where the electric potential is zero.   *(4 marks)*
   (c)   Point P in the diagram is equidistant from the two charges.
      (i)   On a copy of the diagram, draw two arrows at P to represent the directions and relative magnitudes of the components of the electric field at P due to each of the charges.
     (ii)   Hence draw an arrow, labelled R, on your diagram at P to represent the direction of the resultant electric field at P.   *(3 marks)*
                                                        AQA, 2006

2   A small charged sphere of mass 2.1×10^{-4} kg, suspended from a thread of insulating material, was placed between two vertical parallel plates 60 mm apart. When a potential difference of 4200 V was applied to the plates, the sphere moved until the thread made an angle of 6.0° to the vertical, as shown in **Figure 2**.

▲ **Figure 2**

   (a)   Show that the electrostatic force F on the sphere is given by $F = mg \tan 6.0°$ where m is the mass of the sphere.   *(3 marks)*
   (b)   Calculate:
      (i)   the electric field strength between the plates,
     (ii)   the charge on the sphere.   *(3 marks)*
                                                           AQA, 2003

3 **Figure 3** shows some of the equipotential lines that are associated with a point negative charge Q.

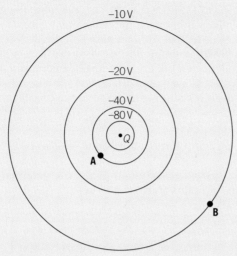

▲ **Figure 3**

(a) **(i)** Explain why the potentials have a negative sign.
  (ii) Draw on a copy of the diagram three electric field lines. Use arrows to show the direction of the field. *(4 marks)*
(b) **(i)** Use data from the diagram, which is full size, to show that the charge Q is about -4.5×10^{-11} C.
  (ii) Calculate the electric field strength at B. *(4 marks)*
(c) **(i)** Calculate the energy, in J, transferred when an electron moves from A to B in the field.
  (ii) State and explain
  • why the kinetic energy of the electron increases as it moves from A to B,
  • how the de Broglie wavelength of the electron changes as it moves from A to B.
  (6 marks)
  AQA, 2003

4 **Figure 4** shows the parallel deflecting plates with some dimensions of the ink-jet cartridge. In order to land in the centre of the gutter the ink droplet must leave the plates at an angle of 35°. On entering the electric field the ink droplet carries a charge of -2×10^{-10} C and travels with a horizontal velocity of $20\,\text{m s}^{-1}$.

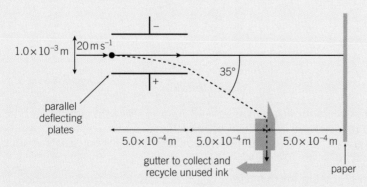

▲ **Figure 4**

(a) **(i)** Draw a vector diagram to show the components and the resultant of the velocity of the charged ink droplet as it leaves the deflecting field. Determine the size of the vertical component.

(ii) Find the time for which the ink droplet is between the deflecting plates and hence calculate its vertical acceleration during this time.

(iii) For an ink droplet of mass 2.9×10^{-10} kg, calculate the electric force acting on the ink droplet whilst it is between the deflecting plates.

(iv) Calculate the electric field strength between the deflecting plates.

(v) Calculate the potential difference between the deflecting plates. *(12 marks)*

(b) The uncharged, undeflected ink droplets travel beyond the deflecting plates towards the paper. With the aid of a suitable calculation, discuss whether or not the printer manufacturer needs to take into consideration the droplet falling under gravity.

(4 marks)

AQA, 2003

5 Dry air ceases to be an insulator if it is subjected to an electric field strength of 3.3 kV mm^{-1} or more.

(a) (i) Show that the electric field strength E and the potential V at the surface of a charged sphere of radius R are related by

$$E = \frac{V}{R}$$

(ii) The dome of a Van de Graaff generator has a radius of 0.20 m. Calculate the maximum potential of this dome in dry air. *(5 marks)*

(b) Two high-voltage conductors are joined together using a small sphere, as shown in **Figure 5**.

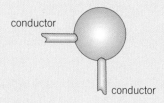

conductor

conductor

▲ **Figure 5**

The conductors are used to transmit alternating current at an rms potential of 100 kV.

(i) Calculate the peak potential of the conductor,

(ii) Calculate the minimum diameter of the sphere necessary to ensure the surrounding air does not conduct. *(3 marks)*

AQA, 2004

6 (a) **Figure 6** shows a small charged metal sphere carrying a charge Q. The potential of the sphere is 8000 V. Copy Figure 6.

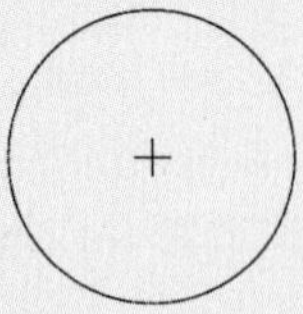

+

▲ **Figure 6**

(i) Draw on your diagram at least six lines to show the direction of the electric field in the region around the charged sphere.

(ii) Draw on your diagram the equipotential lines for potentials of 4000 V and 2000 V.

(iii) The equation for the field strength at a distance r from the sphere is $\dfrac{Q}{4\pi\varepsilon_0 r^2}$. State the name of the quantity represented by ε_0.

(5 marks)

(b) **Figure 7** shows an arrangement for determining the charge on a small sphere.

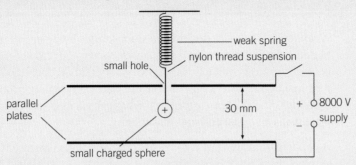

▲ **Figure 7**

The sphere is suspended from a spring of spring constant $0.18\,\text{N\,m}^{-1}$. It hangs between two parallel plates which can be connected to a high-voltage supply.
  (i)   Explain why nylon thread is used for the suspension.
  (ii) Calculate the extension of the spring when a sphere of mass $1.5\,\text{g}$ is suspended from it.
  (iii) Calculate the magnitude of the electric field strength between the plates when the $8000\,\text{V}$ supply is switched on.
  (iv) When the $8000\,\text{V}$ supply is switched on, the sphere moves down a further $4.5\,\text{mm}$. Calculate the charge on the sphere.   *(8 marks)*
(c) One problem with this arrangement is the oscillations of the sphere that occur when the switch is closed.
  (i)   Show that the period of the oscillations produced is about $0.6\,\text{s}$.
  (ii) In practice the oscillations are damped. Sketch a graph showing how the amplitude of the oscillations changes with time for the damped oscillation.
                                                                                      (5 marks)
                                                                                      AQA, 2007

7   The Earth has an electric charge. The electric field strength outside the Earth varies in the same way as if this charge were concentrated at the centre of the Earth. The axes in **Figure 8** represent the electric field strength, E, and the distance from the centre of the Earth, r. The electric field strength at **A** has been plotted.

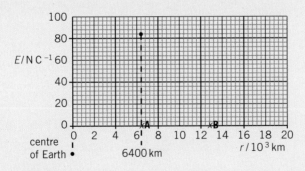

▲ **Figure 8**

(a) (i)   Determine the electric field strength at **B** and then complete the graph to show how the electric field strength varies with distance from the centre of the Earth for distances greater than $6400\,\text{km}$.
     (ii) State how you would use the graph to find the electric potential difference between the points **A** and **B**.   *(4 marks)*
(b) (i)   Calculate the total charge on the Earth.
     (ii) The charge is distributed uniformly over the Earth's surface. Calculate the charge per square metre on the Earth's surface.   *(4 marks)*
                                                                                      AQA, 2002

Learning objectives:

→ Describe in terms of electron flow what is happening when a capacitor charges up.

→ Relate the potential difference (pd) across the plates of a capacitor to the charge on its plate.

→ Discuss what capacitors are used for.

Specification reference: 3.7.4.1

A capacitor is a device designed to store charge. Two parallel metal plates placed near each other form a capacitor. When the plates are connected to a battery, electrons move through the battery and are forced from its negative terminal of the battery onto one of the plates. An equal number of electrons leave the other plate to return to the battery via its positive terminal. So each plate gains an equal and opposite charge.

A capacitor consists of two conductors insulated from each other. The symbol for a capacitor is shown in Figure 1. As explained above, when a capacitor is connected to a battery, one of the two conductors gains electrons from the battery, and the other conductor loses electrons to the battery. When we say that the charge stored by the capacitor is Q, we mean that one conductor stores charge $+Q$ and the other conductor stores charge $-Q$.

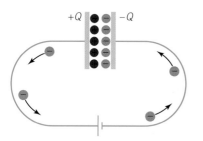

▲ **Figure 1a** *Storing charge on a capacitor*

▲ **Figure 1b** *The capacitor circuit symbol*

Charging a capacitor at constant current

Figure 2 shows how this can be achieved using a variable resistor, a switch, a microammeter, and a cell in series with the capacitor. When the switch is closed, the variable resistor is continually adjusted to keep the microammeter reading constant. At any given time t after the switch is closed, the charge Q on the capacitor can be calculated using the equation $Q = It$, where I is the current.

By using a high-resistance voltmeter connected in parallel with the capacitor, you can measure the capacitor potential difference (pd). To investigate how the capacitor pd changes with time for a constant current, use the variable resistor to keep the current constant, and either

- use a stopwatch and measure the voltmeter reading at measured times, or
- use a data logger as shown in Figure 2.

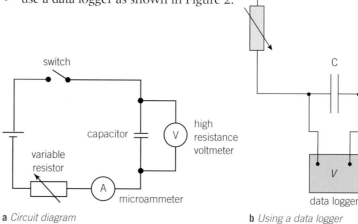

a *Circuit diagram*   **b** *Using a data logger*

▲ **Figure 2** *Investigating capacitors*

Typical readings for a current of 15 µA are shown in Table 1. The charge Q has been calculated using $Q = It$.

▼ **Table 1** *Current = 15 μA*

time t/s	0	20	40	60	80	100
pd V/volts	0	0.29	0.62	0.90	1.22	1.50
charge Q/μC	0	300	600	900	1200	1500

The graph of charge stored, Q, against pd, V, for these measurements is shown in Figure 3. The measurements give a straight line passing through the origin. Therefore, the charge stored, Q, is proportional to the pd, V. In other words, the charge stored per volt is constant.

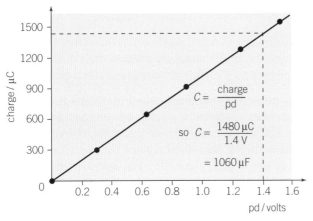

▲ **Figure 3** *Graph of results*

The capacitance C of a capacitor is defined as the charge stored per unit pd.

The unit of capacitance is the farad (F), which is equal to one coulomb per volt. Note that $1.0\,\mu F = 1.0 \times 10^{-6}\,F$.

For a capacitor that stores charge Q at pd V, its capacitance can be calculated using the equation

$$C = \frac{Q}{V}$$

Note:
Rearranging this equation gives $Q = CV$ or $V = \frac{Q}{C}$.

Capacitor uses

Capacitors are used in:

- smoothing circuits (i.e., circuits that smooth out unwanted variations in voltage)
- back-up power supplies (i.e., circuits that take over when the mains supply is interrupted)
- timing circuits, (i.e., circuits that switch on or off automatically after a preset delay)
- pulse-producing circuits (i.e., circuits that switch on and off repeatedly)
- tuning circuits (i.e., circuits that are used to select radio stations and TV channels)
- filter circuits (i.e., circuits that remove unwanted frequencies).

Study tip

The farad is a very large unit. In practice, capacitance is measured in μF, nF, or pF.

When you use $C = \frac{Q}{V}$, remember that charge in μC and pd in V gives capacitance in μF.

Summary questions

1 Complete the following table.

	(a)	(b)	(c)	(d)
charge /μC	60	330		6.30
pd /V	12		9.0	4.5
capacitance /μF		150	1100	

2 A 22 μF capacitor is charged by using a constant current of 2.5 μA to a pd of 12.0 V. Calculate:

   a the charge stored on the capacitor at 12.0 V,

   b the time taken.

3 A capacitor is charged by using a constant current of 0.5 μA to a pd of 5.0 V in 55 s. Calculate:

   a the charge stored,

   b the capacitance of the capacitor.

4 A capacitor is charged by using a constant current of 24 μA to a pd of 4.2 V in 38 s. The capacitor is then charged from 4.2 V by using a current of 14 μA in 50 s. Calculate:

   a charge stored at a pd of 4.2 V,

   b the capacitance of the capacitor,

   c the extra charge stored at a current of 14 μA,

   d the pd after the extra charge was stored.

23.2 Energy stored in a charged capacitor

Learning objectives:

→ Explain why a capacitor stores energy as it is being charged.

→ Describe the form of energy that is stored by a capacitor.

→ Describe what happens to the amount of energy stored if the charge stored is doubled.

Specification reference: 3.7.4.3

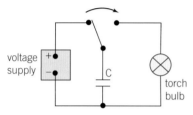

▲ **Figure 1** *Releasing stored energy*

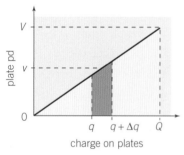

▲ **Figure 2** *Energy stored in a capacitor*

When a capacitor is charged, energy is stored in it because electrons are forced onto one of its plates and taken off the other plate. The energy is stored in the capacitor as electric potential energy. A charged capacitor discharged across a torch bulb will release its energy in a brief flash of light from the bulb, as long as the capacitor has first been charged to the operating pd of the bulb. Charge flow is rapid enough to give a large enough current to light the bulb, but only for a brief time. You could replace the bulb with a miniature electric motor. The motor would briefly spin when the capacitor is discharged through it.

How much energy is stored in a charged capacitor? The charge is forced onto the plates by the battery. In the charging process, the pd across the plates increases in proportion to the charge stored, as shown in Figure 2.

Consider one step in the process of charging a capacitor of capacitance C when the charge on the plates increases by a small amount Δq from q to $q + \Delta q$. The energy stored ΔE in the capacitor is equal to the work done to force the extra charge Δq on to the plates and is given by $\Delta E = v\Delta q$, where v is the average pd during this step. In Figure 2, $v\Delta q$ is represented by the area of the vertical strip of width Δq and height v under the line. Therefore, the area of this strip represents the work done ΔE in this small step.

Now consider all the small steps from zero pd to the final pd V. The total energy stored E is obtained by adding up the energy stored in each small step. In other words, E is represented by the total area under the line from zero pd to pd V. Because this area is a triangle of height V and base length Q $(= CV)$, the total energy stored E = triangle area $= \frac{1}{2} \times$ height $\times$ base $= \frac{1}{2}VQ$.

$$\text{Energy stored by the capacitor } E = \frac{1}{2}QV$$

Notes:

1. Using $Q = CV$ or $V = \frac{Q}{C}$, you can write the above equation as
    $$E = \frac{1}{2}CV^2 \text{ or } E = \frac{1}{2}\frac{Q^2}{C}$$

2. In the charging process, the battery forces charge Q through pd V round the circuit and therefore transfers energy QV to the circuit. Thus, 50% of the energy supplied by the battery $(= \frac{1}{2}QV)$ is stored in the capacitor. The other 50% is wasted due to resistance in the circuit as it is transferred to the surroundings when the charge flows in the circuit.

Measuring the energy stored in a charged capacitor

A joulemeter can measure the energy transfer from a charged capacitor to a light bulb when the capacitor discharges (Figure 3). Before the discharge starts, the capacitor pd V is measured and the joulemeter reading recorded. When the capacitor has discharged, the joulemeter reading is recorded again. The difference between these two joulemeter readings is the energy transferred from the capacitor during the discharge process. This is the total energy stored in the capacitor before it discharged. You can compare this with the calculation of the energy stored by using the equation $E = \frac{1}{2}CV^2$.

Study tip

Don't forget that doubling the charge also doubles the voltage.

You can see the effect of this on energy if you use the equation $E = \frac{1}{2}CV^2$ instead of $E = \frac{1}{2}QV$.

The energy stored in a thundercloud

Imagine that a thundercloud and the Earth below it are like a pair of charged parallel plates. Because the thundercloud is charged, a strong electric field exists between the thundercloud and the ground. The potential difference between the thundercloud and the ground is $V = Ed$, where E is the electric field strength and d is the height of the thundercloud above the ground.

- For a thundercloud carrying a constant charge Q, the energy stored $= \frac{1}{2}QV = \frac{1}{2}QEd$.
- If the thundercloud is forced by winds to rise up to a new height d', then the energy stored $= \frac{1}{2}QEd'$.
- Because the electric field strength is unchanged (because it depends on the charge per unit area; see Topic 22.2), then the increase in the energy stored $= \frac{1}{2}QEd' - \frac{1}{2}QEd = \frac{1}{2}QE\Delta d$, where $\Delta d = d' - d$.

The energy stored increases because work is done by the force of the wind. It has to overcome the electrical attraction between the thundercloud and the ground to make the charged thundercloud move away from the ground.

The insulating property of air breaks down if it is subjected to an electric field stronger than about $300\,\text{kV m}^{-1}$. Prove for yourself that, for every metre rise of the thundercloud carrying a charge of $20\,\text{C}$, the energy stored would increase by $3\,\text{MJ}$. At a height of $500\,\text{m}$, the energy stored would be $1500\,\text{MJ}$.

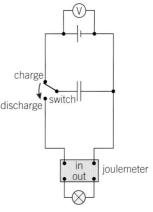

▲ **Figure 3** *Measuring energy stored*

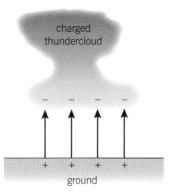

▲ **Figure 4** *Energy in a thundercloud*

Summary questions

1 Calculate the charge and energy stored in a $10\,\mu\text{F}$ capacitor charged to a pd of

   a $3.0\,\text{V}$,   **b** $6.0\,\text{V}$.

2 A $50\,000\,\mu\text{F}$ capacitor is charged from a $9\,\text{V}$ battery then discharged through a light bulb in a flash of light lasting $0.2\,\text{s}$. Calculate:

   a the charge and energy stored in the capacitor before discharge,

   b the average power supplied to the light bulb.

3 An uncharged $2.2\,\mu\text{F}$ capacitor is connected to a $3.0\,\text{V}$ battery. Calculate:

   a the charge and energy

      i stored in the capacitor,

      ii supplied by the battery.

   b Explain the difference between the energy supplied by the battery and the energy stored in the capacitor.

4 In Figure 5, a $4.7\,\mu\text{F}$ capacitor is charged from a $12.0\,\text{V}$ battery by connecting the switch to X. The switch is then reconnected to Y to charge a $2.2\,\mu\text{F}$ capacitor from the first capacitor, causing the charge to be shared in proportion to the capacitance of each capacitor.

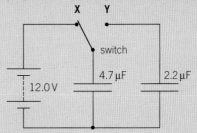

▲ **Figure 5**

Calculate:

   a the initial charge and energy stored in the $4.7\,\mu\text{F}$ capacitor,

   b **i** the final charge stored by each capacitor,

      ii the final pd across the two capacitors,

   c the final energy stored in each capacitor.

   Explain the loss of energy stored.

Learning objectives:

→ Describe and interpret the shape of the Q–t charging curves and the shape of the Q–t discharging curves.

→ Explain which circuit components you would change to make the charge/discharge slower.

→ Define the time constant of a capacitor–resistor circuit.

Specification reference: 3.7.4.4

Capacitor discharge through a fixed resistor

When a capacitor discharges through a fixed resistor, the discharge current decreases gradually to zero. Figure 1a shows a circuit in which a capacitor is discharged through a resistor when the switch is changed over. The reason why the current decreases gradually is that the pd across the capacitor decreases as it loses charge. Because the resistor is connected directly to the capacitor, the resistor current $\left(= \dfrac{\text{pd}}{\text{resistance}}\right)$ decreases as the pd decreases.

The situation is a bit like water emptying through a pipe at the bottom of a container. When the container is full, the flow rate out of the pipe is high because the water pressure at the pipe is high. As the container empties, the water level falls, so the water pressure at the pipe falls, and the flow rate decreases.

The graphs in Figure 1 show how the current and the charge decrease with time. Both curves have the same shape because both the current and the charge (and pd) decrease **exponentially**. This means that any of these quantities decreases by the same factor in equal intervals of time. For example, for initial charge Q_0, if the charge is $0.9Q_0$ after a particular time t_1, the charge will be

- $0.9 \times 0.9Q_0$ after time $2t_1$,
- $0.9 \times 0.9 \times 0.9Q_0$ after time $3t_1$ …
- $0.9^n Q_0$ after time nt_1

To understand why the decrease is exponential, consider one small step in the discharge process of a capacitor C through a resistor R when the charge decreases from Q to $Q - \Delta Q$ in the time interval Δt.

At this stage, the current $I = \dfrac{\text{pd across the plates, } V}{\text{resistance, } R} = \dfrac{Q}{CR}$, because $V = \dfrac{Q}{C}$. The current is proportional to the charge, which is proportional to the pd. So the curves all have the same shape. Assuming that Δt is small enough, the decrease of charge $\Delta Q = -I\Delta t$ (with a minus sign because Q decreases).

Therefore, $\Delta Q = \dfrac{-Q}{CR}\Delta t$, which gives

$$\frac{\Delta Q}{Q} = -\frac{\Delta t}{CR}$$

The equation tells you that the fractional drop of charge $\dfrac{\Delta Q}{Q}$ is the same in any short interval of time Δt during the discharge process. For example, suppose $\Delta t = 10\,\text{s}$ and $CR = 100$. Therefore, $\dfrac{\Delta Q}{Q} = -0.1$ $\left(= -\dfrac{\Delta t}{CR}\right)$. So the charge decreases to 0.9 of its initial value at the start of the $10\,\text{s}$ interval. So if the initial charge is Q_0, then the charge remaining on the plates will be

- $0.9Q_0$ after $10\,\text{s}$,
- $0.9 \times 0.9Q_0$ after a further $10\,\text{s}$,
- $0.9 \times 0.9 \times 0.9Q_0$ after a further $10\,\text{s}$, and so on.

In theory, the charge on the plates never becomes zero.

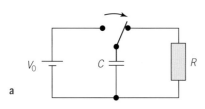

a

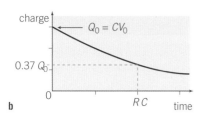

b

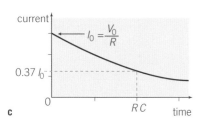

c

▲ **Figure 1** *Capacitor discharging*

Study tip

An exponential decrease graph starts at an intercept with the vertical axis and is asymptotic with the time axis (i.e., the curve approaches but never cuts the time axis).

Exponential changes occur whenever the rate of change of a quantity is proportional to the quantity itself.

Rearranging the equation $\frac{\Delta Q}{Q} = -\frac{\Delta t}{CR}$ gives

$$\frac{\Delta Q}{\Delta t} = -\frac{Q}{CR}$$

For very short intervals of time (i.e., $\Delta t \to 0$), $\frac{\Delta Q}{\Delta t}$ represents the rate of change of charge and is written $\frac{dQ}{dt}$.

Therefore, $\frac{dQ}{dt} = -\frac{Q}{CR}$.

The graphical solution to this equation is shown in Figure 1b. The mathematical solution is

$$Q = Q_0 \, e^{\frac{-t}{RC}}$$

where Q_0 is the initial charge, and e is the exponential function (sometimes written 'exp').

The quantity RC is called the **time constant** for the circuit. At time $t = RC$ after the start of the discharge, the charge falls to 0.37 ($= e^{-1}$) of its initial value.

Time constant = RC

where R is the circuit resistance, and C is the capacitance.

The unit of RC is the second. This is because one ohm $= \dfrac{\text{one volt}}{\text{one ampere}}$,

and one farad $= \dfrac{\text{one coulomb}}{\text{one volt}}$, so

unit of $RC = \dfrac{\text{volts}}{\text{amperes}} \times \dfrac{\text{coulombs}}{\text{volts}} = \dfrac{\text{coulombs}}{\text{amperes}} = \text{seconds}$

Worked example

A $2200\,\mu\text{F}$ capacitor is charged to a pd of $9.0\,\text{V}$ then discharged through a $100\,\text{k}\Omega$ resistor using a circuit as shown in Figure 1.

a   Calculate:
   i   the initial charge stored by the capacitor,
   ii   the time constant of the circuit.
b   Calculate the pd after a time:
   i   equal to the time constant,
   ii $300\,\text{s}$.

Solution

a   i $Q_0 = CV_0 = 2200 \times 10^{-6} \times 9.0 = 2.0 \times 10^{-2}\,\text{C}$.
   ii   Time constant $= RC = 100 \times 10^3 \times 2.2 \times 10^{-3} = 220\,\text{s}$.
b   i   When $t = RC$, $V = V_0 e^{-1} = 0.37 \times 9.0 = 3.3\,\text{V}$.
   ii   When $t = 300\,\text{s}$, $\dfrac{t}{RC} = \dfrac{300}{220} = 1.36$

   Therefore, $V = V_0 e^{\frac{-t}{RC}} = 9.0e^{-1.36} = 2.3\,\text{V}$.

Study tip

You do *not* need to explain why $Q = Q_0 e^{\frac{-t}{CR}}$ is the solution of the equation

$$\frac{dQ}{dt} = -\frac{Q}{CR}$$

Notes

1   The current, the pd, and the charge are all proportional to each other. All three quantities decrease exponentially during capacitor discharge in accordance with the equation $x = x_0 e^{\frac{-t}{CR}}$, where x represents either the current, or the charge, or the pd.

2   The inverse function of e^x is $\ln x$, where $\ln$ is the natural logarithm. To calculate t, given $x, x_0, R,$ and C, using the inverse function of e^x gives $\ln x = \ln x_0 - \left(\dfrac{t}{RC}\right)$.

3   The exponential function

$$e^z = 1 + z + \frac{z^2}{2 \times 1} + \frac{z^3}{3 \times 2 \times 1} + \ldots, \text{ and so on.}$$

It can be shown mathematically that the rate of change of this function in relation to z is the same function. This is why the function appears whenever the rate of change of a quantity is proportional to the quantity itself.

Note that $z = 1$ gives

$$e = 1 + 1 + \frac{1}{2} + \frac{1}{6} + \frac{1}{24}, \text{ etc}$$
$$= 2.718.$$

You can check this on your calculator by keying in 'e^x' then pressing 1 to give 2.718. Keying in -1 instead of 1 gives 0.37 for e^{-1}.

The significance of the time constant

The time constant RC is the time taken, in seconds, for the capacitor to discharge to 37% of its initial charge. Given values of R and C, the time constant can be quickly calculated and used as an approximate measure of how quickly the capacitor discharges. However, as $5RC$ is the time taken to discharge by over 99%, $5RC$ is a better rule of thumb estimate of the time taken for the capacitor to effectively discharge.

Q: Demonstrate that $t = 5RC$ gives a value that is less than 1% of the initial value.

$$\text{A: } 0.37^5 = 0.00690 \ (> 0.010)$$

Synoptic link

Wherever a quantity decreases at a rate that is proportional to the quantity, the decrease is exponential. You will meet exponential decrease again in radioactive decay in Topic 26.6, The theory of radioactive decay, where you will study what is meant by the half-life $T_{\frac{1}{2}}$ of a radioactive isotope. This is the time taken for the number of atoms of a radioactive isotope to decrease to 50% of its initial value. It can be shown that the half-life of a capacitor discharge (i.e., the time taken for the charge (or voltage) to decrease to 50% of the initial value) is equal to 0.69(3) RC.

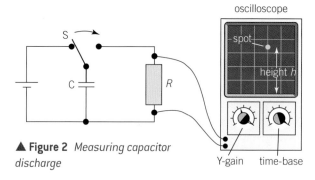

▲ **Figure 2** *Measuring capacitor discharge*

Investigating capacitor discharge

Figure 2 shows how to measure the pd across a capacitor as it discharges through a fixed resistor. An oscilloscope is used because it has a very high resistance so that the discharge current from the capacitor passes only through the fixed resistor. The oscilloscope is used to measure the capacitor pd at regular intervals. You could use a data logger or a digital voltmeter instead of the oscilloscope.

You can then use the measurements to plot a graph of voltage against time. The time taken for the voltage to decrease to 37% $\left(= \frac{1}{e}\right)$ of the initial value can be measured from the graph and compared with the calculated value of RC.

Charging a capacitor through a fixed resistor

When a capacitor is charged by connecting it to a source of constant pd, the charging current decreases as the capacitor charge and pd increase. When the capacitor is fully charged, its pd is equal to the source pd, and the current is zero because no more charge flows in the circuit. The graphs in Figure 3 show how the capacitor charge and current change with time.

- The capacitor charge builds up until the capacitor pd V is equal to the source pd V_0, as shown in Figure 3b. The charge Q_0 on the capacitor is then equal to CV_0. The charge curve is an inverted exponential decrease curve, which flattens out at $Q_0 = CV_0$.

- The time constant for the circuit, RC, is the time taken for the charge to reach 63% of the final charge (i.e., 37% more charge needed to be fully charged). A graph of the capacitor pd V against time has exactly the same shape as the charge curve because $V = \frac{Q}{C}$.

- The capacitor current I decreases exponentially to zero from its initial value I_0, as shown in Figure 3c. The current is always equal to the rate of change of charge. Therefore, the current is given by the gradient of the charge–time graph, and so it decreases exponentially.

1 At any instant during the charging process, the source pd V_0 = the resistor pd + the capacitor pd.

   Therefore, $V_0 = IR + \dfrac{Q}{C}$ at any instant.

2 The initial current $I_0 = \dfrac{V_0}{R}$, assuming that the capacitor is initially uncharged.

3 At time t after charging starts, $I = I_0 \, e^{\frac{-t}{RC}}$.

4 If you combine the three equations above to eliminate I and I_0, you get $V_0 = V_0 \, e^{\frac{-t}{RC}} + \dfrac{Q}{C}$.

So, $\dfrac{Q}{C} = V_0 - V_0 \, e^{\frac{-t}{RC}} = V_0 \left(1 - e^{\frac{-t}{RC}} \right)$.

Because $V = \dfrac{Q}{C}$, then $V = V_0 \left(1 - e^{\frac{-t}{RC}} \right)$ and $Q = CV_0 \left(1 - e^{\frac{-t}{RC}} \right)$.

Figure 4b shows how Q (and V) vary with time. For example,

- at time $t = 0$, $e^{\frac{-t}{RC}} = 1$, so $Q = 0$ and $V = 0$.

- as time $t \rightarrow \infty$, $e^{\frac{-t}{RC}} \rightarrow 0$ so $Q \rightarrow Q_0$ and $V \rightarrow V_0$

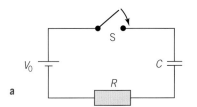

a

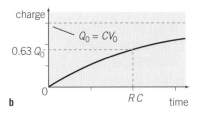

b

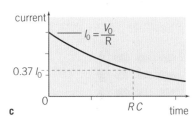

c

▲ **Figure 3** *Capacitor charging*

Summary questions

1 A 50 µF capacitor is charged by connecting it to a 6.0 V battery then discharged through a 100 kΩ resistor.

Calculate:

a i the charge stored in the capacitor immediately after it has been charged,

  ii the time constant of the circuit.

b i Estimate how long the capacitor would take to discharge to about 2 V.

  ii Estimate the resistance of the resistor that you would use in place of the 100 kΩ resistor if the discharge is to be 99% completed within about 5 s.

2 A 68 µF capacitor is charged to a pd of 9.0 V then discharged through a 20 kΩ resistor.

a Calculate:

  i the charge stored by the capacitor at a pd of 9.0 V,

  ii the initial discharge current.

b Calculate the pd and the discharge current 5.0 s after the discharge started.

3 A 2.2 µF capacitor is charged to a pd of 6.0 V and then discharged through a 100 kΩ resistor. Calculate:

a the charge and energy stored in this capacitor at 6.0 V,

b the pd across the capacitor 0.5 s after the discharge started,

c the energy stored at this time.

4 An uncharged 4.7 µF capacitor is charged to a pd of 12.0 V through a 200 Ω resistor and then discharged through a 220 kΩ resistor. Calculate:

a i initial charging current,

  ii the energy stored in the capacitor at 12.0 V,

b the time taken for the pd to fall from 12.0 V to 3.0 V,

c the energy lost by the capacitor in this time.

Synoptic link

You can find out how to model rates of flow of charge in spreadsheets on a computer in Topic 30.6, Graphical and computational modelling.

Learning objectives:
→ Explain how a dielectric affects a capacitor.
→ Define relative permittivity and dielectric constant.
→ Describe the action of a simple polar molecule rotating in an electric field.

Specification reference: 3.7.4.2

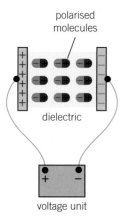

polarised molecules

dielectric

voltage unit

▲ **Figure 1** *Dielectric action*

Dielectric action

A capacitor is a device designed to store charge. The simplest type of capacitor is made of two parallel metal plates opposite each other as shown in Figure 1a in Topic 23.1. When a battery is connected to the plates, electrons from the negative terminal of the battery move onto the plate connected to that terminal. At the same time, electrons move from the other plate to the positive terminal of the battery, leaving this plate with a positive charge. The two plates store equal and opposite amounts of charge.

The charge stored on the plates can be increased by inserting a **dielectric** between the plates. Dielectrics are electrically insulating materials that increase the ability of a parallel-plate capacitor to store charge when a dielectric is placed between the plates of the capacitor. Polythene and waxed paper are examples of dielectrics.

Consider what happens when a dielectric is placed between two oppositely charged parallel plates connected to a battery. Each molecule of the dielectric becomes **polarised**. This means that its electrons are pulled slightly towards the positive plate as shown in Figure 1. So the surface of the dielectric facing the positive plate gains negative charge at the expense of the other side of the dielectric that faces the negative plate. The other surface of the dielectric loses negative charge, so some positive charge is left on its surface.

In some dielectric substances, the molecules are already polarised, but they lie in random directions. These molecules, called polar molecules, turn when the dielectric is placed between the charged plates because their electrons are attracted slightly to the positive plate. The effect is the same as with non-polar molecules – the surface of the dielectric near the positive plate gains negative charge, and the other surface gains positive charge.

As a result, more charge is stored on the plates because

- the positive side of the dielectric attracts more electrons from the battery onto the negative plate.

- the negative side of the dielectric pushes electrons back to the battery from the positive plate.

The effect of a dielectric is to increase the charge stored in a capacitor for any given pd across the capacitor terminals. In other words, its effect is to increase the capacitance of the capacitor.

Relative permittivity

For a fixed pd across a parallel-plate capacitor with an empty space between its plates, the charge stored is increased by inserting a dielectric substance between the plates. The ratio of the charge stored with the dielectric to the charge stored without the dielectric may be defined as the **relative permittivity**, ε_r, of the dielectric substance.

$$\text{relative permittivity } \varepsilon_r = \frac{Q}{Q_0}$$

where Q = charge stored by a parallel-plate capacitor when the space between the plates of the capacitor is completely filled with the dielectric substance, and Q_0 = charge stored at the same pd when the space is completely empty.

For a fixed pd V between the plates, $Q = CV$ and $Q_0 = C_0V$, where C is the capacitance of the parallel-plate capacitor with the dielectric completely between the plates and C_0 is the capacitance when the space is completely empty. Therefore, $\dfrac{Q}{Q_0} = \dfrac{C}{C_0}$, so the relative permittivity may be defined by the equation

$$\varepsilon_r = \frac{C}{C_0}$$

The relative permittivity ε_r of a substance is also called its **dielectric constant**. Typical values for ε_r are 2.3 for polythene, 2.5 for waxed paper, 7 for mica and 81 for water! The large value of ε_r for water is the reason why ionic crystals such as sodium chloride (common salt) dissolve in water. See Topic 22.4.

Capacitor design

For a parallel-plate capacitor with dielectric filling the space between the plates, its capacitance $C = \dfrac{A\varepsilon_0\varepsilon_r}{d}$, where A is the surface area of each plate and d is the spacing between the plates.

This equation shows that a large capacitance can be achieved by

- making the area A as large as possible,
- making the plate spacing d as small as possible,
- filling the space between the plates with a dielectric which has a relative permittivity as large as possible.

Most capacitors consist of two strips of aluminium foil separated by a layer of dielectric, all rolled up as shown in Figure 2. This arrangement makes the capacitance as large as possible because the area A is as large as possible and the spacing d is as small as possible.

Capacitor factors

In Topic 22.2, you saw that for an 'empty' parallel-plate capacitor, the charge stored per unit surface area $\dfrac{Q_0}{A} = \varepsilon_0 E$, where E is the electric field strength and ε_0 is the absolute permittivity of free space. This equation may be written as $\dfrac{Q_0}{A} = \varepsilon_0\dfrac{V}{d}$ because $E = \dfrac{V}{d}$, where V is the pd between the plates and d is the spacing between the plates. Rearranging this equation gives the 'empty' capacitor's capacitance $C_0 = \dfrac{Q_0}{V} = \dfrac{A\varepsilon_0}{d}$, where A is the surface area of each plate. If the space between the plates had been filled completely with a dielectric, the capacitance would have been increased from C_0 to $C = \varepsilon_r C_0$.

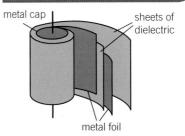

metal cap

sheets of dielectric

metal foil

▲ **Figure 2** *Capacitor design*

Measuring the relative permittivity of a dielectric substance

In this investigation, a capacitor is formed by placing a dielectric sheet of uniform thickness between two parallel metal plates insulated from each other by small pieces of suitable insulation. A potential divider is used to apply a constant pd between the two plates. The capacitor is repeatedly charged and discharged through a microammeter using a reed switch operating at a constant frequency. Figure 3 shows the circuit.

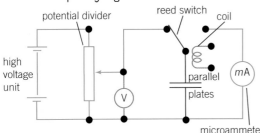

potential divider

reed switch   coil

high voltage unit

V

parallel plates

mA

microammeter

▲ **Figure 3** *Measuring relative permittivity*

The microammeter current reading is proportional to the charge gained by the capacitor each time the switch charges it. The current is measured with the dielectric in place and with the dielectric replaced by small insulating spacers of the same thickness as the dielectric sheet.

The relative permittivity of the dielectric $= \dfrac{I}{I_0}$, where I is the current with the dielectric present and I_0 is the current without the dielectric present.

Note that a capacitance meter may be used to measure the capacitance of the arrangement directly with and without the dielectric between the plates. The ratio $\dfrac{C}{C_0}$ gives the relative permittivity.

Polarisation mechanisms

The relative permittivity of a dielectric in a constant electric field is due to three different polarisation mechanisms.

1 **Orientation polarisation** occurs in substances which contain molecules where covalent bonds are formed between atoms of different elements. The electrons in each covalent bond are shared unequally between the two atoms joined by the bond. The two atoms form a permanent electric **dipole** in which one atom is positively charged and the other is negatively charged. When the electric field is applied, the two atoms in each bond are displaced in opposite direction so the dipole tends to align with the field by turning slightly. However, when the field is absent, the dipoles lie in random directions.

2 **Ionic polarisation** occurs in substances where ions are held together by ionic bonds. The oppositely-charged ions of each ionic bond are displaced in opposite directions when an electric field is applied. As a result, the ions of each bond form a dipole that tends to align with the field.

3 **Electronic polarisation** occurs where the electrons of each atom are displaced relative to the nucleus of the atom when an electric field is applied. The centre of the electron distribution therefore no longer coincides with the nucleus. As a result, the electron distribution and the nucleus form a dipole that tends to align with the field.

In an alternating electric field, the polar dipoles rotate and the non-polar dipoles oscillate one way then the opposite way as the field strength increases and decreases. At low frequencies, the three polarisation mechanisms alternate in phase with the field at low frequency. However, as the frequency increases, each mechanism ceases to work due to the inertia of the particles involved and the resistive forces that oppose the motion of the dipoles. Figure 4 shows how the relative permittivity changes as the frequency of the applied field increases.

The mass of the particles being moved by the field determines which mechanism ceases first as the frequency increases. The greater the mass of each particle, the greater its inertia is, and so the frequency at which it ceases to contribute is lower. Orientation polarisation therefore ceases first as the frequency increases, then ionic polarisation, then electronic polarisation.

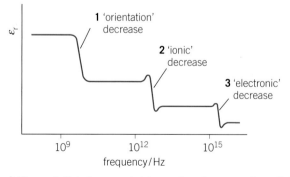

▲ **Figure 4** *Relative permittivity against frequency for a dielectric*

Capacitor applications

1 Any electronic timing circuit or time-delay circuit makes use of capacitor discharge through a fixed resistor.

Figure 5 shows an alarm circuit. In this circuit, the alarm rings if the input voltage to the circuit drops below a particular value after the switch is reset. The time delay between resetting the switch and the alarm ringing can be increased by increasing the resistance R or the capacitance C. This type of change to the circuit would make the discharge of C through R slower. This would then increase the time for the capacitor voltage to decrease enough to make the alarm ring.

2 Capacitor smoothing is used in applications where sudden voltage variations or glitches can have undesirable effects. For example, mains appliances being switched on or off in a building could affect computers connected to the mains supply in the building. Inside a computer there is a large capacitor that will supply current if the mains supply is interrupted. In this way, the computer circuits will continue to function normally.

Q: Explain how the time delay would be affected if the resistance R and the capacitance C in Figure 5 were both doubled.

A: The time constant RC and therefore the time delay would be four times greater.

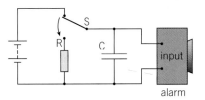

▲ **Figure 5** *A time-delayed alarm circuit*

Summary questions

1 A parallel-plate capacitor consists of two insulated metal plates separated by an air gap. A battery in series with a switch is connected to the plates. The capacitor is charged by closing the switch to charge the capacitor to a constant pd. A sheet of dielectric is then inserted between the plates.

   a When the sheet of dielectric is inserted, state the change that takes place to:

      i the capacitance C of the capacitor,

      ii the charge Q stored by the capacitor.

   b State and explain the change that takes place to the energy stored by the capacitor.

2 In the test in Q1, the switch remains open and the sheet of dielectric is then removed. State and explain how the energy stored by the capacitor changes when the dielectric is removed.

3 An air-filled parallel-plate capacitor has a capacitance of 1.4 pF.

   a The space between the plates is completely filled with a sheet of dielectric that has a relative permittivity of 7.0. Calculate the capacitance of the capacitor with the dielectric present.

   b Calculate the energy stored by the capacitor in (a) when the pd across it is 15.0 V.

4 An electrolytic capacitor contains a very thin layer of dielectric formed when the capacitor is first charged. The insulating property of the dielectric in a certain 100 mF electrolytic capacitor breaks down if the electric field strength across it exceeds $700\,MV\,m^{-1}$. The maximum pd that can be applied to the capacitor is 100 V.

   a Calculate the thickness of the dielectric layer,

   b The effective surface area of each capacitor plate is $1.6\,m^2$. Estimate the energy stored per unit volume in the capacitor at 100 V.

1   In an experiment to measure the capacitance C of a capacitor, the circuit in **Figure 1** was used to charge the capacitor then discharge it through a resistor of known resistance R.

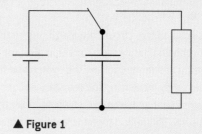

▲ Figure 1

(a) The capacitor pd V at time t after the discharge commenced is given by $V = V_0 e^{\frac{-t}{CR}}$. Show that this equation can be rearranged into an equation of the form $\ln V = a - bt$, where a and b are constants, and determine expressions for a and b.   *(4 marks)*

(b) As the capacitor discharged, its pd was measured every 30 seconds using a digital voltmeter. The measurements were taken three times as shown in Table 1.

▼ Table 1

t/s	0	30	60	90	120	150	180	210	240	270	300
	4.50	3.82	3.26	2.78	2.33	2.00	1.70	1.43	1.23	1.04	0.89
V/V	4.51	3.81	3.25	2.77	2.35	2.10	1.72	1.43	1.25	1.02	0.90
	4.50	3.83	3.25	2.76	2.34	1.98	1.69	1.42	1.22	1.04	0.87
mean V/V	4.503	3.820	3.253	2.760	2.340	2.027	1.703				
$\ln V$	1.505	1.340	1.180	1.017	0.850	0.707	0.532				

(i)   Complete the missing entries in Table 1.
(ii)   Use the measurements to plot a graph of $\ln V$ on the y-axis against t on the x-axis.
(iii) Use your graph to determine the time constant of the discharge circuit.
(iv)   The resistance R of the resistor was $68\,\text{k}\Omega$. Determine the capacitance C of the capacitor.   *(10 marks)*

(c) (i)   Discuss the reliability of the measurements.
(ii) Estimate the accuracy of your value of capacitance, given the resistor value is accurate to within 1%.   *(4 marks)*

2   **Figure 2** shows a $2.0\,\mu\text{F}$ capacitor connected to $150\,\text{V}$ supply.

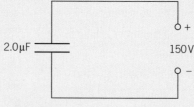

▲ Figure 2

(a) Calculate the charge on the capacitor.   *(2 marks)*

(b) (i)   Suggest a graph that could be drawn in order to calculate the energy stored in the capacitor by finding the area under the graph.
(ii) Calculate the energy stored by the capacitor when it has a pd of $150\,\text{V}$ across it.   *(3 marks)*

(c) The charged capacitor is removed from the power supply and discharged by connecting a 220 kΩ resistor across it.
  (i) Calculate the maximum discharge current.
  (ii) Show that the current will have fallen to 10% of its maximum value in a time of approximately 1 s.   (*5 marks*)

AQA, 2002

3 A capacitor of capacitance 330 μF is charged to a potential difference of 9.0 V. It is then discharged through a resistor of resistance 470 kΩ.
  (a) Calculate:
    (i) the energy stored by the capacitor when it is fully charged,   (*2 marks*)
    (ii) the time constant of the discharging circuit,   (*1 mark*)
    (iii) the pd across the capacitor 60 s after the discharge has begun.   (*3 marks*)
  (b) The capacitor is charged using a 9.0 V battery and negligible internal resistance in series with a 1.0 kΩ resistor. Calculate:
    (i) the time constant for this circuit, and sketch graphs to show how the capacitor pd and the current changed with time during charging,   (*5 marks*)
    (ii) the energy supplied by the battery and the energy supplied to the capacitor during the charging process, and explain the difference.   (*3 marks*)

AQA, 2004

4 A student used a voltage sensor connected to a data logger to plot the discharge curve for a 4.7 μF capacitor. **Figure 3** shows the graph she obtained.

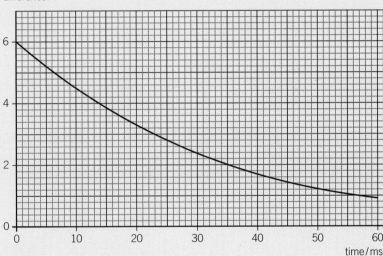

▲ **Figure 3**

Use data from the graph to calculate:
  (a) the initial charge stored,   (*2 marks*)
  (b) the energy stored when the capacitor had been discharging for 35 ms,   (*3 marks*)
  (c) the time constant for the circuit,   (*3 marks*)
  (d) the resistance of the circuit through which the capacitor was discharging.   (*2 marks*)

AQA, 2002

5 **Figure 4** shows a circuit that may be used to investigate the capacitance of a capacitor. The switch moves rapidly between **X** and **Y**, making contact with each terminal 400 times per second. When it makes contact with **X** the capacitor **C** charges, and when it makes contact with **Y** the capacitor discharges through the resistor **R**.

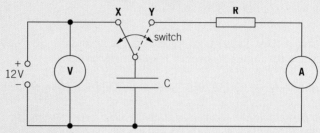

▲ Figure 4

(a) **R** has a value of $220\,\Omega$. The time constant for the circuit is 2.2×10^{-4} s. Calculate the value of capacitor **C**. *(1 mark)*

(b) Calculate the periodic time, T, for the oscillation of the switch. *(2 marks)*

(c) The switch makes contact with **Y** for time $\frac{T}{2}$. The capacitor discharges from 12 V during this time.
   (i) Calculate the voltage across the charged capacitor after a time $\frac{T}{2}$.

   (ii) Explain whether or not it is reasonable to assume that the capacitor has completely discharged in the time $\frac{T}{2}$. *(4 marks)*

AQA, 2007

6 (a) Explain what is meant by a capacitance of 1 farad (F). *(1 mark)*

(b) The capacitance of a capacitor is 2.3×10^{-11} F. When the potential difference across it is 6.0 V, calculate:
   (i) the charge it stores,   (ii) the energy it stores. *(4 marks)*

(c) A student charged the capacitor and then tried to measure the potential difference between the plates using an oscilloscope. The student observed the trace shown in **Figure 5** and concluded that the capacitor was discharging through the oscilloscope.

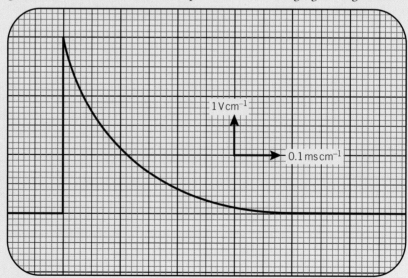

▲ Figure 5

Calculate the resistance of the oscilloscope. *(3 marks)*

AQA, 2003

7 A torch bulb produces a flash of light when a 270 μF capacitor is discharged across it.

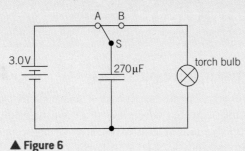

▲ **Figure 6**

(a) The capacitor is charged to a pd of 3.0 V from the battery, as shown in **Figure 6**.
Calculate:
   (i)   the energy stored in the capacitor,
   (ii) the work done by the battery.   *(3 marks)*

(b) The capacitor is discharged by moving switch **S** in the diagram from **A** to **B**.
The discharge circuit has a total resistance of 1.5 Ω.
   (i)   Show that almost all of the energy stored in the capacitor is released when the capacitor pd has decreased from 3.0 V to 0.3 V.
   (ii) Emission of light from the torch bulb ceases when the pd falls below 2.0 V. Calculate the duration of the light flash.
   (iii) Assuming that the torch bulb produces photons of average wavelength 500 nm, estimate the number of photons released during the light flash.   *(8 marks)*

                                                     AQA, 2006

8 A student uses a system shown in **Figure 7** to measure the contact time of a metal ball when it bounces on a metal block.

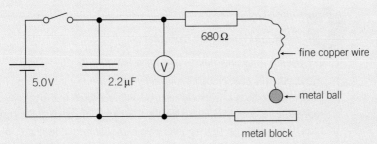

▲ **Figure 7**

The student charges the capacitor by closing the switch, records the voltmeter reading and then opens the switch. The student then releases the ball and measures the voltage after the ball has rebounded from the metal block.
In one test the student records an initial voltage of 5.0 V and a final voltage of 2.2 V.

(a) Calculate the time for which the ball is in contact with the block.   *(3 marks)*
(b) (i)   Calculate the energy lost by the capacitor during the discharge.
    (ii) State where this energy is dissipated and the form it will take.   *(4 marks)*

                                                     AQA, 2002

Learning objectives:

→ Measure the strength of a magnetic field.

→ State the factors that the magnitude of the force on a current-carrying wire depends on.

→ Determine the direction of the force on a current-carrying wire in a magnetic field.

Specification reference: 3.7.5.1

Learning objectives:

→ Measure the strength of a magnetic field.

→ State the factors that the magnitude of the force on a current-carrying wire depends on.

→ Determine the direction of the force on a current-carrying wire in a magnetic field.

Specification reference: 3.7.5.1

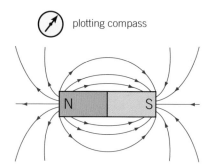

▲ **Figure 1** *The magnetic field near a bar magnet*

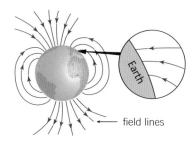

▲ **Figure 2** *The Earth's magnetic field*

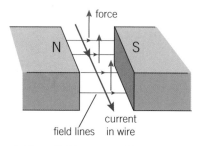

▲ **Figure 3** *The motor effect*

Magnetic field patterns

Magnetism is a topic with a long scientific history stretching back thousands of years when lodestone was used by explorers as a navigational aid. Scientific research over the past 50 years or so has led to many applications such as particle accelerators, powerful microwave transmitters, magnetic discs and tape, superconducting magnets, and magnetic resonance scanners. Magnetism is a valuable scientific tool used by archaeologists, astronomers, and geologists. In short, magnetism has always been a fascinating topic and continues to be so.

A magnetic field is a force field surrounding a magnet or current-carrying wire which acts on any other magnet or current-carrying wire placed in the field. The magnetic field of a bar magnet is strongest at its ends which are referred to as **north-seeking** and **south-seeking** 'poles' according to which direction, north or south, each end points when the magnet is free to align itself with the horizontal component of the Earth's magnetic field. A **line of force** (or magnetic field line) of a magnetic field is a line along which a north pole would move in the field.

➕ The Earth's magnetism

The Earth's magnetic field is not unlike the field of a giant bar magnet inside the Earth. It is caused by circulation currents in the molten iron in the Earth's core. The Earth's magnetic poles are known to drift gradually and a magnetic compass points to the Earth's magnetic north pole. At the present time, the Earth's north magnetic pole is in Northern Canada. As a result of studying the magnetism of rocks, scientists discovered that the continents are drifting across the Earth's surface on giant 'tectonic plates' and that the Earth's magnetic field can reverse quite suddenly!

Q: Describe the direction of the Earth's magnetic field at the magnetic north pole.

A: it is vertically downwards

The force on a current-carrying wire in a magnetic field

A current-carrying wire placed at a non-zero angle to the lines of force of an external magnetic field experiences a force due to the field. This effect is known as the **motor effect**. The force is perpendicular to the wire and to the lines of force.

The motor effect can be tested using the simple arrangement shown in Figure 3. The wire is placed between opposite poles of a U-shaped magnet so it is at right angles to the lines of force of the magnetic field.

When a current flows, the section of the wire in the magnetic field experiences a force that pushes it out of the field. The magnitude of the force depends on the current, the strength of the magnetic field, the length of the wire, and the angle between the lines of force of the field and the current direction.

The force is:

- greatest when the wire is at right angles to the magnetic field,
- zero when the wire is parallel to the magnetic field.

The direction of the force can be related to the direction of the field and to the direction of the current using **Fleming's left-hand rule** shown in Figure 4. If the current is reversed or if the magnetic field is reversed, the direction of the force is reversed.

The magnitude of the force on a current-carrying wire in a magnetic field can be investigated using the arrangement shown in Figure 5. The stiff wire frame is connected in series with a switch, an ammeter, a variable resistor, and a battery. When the switch is closed, the magnet exerts a force on the wire which can be measured from the change of the top pan balance reading.

The tests above show that the force F on the wire is proportional to:

1 the current I,   2 the length l of the wire.

The **magnetic flux density** B of the magnetic field, sometimes referred to as the strength of the magnetic field, is defined as the force per unit length per unit current on a current-carrying conductor at right angles to the magnetic field lines.

Therefore, for a wire of length l carrying a current I in a uniform magnetic field B **at 90° to the field lines**, the force F on the wire is given by

$$F = BIl$$

1 The unit of B is the tesla (T), equal to $1\,\text{N}\,\text{m}^{-1}\,\text{A}^{-1}$.
2 The direction of the force is given by Fleming's left-hand rule. See Figure 4.

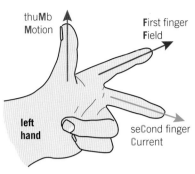

▲ **Figure 4** *Fleming's left-hand rule*

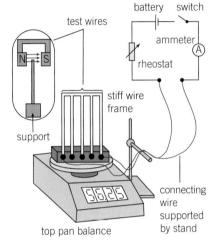

▲ **Figure 5** *Measuring the force on a current-carrying wire in a magnetic field*

> **Study tip**
>
> $F = BIl$ applies only when B and I are at right angles.

Worked example

A straight horizontal wire XY of length 5.0 m is in a uniform horizontal magnetic field of magnetic flux density 120 mT. The wire is at an angle of 90° to the field lines which are due north in direction. The wire conducts a current of 14 A from east to West. Calculate the magnitude of the force on the wire and state its direction.

Solution

$B = 120\,\text{mT} = 0.12\,\text{T}$

$F = BIl = 0.12 \times 14 \times 5.0 = 8.4\,\text{N}$

The force on the wire is vertically downwards.

Note, for a straight wire at angle θ to the magnetic field lines, the force on the wire is due to the component of the magnetic field perpendicular to the wire, $B\sin\theta$. Therefore the magnitude of the force on the wire $F = BIl\sin\theta$. This extension of $F = BIl$ is not required for this specification.

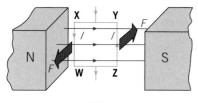

a *top view of coil parallel to field*

b *top view of coil at angle α to field*

▲ **Figure 6** *Couple on a coil*

The couple on a coil in a magnetic field

Consider a rectangular current-carrying coil in a uniform horizontal magnetic field, as shown in Figure 6. The coil has n turns of wire and can rotate about a vertical axis.

- The long sides of the coil are vertical. Each wire down each long side experiences a force BIl where l is the length of each long side. Each long side therefore experiences a horizontal force $F = (BIl)n$ in opposite directions at right angles to the field lines.

- The pair of forces acting on the long sides form a couple as the forces are not directed along the same line. The torque of the couple $= Fd$, where d is the perpendicular distance between the line of action of the forces on each side. See Topic 24.4. If the plane of the coil is at angle α to the field lines, then $d = w\cos\alpha$ where w is the width of the coil.

- Therefore, the torque $= Fw\cos\alpha = BIlnw\cos\alpha = BIAn\cos\alpha$, where the coil area $A = lw$. If $\alpha = 0$ (i.e., the coil is parallel to the field), the torque $= BIAn$ as $\cos 0 = 1$.

- If $\alpha = 90°$ (i.e., the coil is perpendicular to the field), the torque $= 0$ as $\cos 90° = 0$.

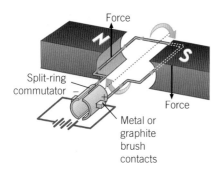

Split-ring commutator

Force

Force

Metal or graphite brush contacts

▲ **Figure 7** *In an electric motor*

The electric motor

The simple electric motor consists of a coil of insulated wire which spins between the poles of a U-shaped magnet.

When a direct current passes round the coil,

- the wires at opposite edges of the coil are acted on by forces in opposite directions,

- the force on each edge makes the coil spin about its axis.

Current is supplied to the coil via a split-ring commutator. The direction of the current round the coil is reversed by the split-ring commutator each time the coil rotates through half a turn. This ensures the current along an edge changes direction when it moves from one pole face to the other. As shown in Figure 7, the result is that the force on each edge continues to turn the coil in the same direction.

In a practical electric motor, several evenly spaced 'armature' coils are wound on an iron core. Each coil is connected to its own section of the commutator. The result is that each coil in sequence experiences a torque when it is connected to the voltage supply so the armature is repeatedly pushed round. Because the iron core makes the field radial, each coil is in the plane of the field (i.e., $\alpha = 0$) for most of the time. As a result, the torque is steady and the motor runs more smoothly. In addition, the iron core makes the field much stronger so the torque of the motor is much greater.

By using an electromagnet connected to the same voltage supply as the coils, an electric motor can operate with alternating current or with direct current. This is because the magnetic field reverses each time the armature current reverses when an ac supply is used. So the turning effect on the armature is unchanged in direction.

▲ **Figure 8** *A practical electric motor*

Summary questions

1   The table below relates the force on a current-carrying wire which is at right angles to the lines of force of a magnetic field and the current. Complete the table below by working out the missing data in each column.

▼ Table 1

	(a)	(b)	(c)	(d)
B / T	0.20 T vertically down	0.20 T vertically down	?	0.1 T horizontal due ?
I / A	3.0 A horizontal due north	?	3.0 A horizontal due north	2.0 A vertically up
l / m	0.040 m	0.040 m	0.040 m	0.040 m
F / N	?	0.036 N horizontal due south	0.024 N horizontal due west	? horizontal due east

2   a   A straight vertical wire of length 0.10 m carries a downward current of 4.0 A in a uniform horizontal magnetic field of flux density 55 mT that acts due north. Determine the magnitude and direction of the force on the wire.

   b   A straight horizontal wire of length 50 mm carrying a constant current is in a uniform magnetic field of flux density 140 mT which acts vertically downwards. The wire experiences a force of 28 mN in a direction which is due north. Determine the magnitude and the direction of the current in the wire.

3   A rectangular coil of width 60 mm and of length 80 mm has 50 turns. The coil was placed horizontally in a uniform horizontal magnetic field of flux density 85 mT with its shorter side parallel to the field lines. A current of 8.0 A was passed through the coil. Sketch the arrangement and determine the force on each side of the coil.

4   The Earth's magnetic field at a certain position on the Earth's surface has a horizontal component of 18 µT due north and a downwards vertical component of 55 µT. Calculate:

   a   the magnitude of the Earth's magnetic field at this position,

   b   the magnitude and direction of the force on a vertical wire of length 0.80 m carrying a current of 4.5 A downwards.

Learning objectives:

→ Describe what happens to charged particles in a magnetic field.

→ Explain why a force acts on a wire in a magnetic field when a current flows along the wire.

→ State the equation used to find the force on a moving charge.

Specification reference: 3.7.5.2

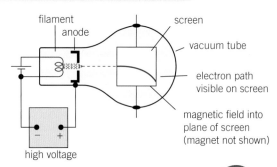

▲ **Figure 1** *An electron deflection tube*

Synoptic link

The positron was the first antimatter particle to be discovered when a β particle track in a magnetic field was found that curved in the opposite direction to the β⁻ tracks. See Topic 1.4, Particles and antiparticles.

Electron beams

Figure 1 shows a vacuum tube designed to show the effect of a magnetic field on an electron beam. The production of the electron beam is explained in Topic 24.3. The path of the beam can be seen where it passes over the fluorescent screen in the tube. The beam is deflected downwards when a magnetic field is directed into the plane of the screen. Each electron in the beam experiences a force due to the magnetic field. The beam follows a circular path because the direction of the force on each electron is perpendicular to the direction of motion of the electron (and to the field direction). The direction of the force on an electron in the beam can also be worked out using Fleming's left-hand rule, provided we remember the convention that the current direction is opposite to the direction in which the electrons move.

The reason why a current-carrying wire in a magnetic field experiences a force is that the electrons moving along the wire are pushed to one side by the force of the field. If the electrons in Figure 1 had been confined to a wire, the whole wire would have been pushed downwards.

A high-energy collision

Magnetic fields are used in particle physics detectors to separate different charged particles out and, as explained in Topic 24.3, to measure their momentum from the curvature of the tracks they create. All charged particles moving across the lines of a magnetic field are acted on by a force due to the field. Positively charged particles such as

▲ **Figure 2** *Charged particles in a magnetic field*

protons are pushed in the opposite direction to negatively charged particles such as electrons. Figure 2 shows charged particles curving across a magnetic field. The positively charged particles curve in the opposite direction to the negatively charged particles. The particles were created by a collision between a fast-moving incoming particle and the nucleus of an atom.

Q: What type of event created the two oppositely-curved tracks near the top of the picture?

A: a pair production event

Force on a moving charge in a magnetic field

A beam of charged particles crossing a vacuum tube is an electric current across the tube. Suppose each charged particle has a charge Q and moves at speed v. In a time interval t, each particle travels a distance vt. Its passage is equivalent to current $I = \dfrac{Q}{t}$ along a wire of length $l = vt$.

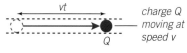

▲ **Figure 3** *Force on a moving charge*

If the particles pass through a uniform magnetic field in a direction at right angles to the field lines, each particle experiences a force F due to the field. If the particles were confined to a wire, the force would be given by $F = BIl$. For moving charges, the same equation applies where $I = \dfrac{Q}{t}$ and $l = vt$.

Therefore, for a charged particle moving across a uniform magnetic field in a direction at right angles to the field, $F = BIl = B\left(\dfrac{Q}{t}\right)(vt) = BQv$

For a particle of charge Q moving through a uniform magnetic field at speed v in a perpendicular direction to the field, the force on the particle is given by

$$F = BQv$$

If the direction of motion of a charged particle in a magnetic field is at angle θ to the lines of the field, then the component of B perpendicular to the direction of motion of the charged particle, $B\sin\theta$, is used to give $F = BQv\sin\theta$.

- If the velocity of the charged particle is perpendicular to the direction of the magnetic field, $\theta = 90°$, so the equation becomes $F = BQv$ because $\sin 90° = 1$.

- If the velocity of the charged particle is parallel to the direction of the magnetic field, $\theta = 0$, so $F = 0$ because $\sin 0 = 0$.

Note:
The equation $F = BQv\sin\theta$ is not required for this specification.

Study tip

Stationary charges in a magnetic field experience no **magnetic force**. Also when applying Fleming's left-hand rule to charged particles, the current direction for negative particles is in the opposite direction to the direction of motion of the particles.

Note

The Hall probe is not on the specification and is presented here as an application of the equations for the electric and magnetic force on a charged particle.

The Hall probe

Hall probes are used to measure magnetic flux density (Figure 4). A Hall probe contains a slice of semiconducting material. Figure 5 shows the slice in a magnetic field with the field lines perpendicular to the flat side of the slice. A constant current passes through the slice as shown. The charge carriers (which are electrons in an n-type semiconductor) are deflected by the field. As a result, a potential difference is created between the top and bottom edges of the slice. This effect is known as the Hall effect after its discoverer.

▲ **Figure 4** Using a Hall probe

The pd, referred to as the Hall voltage, is proportional to the magnetic flux density, provided the current is constant. This is because each charge carrier passing through the slice is subjected to a magnetic force $F_{mag} = BQv$, where v is the speed of the charge carrier. Once the Hall voltage has been created, the magnetic deflection of a charge carrier entering the slice is opposed by the force on it due to the electric field created by the Hall voltage. The electric field force $F_{elec} = \dfrac{QV_h}{d}$, where V_h represents the Hall voltage and d is the distance between the top and bottom sides of the slice. See Topic 22.2. Therefore, $\dfrac{QV_h}{d} = BQv$ gives $V_h = Bvd$. For constant current, v is constant so V_h is proportional to B.

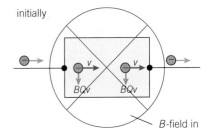

initially

B-field in

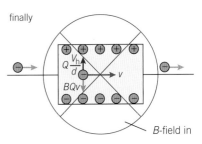

finally

B-field in

▲ **Figure 5** The Hall voltage

Summary questions

$e = 1.6 \times 10^{-19}$ C

1 a In Figure 1, how would the force on the electrons in the magnetic field differ if:

   i   the magnetic field was reversed in direction,

   ii the magnetic field was reduced in strength,

   iii the speed of the electrons was increased.

   b Calculate the force on an electron that enters a uniform magnetic field of flux density 150 mT at a velocity of 8.0×10^6 m s^{-1} at an angle of

   i   90°,   ii 0° to the field.

2 Electrons in a vertical wire move upwards at a speed of 2.5×10^{-3} m s^{-1} into a uniform horizontal magnetic field of magnetic flux density 95 mT. The field is directed along a line from south to north as shown in Figure 6. Calculate the force on each electron and determine its direction.

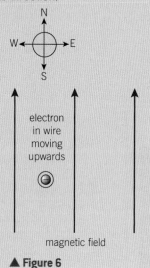

▲ Figure 6

3 A beam of protons and π^+ mesons moving at the same speed is directed into a uniform magnetic field in the same direction as the field.

   a Explain why the beam is not deflected by the field.

   b If the particles had been directed into the field in a direction at right angles to the field lines at the same speed, state and explain what effect this would have had on the beam.

4 In a Hall probe, electrons passing through the semiconductor experience a force due to a magnetic field.

   a Explain why a potential difference is created across the semiconductor as a result of the application of the magnetic field.

   b When the magnetic flux density was 90 mT, each electron moving through the slice experiences a force of 6.4×10^{-20} N due to the magnetic field. Calculate:

   i   the mean speed of the electrons passing through the slice,

   ii the magnetic force on each electron if the magnetic flux density is increased to 120 mT.

24.3 Charged particles in circular orbits

Magnetic fields are used to control beams of charged particles in many devices, from television tubes to high-energy accelerators. The force of the magnetic field on a moving charged particle is at right angles to the direction of motion of the particle.

- No work is done by the magnetic field on the particle as the force always acts at right angles to the velocity of the particle. Its direction of motion is changed by the force but not its speed. The kinetic energy of the particle is unchanged by the magnetic field.
- In accordance with Fleming's left-hand rule, the magnetic force is always perpendicular to the velocity at any point along the path. The particle therefore moves on a circular path with the force always acting towards the centre of curvature of the circular path. See Topic 17.2.
- The force causes a centripetal acceleration because it is perpendicular to the velocity. Figure 1 shows the deflection of a beam of electrons in a uniform magnetic field. The path is a complete circle because the magnetic field is uniform and the particle remains in the field.

The radius, r, of the circular orbit in Figure 1 depends on the speed v of the particles and the magnetic flux density B.

At any point on the orbit, the particle is acted on by a magnetic force $F = BQv$ and it experiences a centripetal acceleration $a = \dfrac{v^2}{r}$ towards the centre of the circle.

Applying Newton's second law in the form $F = ma$ gives

$$BQv = \frac{mv^2}{r}$$

Rearranging this equation gives

$$r = \frac{mv}{BQ}$$

This equation shows that r decreases (so the path is more curved):

1 if B is increased or if v is decreased,
2 if particles with a larger specific charge, $\dfrac{Q}{m}$, are used.

Thermionic devices

In Figure 1, the beam of electrons is produced by an 'electron gun'. This consists of an electrically heated filament wire near a positively charged metal anode which attracts electrons emitted by the hot filament wire. This emission process is called thermionic emission. The electrons pass through a small hole in the anode to form the beam. The greater the potential difference between the anode and the filament wire, the higher the speed of the electrons when they reach the anode so the faster they are in the beam. The oscilloscope, the cathode ray television tube and the magnetron valve used in microwave cookers and radar systems all rely on thermionic emission.

Learning objectives:

→ Describe what happens to the direction of the magnetic force when electrons are deflected by a magnetic field.

→ Explain why the moving charges move in a path that is circular.

→ State the factors that affect the radius of the circular path.

Specification reference: 3.7.5.1

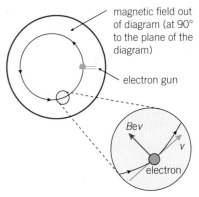

▲ **Figure 1** *A circular orbit in a magnetic field*

Synoptic link

We can apply the centripetal acceleration formula $a = \dfrac{v^2}{r}$ because the particle is in uniform circular motion. See Topic 17.2, Centripetal acceleration.

Study tip

Check whether a question refers to radius or diameter. Remember that equipment must be evacuated to prevent loss of speed.

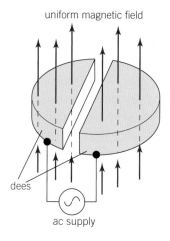

uniform magnetic field

dees

ac supply

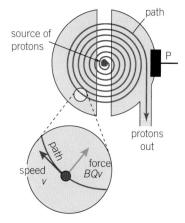

source of protons

path

P

protons out

path

speed
v

force
BQv

▲ **Figure 2** *The cyclotron*

The cyclotron

The cyclotron is used in hospitals to produce high-energy beams for radiation therapy. It consists of two hollow D-shaped electrodes (referred to as 'dees') in a vacuum chamber. With a uniform magnetic field applied perpendicular to the plane of the dees, a high-frequency alternating voltage is applied between the dees.

Charged particles are directed into the one of the dees near the centre of the cyclotron. The charged particles are forced on a circular path by the magnetic field, causing them to emerge from the dee they were directed into. As they cross into the other dee, the alternating voltage reverses so they are accelerated into the other dee where they are once again forced on a circular path by the magnetic field. On emerging from this dee, the voltage reverses again and accelerates the particles into the first dee where the process is repeated. This occurs because the time taken by a particle to move round its semi-circular path in each dee does not depend on the particle speed (provided the speed stays much less than the speed of light c). This is because $r = \frac{mv}{BQ}$ so the time taken to complete the semi-circle $= \frac{\pi r}{v} = \frac{m\pi}{BQ}$, which is independent of the particle's speed.

1   Each time a particle crosses from one dee to the other it gains speed and its radius of orbit increases. The particles emerge from the cyclotron when the radius of orbit is equal to the dee radius R.
    Using the equation $r = \frac{mv}{BQ}$, it follows that the speed v of the particles on exit from the cyclotron is given by $v = \frac{BQR}{m}$.

2   The time T for one full cycle of the alternating voltage must be equal to the time taken by a particle to complete one full circle.
    Hence $T = \frac{2m\pi}{BQ}$. Therefore the frequency f of the alternating voltage must be set at a value given by the equation $f = \frac{1}{T} = \frac{BQ}{2\pi m}$.

The mass spectrometer

The mass spectrometer is used to analyse the type of atoms present in a sample. The atoms of the sample are ionised and directed in a narrow beam at the same velocity into a uniform magnetic field. Each ion is deflected in a semi-circle by the magnetic field onto a detector, as shown in Figure 3. The radius of curvature of the path of each ion depends on the specific charge $\frac{Q}{m}$ of the ion in accordance with the equation $r = \frac{mv}{BQ}$. Each type of ion is deflected by a different amount onto the detector. The detector is linked to a computer which is programmed to show the relative abundance of each type of ion in the sample.

The ions in the beam enter the magnetic field at the same velocity because they pass through a velocity selector,

as shown in Figure 3. The velocity selector consists of a magnet and a pair of parallel plates at spacing d and voltage V_p due to a high voltage supply. The magnet and the plates are aligned so each ion passing through the velocity selector is acted on by an electric field force, $F_{elec} = \frac{QV_p}{d}$, in the opposite direction to a magnetic field force $F_{mag} = B_S Qv$ where B_S is the magnetic flux density of the magnet in the velocity selector. Ions moving at a certain velocity such that $B_S Qv = \frac{QV_p}{d}$ experience equal and opposite forces so pass through undeflected. All other ions are deflected and do not pass through the collimator slit. So although the beam emerging from the

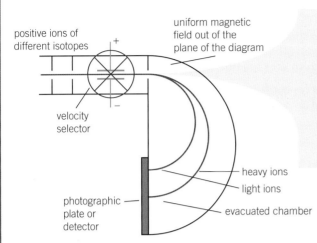

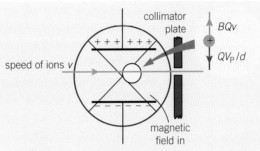

▲ Figure 3 *The mass spectrometer*

Synoptic link

The magnetic field force is cancelled out by the electric field force. See Topic 22.2, Electric field strength, for the force on a charged particle in a uniform electric field.

collimator consists of different types of ions, they all have the same speed $v = \dfrac{V_p}{B_s d}$.

Q: State whether a proton or a singly charged helium ion is deflected more easily in a mass spectrometer.

A: a proton

Summary questions

$e = 1.6 \times 10^{-19}$ C, $\dfrac{e}{m}$ for the electron $= 1.76 \times 10^{11}$ C kg^{-1}

1 A beam of electrons at a speed of 3.2×10^7 m s^{-1} is directed into a uniform magnetic field of flux density 8.5 mT in a direction perpendicular to the field lines. The electrons move on a circular orbit in the field.

   a i Explain why the electrons move on a circular orbit.

   ii Calculate the radius of the orbit.

   b The flux density is adjusted until the radius of orbit is 65 mm. Calculate the flux density for this new radius.

2 A narrow beam of electrons was directed at a speed of 2.9×10^7 m s^{-1} into a uniform magnetic field.

   a The beam followed a circular path of radius 35 mm in the magnetic field. Calculate the flux density of the magnetic field.

   b The speed of the electrons in the beam was halved by reducing the anode voltage. Calculate the new radius of curvature of the beam in the field.

3 The first cyclotron, used to accelerate protons, was 0.28 m in diameter and was in a magnetic field of flux density 1.1 T.

   a Show that protons emerged from this cyclotron at a maximum speed of 1.5×10^7 m s^{-1}.

   b Calculate the maximum kinetic energy, in MeV, of a proton from this accelerator.

   The mass of a proton $= 1.67 \times 10^{-27}$ kg, 1 MeV $= 1.6 \times 10^{-13}$ J

4 In a mass spectrometer, a beam of ions at a speed of 7.6×10^4 m s^{-1} was directed into a uniform magnetic field of flux density 680 mT.

   a An ion was deflected in a semi-circular path of diameter 28 mm on to the detector. Calculate the specific charge of the ion.

   b A different type of ion was deflected onto the same detector when the magnetic flux density was changed to 400 mT. Calculate the specific charge of this ion.

1 (a) The equation $F = BIl$, where the symbols have their usual meanings, gives the magnetic force that acts on a conductor in a magnetic field.
   Give the unit of each of the quantities in the equation: F, B, I, l.
   State the condition under which the equation applies. *(2 marks)*

(b) **Figure 1** shows a horizontal copper bar of 25 mm × 25 mm square cross-section and length l carrying a current of 65 A.

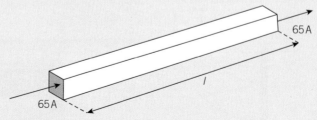

▲ **Figure 1**

(i) Calculate the minimum value of the flux density of the magnetic field in which it should be placed if its weight is to be supported by the magnetic force that acts on it.

density of copper = $8.9 \times 10^3 \, \mathrm{kg \, m^{-3}}$

(ii) Copy the diagram and draw an arrow to show the direction in which the magnetic field should be applied if your calculation in part (i) is to be valid. Label this arrow M. *(5 marks)*

AQA, 2003

2 A 'bus bar' is a metal bar which can be used to conduct a large electric current. In a test, two bus bars, **X** and **Y**, of length 0.83 m are clamped at either end parallel to each other, as shown in **Figure 2**.

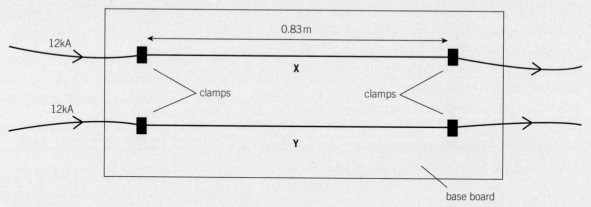

▲ Figure 2

(a) When a constant current of 12 kA is carried by each bus bar, they exert a force of 180 N on each other. This force is due to the magnetic field created by the current carried by each bus bar.
(i) Calculate the magnetic flux density due to the current in one bus bar at the position of the other bus bar.
(ii) The magnetic flux density at any given distance from a straight conductor is proportional to the current through the conductor. Calculate the force on each bus bar if X carried a current of 6 kA and Y carried a current of 12 kA in the same direction. *(6 marks)*

(b) When the same alternating current is passed through the two bus bars, both vibrate strongly.

   (i)   Explain why the bars vibrate.

   (ii) State **one** way the amplitude of the vibrations could be reduced without reducing the current.

   (4 marks)

   AQA, 2007

3 **(a)**

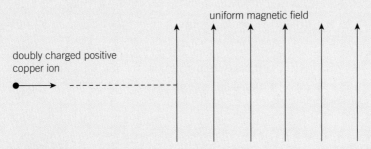

▲ **Figure 3**

Figure 3 shows a doubly charged positive ion of the copper isotope $^{63}_{29}\text{Cu}$ that is projected into a vertical magnetic field of flux density 0.28 T, with the field directed upwards. The ion enters the field at a speed of $7.8 \times 10^5 \,\text{m s}^{-1}$.

   (i)   State the initial direction of the magnetic force that acts on the ion.

   (ii) Describe the subsequent path of the ion as fully as you can. Your answer should include both a qualitative description and a calculation.

   mass of $^{63}_{29}\text{Cu}$ ion = 1.05×10^{-25} kg   *(5 marks)*

(b) State the effect on the path in part (a) if the following changes are made separately.

   (i)   The strength of the magnetic field is doubled.

   (ii) A singly charged positive Cu ion replaces the original one.   *(3 marks)*

   AQA, 2004

4 **Figure 4** shows a diagram of a mass spectrometer.

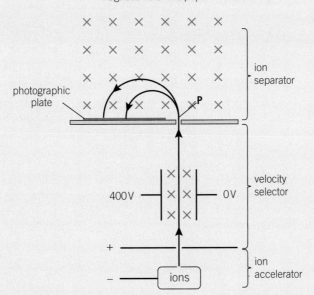

▲ **Figure 4**

(a) The magnetic field strength in the velocity selector is 0.14 T and the electric field strength is 20 000 V m⁻¹.

    (i) Define the unit for magnetic flux density, the tesla.

    (ii) Show that the velocity selected is independent of the charge on an ion.

    (iii) Show that the velocity selected is about 140 km s⁻¹.   *(5 marks)*

(b) A sample of nickel is analysed in the spectrometer. The two most abundant isotopes of nickel are $^{58}_{28}\text{Ni}$ and $^{60}_{28}\text{Ni}$. Each ion carries a single charge of $+1.6 \times 10^{-19}$ C.

The $^{58}_{28}\text{Ni}$ ion strikes the photographic plate 0.28 m from the point **P** at which the ion beam enters the ion separator.

Calculate:

    (i) the magnetic flux density of the field in the ion separator,

    (ii) the separation of the positions where the two isotopes hit the photographic plate.   *(5 marks)*

                                                             AQA, 2003

5   The protons in an accelerator were directed at a solid target, causing antiprotons and negative pions, as well as other particles and antiparticles, to emerge at high speed from the target. A uniform magnetic field was used to separate the negative particles from the uncharged and positive particles, as shown in **Figure 5**.

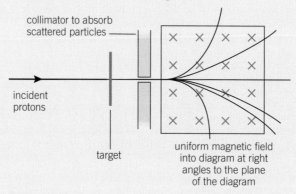

▲ **Figure 5**

(a) Show that the speed, *v*, of a charged particle moving in a circular path of radius *r* in a uniform magnetic field *B* is given by

$$v = \frac{BQr}{m}$$

where *m* is the mass of the particle and *Q* is its charge.   *(1 mark)*

(b) An antiproton and a negative pion follow the same path in the magnetic field. Explain why they have the same momentum but different speeds.   *(3 marks)*

(c) State, in terms of quarks and antiquarks, the composition of each of the following: antiproton, negative pion.   *(3 marks)*

                                                             AQA, 2004

6   **Figure 6** shows the arrangement of an apparatus for determining the masses of ions. In an evacuated chamber, positive ions from an ion source pass through the slit at **P** with the same velocity *v*. After passing **P**, the ions enter a region over which a uniform magnetic field is applied. The ions travel in a semi-circular path of diameter *d* and are detected at points such as **R**.

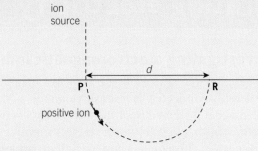

▲ **Figure 6**

(a) (i) State the direction of the applied magnetic field.
   (ii) Explain why the ions travel in a semi-circular path whilst in the magnetic field.
   (iii) By considering the force that acts on an ion of mass m and charge Q, having velocity v, show that the diameter d of the path of the ions is given by

$$d = \frac{2mv}{BQ}$$

   where B is the flux density of the magnetic field. (*7 marks*)

(b) In an experiment using singly ionised magnesium ions travelling at a velocity of $7.5 \times 10^4 \, \mathrm{m\,s^{-1}}$, d was $110 \, \mathrm{mm}$ when B was $0.34 \, \mathrm{T}$. Use this result to calculate the charge to mass ratio of these ions. (*2 marks*)

(c) (i) Some ions of the same element, whilst travelling at the same velocity as each other at **P**, may arrive at a point that is close to, but slightly different from, **R**. Explain why this might happen.
   (ii) Other ions of the same element, also travelling at the same velocity at **P** as all of the others, may travel in a path whose diameter is half that of the others. Explain why this might happen. (*3 marks*)

AQA, 2007

7 **Figure 7** shows the path of protons in a proton synchrotron.

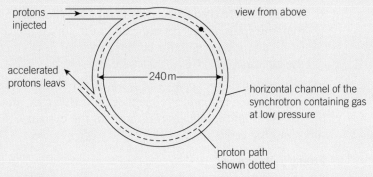

▲ **Figure 7**

The protons are injected at a speed of $1.2 \times 10^5 \, \mathrm{m\,s^{-1}}$ and a magnetic field is applied to make them move in a circular path.

(a) Calculate the magnetic flux density of the field required for protons to move in the circular path when their speed is $1.2 \times 10^5 \, \mathrm{m\,s^{-1}}$.

(b) Explain how the magnetic flux density required to maintain the circular path has to change as the kinetic energy of the protons increases. (*5 marks*)

AQA, 2007

Learning objectives:

→ Describe what must happen to a conductor (or to the magnetic field in which it's placed) for electricity to be generated.

→ State the factors that would cause the induced emf to be greater.

→ Discuss whether an induced emf always causes a current to flow.

Specification reference: 3.7.5.4

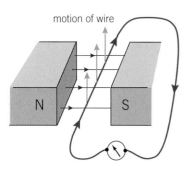

▲ **Figure 1** *Generating an electric current*

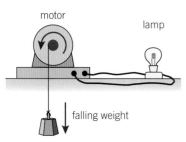

▲ **Figure 2** *A motor as a generator*

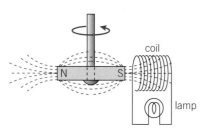

▲ **Figure 3** *A simple dynamo*

Investigating electromagnetic induction ·

To generate electricity, all you need is a magnet and some wire, preferably connected to a sensitive meter, as shown in Figure 1. When the magnet is moved near the wire, a small current passes through the meter. This happens because an electromotive force (emf) is **induced** in the wire. This effect, known as **electromagnetic induction**, occurs whenever a wire cuts across the lines of a magnetic field. If the wire is part of a complete circuit, the **induced emf** forces electrons round the circuit. The induced emf can be increased by:

• moving the wire faster,

• using a stronger magnet,

• making the wire into a coil, as in Figure 3, and pushing the magnet in or out of the coil.

No emf is induced in the wire if the wire is parallel to the magnetic field lines as it moves through the field. The wire must cut across the lines of the magnetic field for an emf to be induced in the wire.

Other methods of generating an induced emf include:

1 **Using an electric motor in reverse,** as in Figure 2. The falling weight makes the motor coil turn between the poles of the magnet in the motor. The emf induced in the coil forces a current round the circuit and so causes the lamp to light. The faster the coil turns, the brighter the lamp is.

2 **Using a cycle dynamo,** as in Figure 3. When the magnet in the dynamo spins, an emf is induced in the coil. If the coil is connected to a lamp, the lamp lights because the emf forces a current round the circuit.

In both examples above, an emf is induced because there is relative motion between coil and the magnet. In the electric motor in reverse, the coil spins and the magnet is fixed. In the dynamo, the magnet spins and the coil is fixed.

Energy changes

When a magnet is moved relative to a conductor (e.g., a wire or a coil), an emf is induced in the conductor. If the conductor is part of a complete circuit which has no other sources of emf, a current passes round the circuit just as if the circuit included a battery. However, unlike the emf of a battery which is constant, the induced emf becomes zero when the relative motion between the magnet and the wires ceases.

An electric current transfers energy from the source of the emf in a circuit to the other components in the circuit. For example, when a dynamo is used to light a lamp, energy is transferred from the dynamo to the lamp. The current through the dynamo coil causes a reaction force on the coil due to the magnet. Work must therefore be done to keep the magnet spinning. The energy transferred from the coil to the lamp is equal to the work done on the coil to keep it spinning, assuming no energy is wasted as sound or due to friction or internal resistance.

The rate of transfer of energy from the source of emf to the other components of the circuit is equal to the product of the induced emf and the current. This is because:

- the induced emf is the energy transferred from the source per unit charge that passes through the source,
- the current is the charge flow per second.

So the induced emf × the current = energy transferred per unit charge from the source × the charge flow per second = energy transferred per second from the source.

The discovery of electromagnetic induction

Electromagnetic induction was discovered by Michael Faraday in 1831 at the Royal Institution, London. Faraday knew that a current passing along a wire produces a magnetic field near the wire and he wanted to know if a magnet could be used to produce a current. Using a magnetic compass near a loop of wire as a detector of current, he showed that the compass deflected whenever the magnet was moved in or out of the wire. He used the term 'electromotive force' (emf) to describe the voltage induced in a wire.

▲ **Figure 4** *Michael Faraday holding a bar magnet — when he demonstrated his discoveries at the Royal Institution, he was asked 'What use is electricity?' He replied, 'What use is a new baby?' No one can tell what can grow from a new discovery*

Understanding electromagnetic induction

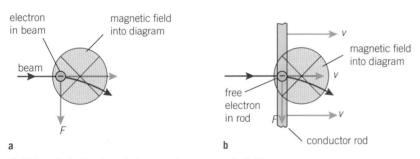

▲ **Figure 5** *Deflection of electrons in a magnetic field*

When a beam of electrons is directed across a magnetic field, each electron experiences a force at right angles to its direction of motion and to the field direction. A metal rod is a tube containing lots of free electrons. If the rod is moved across a magnetic field, as shown in Figure 5, the magnetic field forces the free electrons in the rod to one end away from the other end. So one end of the rod becomes negative and the other end positive. In this way, an emf is induced in the rod. The same effect happens if the magnetic field is moved and the rod is stationary. As long as there is relative motion between the rod and the magnetic field, an emf is induced in the rod. If the relative motion ceases, the induced emf becomes zero because the magnetic field no longer exerts a force on the electrons in the rod. Note that when the rod is part of a complete circuit, the electrons are forced round the circuit. In other words, the induced emf drives a current round the circuit.

Study tip

Think carefully about which of Fleming's rules applies to generators – the riGht-hand rule is for Generators.

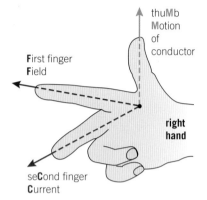

▲ **Figure 6** *Fleming's right-hand rule*

The dynamo rule

In Figure 5, the magnetic field is into the plane of the diagram and the motion of the conductor relative to the rod is rightwards. The electrons in the rod are forced downwards. The direction of the induced current can also be worked out using **Fleming's right-hand rule**, also referred to as the **dynamo rule**, as shown in Figure 6. The direction of the induced current is, in accordance with the current convention, opposite to the direction of the flow of electrons in the conductor.

Summary questions

1   A coil of wire is connected to a sensitive meter.

  a   Explain why the meter shows a brief reading when a magnet is pushed into the coil.

  b   State two ways in which the meter reading could be made larger.

2   An electric motor consists of a coil of wire between the poles of a magnet. The motor is connected to a lamp. A thread wrapped round the motor spindle is used to support a weight, as shown in Figure 2 earlier in this topic.

  a   Explain why the lamp lights when the weight descends.

  b   What difference would have been made if the magnet had been much stronger?

  c   Explain why a lamp connected to a dynamo lights when the dynamo turns.

  d   Why is the dynamo easier to turn when the lamp is disconnected?

3   A horizontal rod aligned along a line from east to west is dropped through a horizontal magnetic field which is directed from south to north.

  a   **i**   What is the direction of the velocity of the rod?

    ii   Determine which end of the rod is positive. Explain your answer.

  b   Explain why no emf is induced in the rod if it is aligned from north to south then dropped in the field.

Coils, currents, and fields

A magnetic field is produced in and around a coil when it is connected to a battery and a current is passed through it. A magnetic compass near the coil is deflected when current passes through it. For a long coil or solenoid, the pattern of the magnetic field lines is like the pattern for a bar magnet – except the magnetic field lines near a bar magnet loop round from the north pole to the south pole of the magnet. Figure 1 shows the magnetic field pattern of a current-carrying solenoid. The field lines pass through the solenoid and loop round outside the solenoid from one end (the north pole) to the other end (the south pole). If each end in turn is viewed from outside the solenoid:

- current passes a**N**ticlockwise (or cou**N**terclockwise) round the '**N**orth pole' end,

- current passes clockwise round the 'south pole' end.

Lenz's law

When a bar magnet is pushed into a coil connected to a meter, the meter deflects. If the bar magnet is pulled out of the coil, the meter deflects in the opposite direction. What determines the direction of the induced current? Consider the north pole of a bar magnet approaching end X of a coil, as shown in Figure 2.

The induced current passing round the circuit creates a magnetic field due to the coil. The coil field must act against the incoming north pole, otherwise it would pull the N-pole in faster, making the induced current bigger, pulling the N-pole in even faster still, etc. Clearly, conservation of energy forbids this creation of kinetic and electrical energy from nowhere. So the induced current creates a magnetic field in the coil which opposes the incoming N-pole. The induced polarity of end X must therefore be a N-pole so as to repel the incoming N-pole. Therefore, the current must go round end X of the coil in an anticlockwise direction, as shown.

If the magnet is removed from inside the coil, the induced current passes round end X of the coil in a clockwise direction. This corresponds to an induced S-pole at end X which therefore opposes the magnet moving away.

> **Lenz's law** states that the direction of the induced current is always such as to oppose the change that causes the current.

The explanation of Lenz's law is that energy is never created or destroyed. The induced current could never be in a direction to help the change that causes it; that would mean producing electrical energy from nowhere, which is forbidden!

Learning objectives:
→ Define the magnetic flux and the magnetic flux linkage.
→ Relate the induced emf in a coil to the magnetic flux linkage through it.
→ State Lenz's law and the conservation law that explains it.

Specification reference: 3.7.5.3; 3.7.5.4

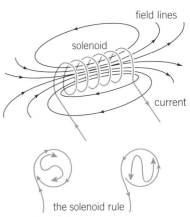

▲ **Figure 1** *The magnetic field near a solenoid*

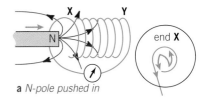

a *N-pole pushed in*

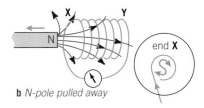

b *N-pole pulled away*

▲ **Figure 2** *Lenz's law*

Regenerative braking

A battery-powered or hybrid electric vehicle contains an alternator that can be used as an electric motor or as a generator. When the alternator is used as an electric motor, it is driven by the batteries. When the brakes are applied, the alternator is used to generate electricity which is used to recharge the battery. Some of the kinetic energy is transferred to electrical energy in the battery. The induced current through the alternator coil creates a magnetic field that acts against the magnetic field of the alternator. So the alternator experiences a braking force which helps to slow the vehicle down.

The fuel consumption of a hybrid vehicle (i.e., a vehicle with a petrol engine and an electric motor) is significantly less than that of a petrol-only vehicle. This is because some of the hybrid vehicle's kinetic energy is converted to chemical energy in its battery when the vehicle brakes. The battery supplies this energy to the electric motor when it takes over from the petrol engine at low speeds.

▲ **Figure 3** *An electric car*

Faraday's law of electromagnetic induction

Consider a conductor of length *l* which is part of a complete circuit cutting through the lines of a magnetic field of flux density *B*.

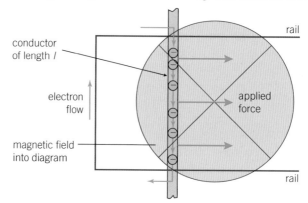

▲ **Figure 4** *Induced emf in a conductor*

An emf ε is induced in the conductor and an induced current *I* passes round the circuit.

The conductor experiences a force $F = BIl$ due to carrying a current in a magnetic field. The force opposes the motion of the conductor and so an equal and opposite force must be applied to the conductor to keep it moving in the field. If the conductor moves a distance Δs in time Δt,

- the work done *W* by the applied force is given by $W = F\Delta s = BIl\Delta s$.
- the charge transfer along the conductor in this time is $Q = I\Delta t$.

Therefore, the induced emf is $\varepsilon = \dfrac{W}{Q} = \dfrac{BIl\Delta s}{I\Delta t} = \dfrac{Bl\Delta s}{\Delta t}$

As $l\Delta s$ is the area *A* 'swept out' by the conductor in time Δt,

the induced emf is $\varepsilon = \dfrac{BA}{\Delta t}$

Note

Think carefully about the energy changes here. The energy transferred by the applied force (i.e., the work done) is transformed to electric energy. The circuit needs to be complete otherwise no induced current flows, no electrical energy is transferred, and the applied force would make the rod move faster and faster.

The product of the **magnetic flux density**, *B*, and the area, *A*, swept out (= *BA*), is called the **magnetic flux**. The concept of magnetic flux is very useful for calculating induced emfs. The example of the conductor cutting across the field lines shows that the induced emf is equal to the magnetic flux swept out by the conductor each second. Michael Faraday was the first person to show how induced emfs could be calculated from magnetic flux changes.

- Magnetic flux $\phi = BA$.
- **Magnetic flux linkage** through a coil of *N* turns $= N\phi = NBA$ where *B* is the magnetic flux density perpendicular to area *A*.
- The unit of magnetic flux is the **weber** (Wb), equal to $1\,T\,m^2$.

Note that flux density *B* (in teslas) is the flux per unit area passing at right angles (i.e., normally) through the area.

Therefore 1 tesla = 1 weber per square metre.

More about flux linkage

1   When the magnetic field is along the normal (i.e., perpendicular) to the coil face, the flux linkage = $N\phi = BAN$.
2   When the coil is turned through 180°, the flux linkage = $-BAN$.
3   When the magnetic field is parallel to the coil area, the flux linkage = 0 as no field lines pass through the coil area.

In general, when the magnetic field is at angle θ to the normal at the coil face, the flux linkage through the coil is $N\phi = BAN \cos\theta$

Faraday's law of electromagnetic induction states that the **induced emf in a circuit is equal to the rate of change of flux linkage through the circuit.**

$$\text{Induced emf } \varepsilon = -N\frac{\Delta\phi}{\Delta t}$$

where $N\dfrac{\Delta\phi}{\Delta t}$ is the change of flux linkage per second.

The minus sign represents the fact that the induced emf acts in such a direction as to opposite the change that causes it (as per Lenz's law).

Examples

1 A moving conductor in a magnetic field

An emf is induced in the conductor provided the conductor cuts across the lines of the magnetic field. The direction of motion of the conductor in Figure 4 is at right angles to the field lines.

As explained earlier, the magnitude of the induced emf $\varepsilon = \dfrac{Bl\Delta s}{\Delta t}$, where l is the length of the conductor and Δs is the distance it moves in time Δt. Note that the change of flux in this time, $\Delta\phi = Bl\Delta s$ so the change of flux per second, $\dfrac{\Delta\phi}{\Delta t}$, is equal to the magnitude of the induced emf.

Because the speed of the conductor, $v = \dfrac{\Delta s}{\Delta t}$, the induced emf

$$\varepsilon = \frac{Bl\,\Delta s}{\Delta t} = Blv$$

$$\text{Induced emf } \varepsilon = Blv$$

2 A fixed coil in a changing magnetic field

Figure 6 shows a small coil on the axis of a current-carrying solenoid. The magnetic field of the solenoid passes through the small coil. If the current in the solenoid changes, an emf is induced in the small coil. This is because the magnetic field through the coil changes so the flux linkage through it changes, causing an induced emf.

The flux linkage through the coil, $N\phi = BAN$, where A is the coil area and N is the number of turns of the coil. Suppose the magnetic flux density changes from B to $B + \Delta B$ in time Δt so the flux linkage changes by an amount $N\Delta\phi$ (= $\Delta B\,AN$).

The magnitude of the induced emf $= \dfrac{N\Delta\phi}{\Delta t} = \dfrac{AN\Delta B}{\Delta t}$

Because B is proportional to the current I in the solenoid, the magnitude of the induced emf is therefore proportional to the rate of change of current in the solenoid.

Note

1   The unit of flux change per second, the weber per second, is the same as the volt. Therefore, the weber is equal to 1 volt second.

2   Whenever the flux linkage through a circuit changes, an emf is induced in the circuit. The flux can be due to a permanent magnet or due to a current-carrying wire.

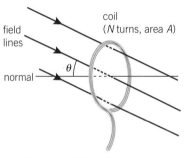

flux linkage = $BAN \cos\theta$

▲ **Figure 5** *Flux linkage*

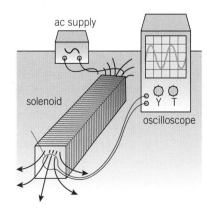

▲ **Figure 6** *A changing magnetic field*

flux linkage

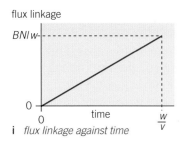

i *flux linkage against time*

induced emf

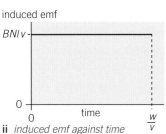

ii *induced emf against time*

▲ **Figure 7** *Flux changes*

3 A rectangular coil moving into a uniform magnetic field

Consider a rectangular coil of N turns, length l and width w moving into a uniform magnetic field of flux density B at constant speed v. Figure 8 shows a similar situation. Suppose the coil enters the field at time $t = 0$.

- The time taken by the coil to enter the field completely
  $$= \frac{\text{coil width}}{\text{speed}} = \frac{w}{v}.$$
  During this time, the flux linkage $N\phi$ increases steadily from 0 to $BNlw$. Therefore, the change of flux linkage per second,
  $$\frac{N\Delta\phi}{\Delta t} = \frac{BNlw}{w/v} = BNlv$$

- When the coil is completely in the field, the flux linkage through it ($=BNlw$) does not change so the induced emf is zero. In other words, the emf induced in the leading side is cancelled by the emf induced in the trailing side once the trailing side has entered the field. Figure 7 shows how the flux linkage and the induced emf change with time.

Summary questions

1 A uniform magnetic field of flux density 72 mT is confined to a region of width 60 mm, as shown in Figure 8. A rectangular coil of length 50 mm and width 20 mm has 15 turns. The coil is moved into the magnetic field at a speed of 10 mm s^{-1} with its longer edge parallel to the edge of the magnetic field.

   a Calculate:

   i the flux linkage through the coil when it is completely in the field,

   ii the time taken for the flux linkage to increase from zero to its maximum value,

   iii the induced emf in the coil as it enters the field.

   b i Sketch a graph to show how the flux linkage through the coil changes with time from the instant the coil enters the field to when it leaves the field completely.

   ii Sketch a graph to show how the induced emf in the coil varies with time.

2 A rectangular coil of length 40 mm and width 25 mm has 20 turns. The coil is in a uniform magnetic field of flux density 68 mT.

   a Calculate the flux linkage through the coil when the coil is at right angles to the field lines.

   b The coil is removed from the field in 60 ms. Calculate the mean value of the induced emf.

3 A circular coil of diameter 24 mm has 40 turns. The coil is placed in a uniform magnetic field of flux density 85 mT with its plane perpendicular to the field lines.

   a Calculate:

   i the area of the coil in m^2,

   ii the flux linkage through the coil.

   b The coil was reversed in a time of 95 ms. Calculate:

   i the change of flux linkage through the coil,

   ii the magnitude of the induced emf.

4 A small circular coil of diameter 15 mm and 25 turns is placed in a fixed position on the axis of a solenoid, as shown in Figure 7. The magnetic flux density of the solenoid at this position varies with current according to the equation $B = kI$, where $k = 1.2 \times 10^{-3}\,\text{T}\,\text{A}^{-1}$.

   a Calculate the flux linkage through the coil when the current in the solenoid is 1.5 A.

   b The current in the solenoid was reduced from 1.5 A to zero in 0.20 s. Calculate the magnitude of the induced emf in the small coil.

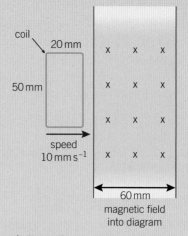

▲ **Figure 8**

The alternating current generator

The simple ac generator consists of a rectangular coil that spins in a uniform magnetic field, as shown in Figure 1. When the coil spins at a steady rate, the flux linkage changes continuously. At an instant when the normal to the plane of the coil is at angle θ to the field lines, the flux linkage through the coil is $N\phi = BAN\cos\theta$, where B is the magnetic flux density, A is the coil area, and N is the number of turns on the coil.

For a coil spinning at a steady frequency, f, $\theta = 2\pi ft$ at time t after $\theta = 0$. So the flux linkage $N\phi$ (= $BAN\cos 2\pi ft$) changes with time as shown in Figure 2.

- The gradient of the graph is the change of flux linkage per second, $\dfrac{N\Delta\phi}{\Delta t}$, so it represents the induced emf. It can be shown mathematically that the induced emf alternates according to the equation

$$\varepsilon = \varepsilon_0 \sin 2\pi ft$$

where f is the frequency of rotation of the coil and ε_0 is the peak emf. The above equation may be written as $\varepsilon = \varepsilon_0 \sin\omega t$ as the angular frequency of the coil is $\omega = 2\pi f$.

- The induced emf is zero when the sides of the coil move parallel to the field lines. At this position, the rate of change of flux is zero and the sides of the coil do not cut the field lines.

- The induced emf is a maximum when the sides of the coil cut the field lines at right angles. At this position, the emf induced in each wire of each side is Blv, where v is the speed of each wire and l is its length. So for N turns and two sides, the induced emf at this position is $\varepsilon_0 = 2NBlv$. This shows the peak emf can be increased by increasing the speed (i.e., the frequency of rotation) or using a stronger magnet, a bigger coil, or a coil with more turns.

Note that $v = \omega\dfrac{d}{2}$, where d is the width of the coil. So $\varepsilon_0 = 2NBlv = BAN\omega$ where the coil area $A = ld$. So $\varepsilon = BAN\omega \sin\omega t$.

Learning objectives:
→ State the two features of the output voltage waveform that change if the coil is turned faster.

→ Explain why the output alternates.

→ Explain why it is preferable for practical generators to have fixed coils and a rotating (electro)magnet.

Specification reference: 3.7.5.4

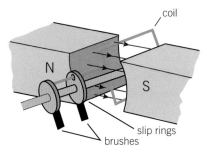

▲ **Figure 1** *The ac generator*

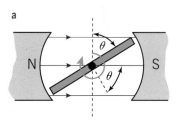

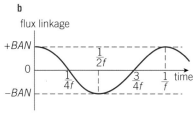

▲ **Figure 2** *Flux linkage in a spinning coil*

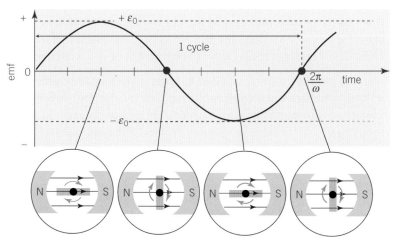

▲ **Figure 3** *Emf against time for an ac generator*

Synoptic link

Remember that the peak value of an alternating pd is $\sqrt{2}$ × the rms value. See Topic 25.4.

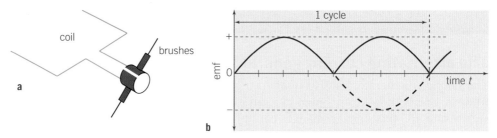

▲ **Figure 4** *The dc generator* **a** *The split-ring commutator* **b** *emf against time*

Note

A dc generator can be made by replacing the two slip rings of the ac generator with a split-ring, as shown in Figure 4. The emf does not reverse its polarity because the connections between the split-ring and the brushes reverse every half cycle.

Study tip

For a rotating coil, the *rate of change* of flux is greatest when the flux through it is zero; the *rate of change* of flux is zero when the flux through it is greatest.

Back emf

An emf is induced in the spinning coil of an electric motor because the flux linkage through the coil changes. The induced emf ε is referred to as a **back emf** because it acts against the pd V applied to the motor in accordance with Lenz's law. At any instant, $V - \varepsilon = IR$, where I is the current through the motor coil and R is the circuit resistance.

Because the induced emf is proportional to the speed of rotation of the motor, the current changes as the motor speed changes.

- At low speed, the current is high because the induced emf is small.
- At high speed, the current is low because the induced emf is high.

Note:
Multiplying the equation $V - \varepsilon = IR$ by I throughout gives $IV - I\varepsilon = I^2R$

Rearranging this equation gives:

| electrical power supplied by the source IV | = | electrical power transferred to mechanical power $I\varepsilon$ | + | electrical power wasted due to circuit resistance I^2R |

➕ The efficiency of an electric motor

When the motor spins without driving a load, it spins at high speed so the current is very small because the back emf is large. Its speed is limited by friction in the bearings and by air resistance. So it uses very little power.

When the motor is used to drive a load, its speed is much less than when off-load so the back emf is smaller and the current is larger. The power it uses from the voltage source that is not transferred as mechanical power to the load is wasted due to the resistance heating effect of the electrical current.

$$\text{The efficiency of an electric motor} = \frac{\text{mechanical power output}}{\text{electrical power supplied}} \; (\times\,100\%)$$

Q An electric car is powered by an electric motor connected to a battery. Describe how the back emf changes when the car speed up.

A: it increases in proportion to the speed.

Power station alternators

A power station alternator has three sets of coils at 120° to one another (Figure 5). Each set of coils produces an alternating emf 120° out of phase with each of the other two emfs. The coils are called the 'stators' because they are stationary, so they don't need slip ring connectors and an electromagnet called the 'rotor' spins between them.

The electromagnet is supplied with current from a dc generator. So the turning of the rotor induces an alternating emf in each set of stator coils. The three phases are distributed via transformers and power lines to factories and local sub-stations. A local sub-station supplies mains electricity to local premises, a third on each of the three phases. This is why your home can sometimes suffer a blackout when other homes nearby still have electricity. This happens when a fault in the local sub-station cuts out one phase but not the others.

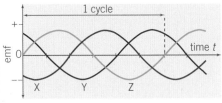

▲ **Figure 5** *Three-phase alternator*

Summary questions

1 **a** An ac generator produces an alternating emf with a peak value of 8.0 V and a frequency of 20 Hz. Sketch a graph to show how the emf varies with time.

   b The frequency of rotation of the ac generator in **a** is increased to 30 Hz. On the same axes, sketch a graph to show how the emf varies with time at 30 Hz.

2 A rectangular coil of N turns and area A spins at a constant frequency f in a uniform magnetic field of flux density B. Complete the table to show how the flux linkage and induced emf relate to the orientation of the coil during one cycle of rotation.

time	orientation of coil	flux linkage	induced emf
0	parallel to field	?	$+\varepsilon_0$
$\dfrac{0.25}{f}$	perpendicular to field	$+BAN$	?
$\dfrac{0.50}{f}$	parallel to field	?	?
$\dfrac{0.75}{f}$	perpendicular to field	?	?

3 The coil of an ac generator has 80 turns, a length of 65 mm and a width of 38 mm. It spins in a uniform magnetic field of flux density 130 mT at a constant frequency of 50 Hz.

   a Calculate the maximum flux linkage through the coil.

   b **i** Show each side of the coil moves at a speed of $6.0\,\text{m s}^{-1}$,

   ii Show that the peak voltage is 8.1 V.

4 An electric motor is to be used to move a variable load. The motor is connected in series with a battery and an ammeter.

   a Explain why the motor current is very small when the load is zero.

   b Explain why the motor current increases when the load is increased.

Learning objectives:

→ Define an alternating current.

→ Explain what is meant by the rms value of an alternating current.

→ Calculate the power supplied by an alternating current.

Specification reference: 3.7.5.5

Alternating current measurements

An alternating current (ac) is a current that repeatedly reverses its direction. In one cycle of an alternating current, the charge carriers move one way in the circuit, then reverse direction, then re-reverse direction.

The frequency f of an alternating current is the number of cycles it passes through each second. The unit of frequency is the hertz (Hz), equal to one cycle per second. Mains electricity has a frequency of 50 Hz. In a mains circuit, each cycle therefore takes 0.020 s ($= \frac{1}{50}$ s). Note that the time for one full cycle, the time period T, is given by $T = \frac{1}{f}$.

The peak value of an alternating current (or pd) is the maximum current (or pd) in either direction. The peak current in a circuit depends on the peak pd of the alternating current source and on the components in the circuit. For example, the peak pd in a mains circuit is 325 V. If this pd is applied to a 100 Ω heating element, the peak current through the heating element would be 3.25 A $\left(= \frac{325\,\text{V}}{100\,\Omega} \right)$.

Note that the peak-to-peak value is the difference between the peak value one way and the peak value in the opposite direction (i.e., twice the peak value).

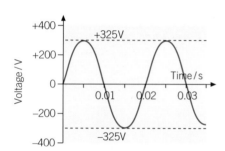

▲ **Figure 1** *The variation of mains pd with time*

Figure 1 shows how the mains pd varies with time. This type of variation is described as sinusoidal because its shape is that of a sine wave. Using transformers, we can change the peak pd of an alternating current but not its frequency.

Observing alternating current

1   Use an oscilloscope to display the waveform (i.e., variation with time) of the alternating pd from a signal generator.

   • Increasing the output pd from the signal generator makes the oscilloscope trace taller. This shows the peak value of the alternating pd has been made larger.

   • Increasing the frequency of the signal generator increases the number of cycles on the screen. This is because the number of cycles per second of the alternating pd has increased.

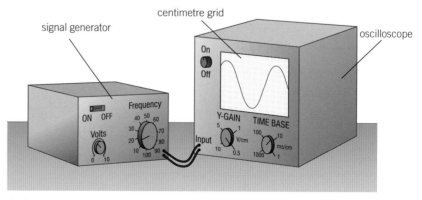

▲ **Figure 2** *Using an oscilloscope*

2   Connect the signal generator to an LED and make the frequency low enough so you can see the brightness of the lamp vary.
   • At very low frequency, you can see the LED light up and then fade out repeatedly. The LED is at its brightest each time the current is at its peak value. This happens twice each cycle corresponding to the current at its peak value in each direction.
   • If the frequency is raised gradually, the L flickers faster and faster until the variation of brightness is too fast to notice. The mains frequency is too high to cause flickering.

The heating effect of an alternating current

Imagine an electric heater supplied with alternating current at a very low frequency. The heater would heat up, then cool down, then heat up, and so on. The heater would repeatedly heat up then go cold.

Recall that the heating effect of an electric current varies according to the square of the current (see Topic 13.2, More about resistance). This is because the electrical power P supplied to the heater for a current I is given by

$$P = IV = I^2R$$

where R is the resistance of the heater element.

Figure 3b shows how the power ($= I^2R$) varies with time.

   • At peak current I_0, maximum power is supplied equal to I_0^2R.
   • At zero current, zero power is supplied.

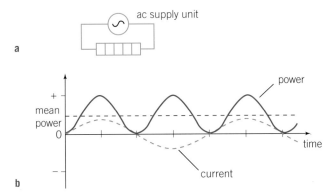

▲ **Figure 3** *Variation of power with time for an alternating current*

For a sinusoidal current, the *mean power* over a full cycle is half the peak power. You can see this from the symmetrical shape of the power curve in Figure 3b about the mean power. The mean power is therefore $\frac{1}{2}I_0^2R$.

The direct current that would give the same power as the mean power is called the **root mean square value** of the alternating current, I_{rms}.

The root mean square value of an alternating current is the value of direct current that would give the same heating effect as the alternating current in the same resistor.

Therefore,

$$(I_{rms})^2R = 0.5I_0^2R$$

Cancelling R from this equation and rearranging the equation gives

$$(I_{rms})^2 = 0.5I_0^2$$

Therefore,

$$I_{rms} = \frac{1}{\sqrt{2}} I_0$$

Also, the root mean square value of an alternating pd is given by

$$V_{rms} = \frac{1}{\sqrt{2}} V_0$$

The root mean square value of an alternating current or pd $= \frac{1}{\sqrt{2}} \times$ the peak value.

Note that for any resistor of known resistance in an alternating circuit, if you know the rms pd or current for the resistor, you can calculate the mean power supplied to it using the rms value.

$$P = (I_{rms})^2 R = \frac{(V_{rms})^2}{R} = I_{rms} V_{rms}$$

For example, if an alternating current of rms value 4 A is passed through a $5\,\Omega$ resistor, the mean power supplied to the resistor is $4^2 \times 5 = 80\,W$.

Study tip

Remember rms $= \dfrac{\text{peak}}{\sqrt{2}}$

Summary questions

1 An alternating current has a frequency of 200 Hz and a peak value of 0.10 A. Calculate:

   a the time for one cycle,

   b the rms current.

2 An alternating current of peak value 3.0 A is passed through a $4.0\,\Omega$ resistor. Calculate

   a i the rms current,   ii the rms pd across the resistor,

   b i the peak power,   ii the mean power supplied to the resistor.

3 A mains electric heater gives a mean output power of 1.0 kW when the rms pd across it is 230 V. Calculate:

   a i the rms current,   ii the peak current, and

      iii the peak power.

   b The heater plug is fitted with a 5 A fuse. Explain why the fuse does not blow even though the peak current through it is more than 5 A.

4 A 12 V, 48 W lamp lights at its normal brightness when it is connected to a source of alternating pd.

   a Calculate:

      i the rms current through it,

      ii the peak current through it, and

      iii the peak pd across it.

   b The cable connecting the lamp to the source has a resistance of $0.5\,\Omega$.

      i Show that the rms pd across the cable is 2.0 V when the lamp is on.

      ii Calculate the peak pd of the source when the lamp is on.

25.5 Transformers

The transformer rule

A transformer changes an alternating pd to a different peak value. Any transformer consists of two coils: the primary coil and the secondary coil. The two coils have the same iron core. When the primary coil is connected to a source of alternating pd, an alternating magnetic field is produced in the core. The field passes through the secondary coil. So an alternating emf is induced in the secondary coil by the changing magnetic field. The symbol for the transformer is shown in Figure 1.

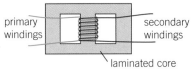
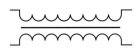

a *practical arrangement*   b *transformer symbol*

▲ **Figure 1** *The transformer*

A transformer is designed so that all the magnetic flux produced by the primary coil passes through the secondary coil.

Let ϕ = the flux in the core passing through each turn at an instant when an alternating pd V_p is applied to the primary coil.

- The flux linkage in the secondary coil = $N_s\phi$, where N_s is the number of turns on the secondary coil. From Faraday's law, the induced emf in the secondary coil, $V_s = N_s\dfrac{\Delta\phi}{\Delta t}$

- The flux linkage in the primary coil = $N_p\phi$, where N_p is the number of turns on the primary coil. From Faraday's law, the induced emf in the primary coil = $N_p\dfrac{\Delta\phi}{\Delta t}$

The induced emf in the primary coil opposes the pd applied to the primary coil, V_p. Assuming the resistance of the primary coil is negligible, so all the applied pd acts against the induced emf in the primary coil, the applied pd is $V_p = N_p\dfrac{\Delta\phi}{\Delta t}$.

Dividing the equation for V_s by the equation for V_p gives

$$\frac{V_s}{V_p} = N_s\frac{\Delta\phi}{\Delta t} \bigg/ N_p\frac{\Delta\phi}{\Delta t}$$

Cancelling $\dfrac{\Delta\phi}{\Delta t}$ from this equation gives the **transformer rule**:

$$\frac{V_s}{V_p} = \frac{N_s}{N_p}$$

- A **step-up transformer** has more turns on the secondary coil than on the primary coil. So the secondary voltage is stepped up compared with the primary voltage (i.e., $N_s > N_p$ so $V_s > V_p$).

- A **step-down transformer** has fewer turns on the secondary coil than on the primary coil. So the secondary voltage is stepped down compared with the primary voltage (i.e., $N_s < N_p$ so $V_s < V_p$).

Learning objectives:
→ Explain the purpose of transformers.
→ Describe the energy changes that take place in a transformer.
→ Discuss how the efficiency of transformers is improved by better design.

Specification reference: 3.7.5.6

Inside a phone charger

A phone charger contains a step-down transformer that steps the mains pd down to a much lower pd (still alternating) then a diode circuit is used to convert the alternating pd to a direct pd. Figure 2 shows how this is done.

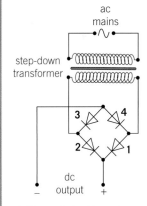

▲ **Figure 2** *Inside a phone charger*

◊ Identify the diodes in Figure 2 that conduct when the transformer output pd is positive at the right of the diode circuit and negative at the left.

A: Diodes 1 and 3.

The changing magnetic flux in the core induces a **back emf** in the primary coil as well as an emf in the secondary coil. The back emf acts against the primary voltage, making the primary current very small when the secondary current is 'off'.

When the secondary current is on, the magnetic field it creates is in the opposite direction to the magnetic field of the primary current. In this situation, the back emf in the primary coil is reduced so the primary current is larger than when the secondary current is off.

Study tip

Transformers only work if the magnetic flux through them is changing; they won't work with steady dc.

A *step-down* transformer steps down the voltage but *increases* the current.

Transformer efficiency

Transformers are almost 100% efficient because they are designed with:

1 low-resistance windings to reduce power wasted due to the heating effect of the current.
2 a laminated core which consists of layers of iron separated by layers of insulator. Induced currents in the core itself, referred to as **eddy currents**, are reduced in this way so the magnetic flux is as high as possible. Also, the heating effect of the induced currents in the core is reduced.
3 a core of '**soft iron**' which is easily magnetised and demagnetised. This reduces power wasted through repeated magnetisation and demagnetisation of the core.

$$\text{The efficiency of a transformer} = \frac{\text{power delivered by the secondary coil}}{\text{power supplied to the primary coil}}$$

$$= \frac{I_s V_s}{I_p V_p} \times 100\%$$

When a device (e.g., a lamp) is connected to the secondary coil, because the efficiency of a transformer is almost equal to 100%,

$$\begin{array}{c}\text{the electrical power supplied} \\ \text{to the primary coil}\end{array} = \begin{array}{c}\text{the electrical power supplied} \\ \text{by the secondary coil}\end{array}$$

Therefore, the current ratio is $\dfrac{I_s}{I_p} = \dfrac{V_p}{V_s} = \dfrac{N_p}{N_s}$

- In a step-up transformer, the voltage is stepped up and the current is stepped down.
- In a step-down transformer, the voltage is stepped down and the current is stepped up.

Worked example

A transformer is used to step down 230 V mains to 12 V. When a 12 V 48 W lamp is connected to the transformer's secondary coil, the lamp lights normally.

a The transformer has 1150 turns on its primary coil. Calculate the number of turns on its secondary coil.
b When the lamp is on, the primary current in the transformer is 0.22 A. Calculate:
  i   the current in the secondary coil,
  ii the efficiency of the transformer.

Solution

a Rearranging $\dfrac{V_s}{V_p} = \dfrac{N_s}{N_p}$ gives $N_s = N_p \dfrac{V_s}{V_p} = \dfrac{1150 \times 12}{230} = 60$ turns

b i   The power supplied to the lamp $= I_s V_s = 48$ W

   therefore $I_s = \dfrac{48}{V_s} = \dfrac{48}{12} = 4.0$ A

   ii Efficiency $= \dfrac{I_s V_s}{I_p V_p} = \dfrac{48}{230 \times 0.22} = 0.95 \ (= 95\%)$

The grid system

Electricity from power stations in the United Kingdom is fed into The National Grid System which supplies electricity to most parts of the country. The National Grid is a network of transformers and cables, underground and on pylons, which covers all regions of the UK Each power station generates alternating current at a precise frequency of 50 Hz at about 25 kV.

Step-up transformers at the power station increase the alternating voltage to 400 kV or more for long-distance transmission via the grid system. Step-down transformers operate in stages, as shown in Figure 3. Factories are supplied with all three phases at either 33 kV or 11 kV. Homes are supplied via a local transformer sub-station with single-phase ac at 230 V.

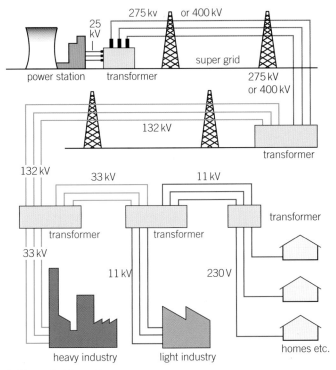

▲ **Figure 3** *The grid system*

Transmission of electrical power over long distances is much more efficient at high voltage than at low voltage. The reason is that the current needed to deliver a certain amount of power is reduced if the voltage is increased. So power wasted due to the heating effect of the current through the cables is reduced. To deliver power P at voltage V, the current required is $I = \dfrac{P}{V}$. If the resistance of the cables is R, the power wasted through heating the cables is $I^2R = \dfrac{P^2R}{V^2}$.

Therefore, the higher the voltage is, the smaller the ratio of the wasted power to the power transmitted is.

For example, for transmission of 1 MW of power through cables of resistance 500 Ω at 25 kV, the current necessary would be 40 A $\left(= \dfrac{1\,\text{MW}}{25\,\text{kV}}\right)$ so the power wasted would be 0.8 MW (= $I^2R = 40^2 \times 500$ W). Prove for yourself that at 400 kV, the power wasted would be about 3 kW. Check your answer using the equation derived above.

1 a Explain why an alternating emf is induced in the secondary coil of a transformer when the primary coil is connected to an alternating voltage supply.

b In terms of electrical power, explain why the current through the primary coil of a transformer increases when a device is connected to the secondary coil.

2 a Explain why a transformer is designed so that as much of the magnetic flux produced by the primary coil of a transformer passes through the secondary coil.

b Explain why a transformer works using alternating current but not using direct current.

3 A transformer has a primary coil with 120 turns and a secondary coil with 2400 turns.

a Calculate the primary voltage needed for a secondary voltage of 230 V.

b A 230 V 60 W lamp is connected to its secondary coil. Calculate the current through

i the secondary coil,

ii the primary coil.

State any assumptions made in this calculation.

4 a Explain why transmission of electrical power over a long distance is more efficient at high voltage than at low voltage.

b A power cable of resistance 200 Ω is to be used to deliver 2.0 MW of electrical power at 120 kV from a power station to an industrial estate. Calculate:

i the current through the cable,

ii the power wasted in the cable.

1   A coil is connected to a centre zero ammeter, as shown in **Figure 1**. A student
    drops a magnet so that it falls vertically and completely through the coil.

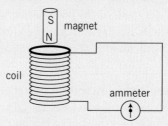

▲ **Figure 1**

   (a) Describe what the student would observe on the ammeter as the magnet
       falls through the coil.   (*2 marks*)
   (b) If the coil were not present the magnet would accelerate downwards at
       the acceleration due to gravity. State and explain how its acceleration in
       the student's experiment would be affected, if at all,
       (i)   as it entered the coil,
       (ii) as it left the coil.   (*4 marks*)
   (c) Suppose the student forgot to connect the ammeter to the coil, therefore leaving
       the circuit incomplete, before carrying out the experiment. Describe and explain
       what difference this would make to your conclusions in part (b).   (*3 marks*)
                                                                     AQA, 2004

2   Faraday's law of electromagnetic induction predicts that the induced emf, E,
    in a coil is given by

$$\frac{\Delta(N\phi)}{t}$$

   (a) (i)   What quantity does the symbol ϕ represent?
       (ii) State the SI unit for ϕ.   (*2 marks*)
   (b) In **Figure 2** the magnet forms the bob of a simple pendulum. The magnet
       oscillates with a small amplitude along the axis of a 240 turn coil that has
       a cross-sectional area of $2.5 \times 10^{-4}\,\mathrm{m}^2$.

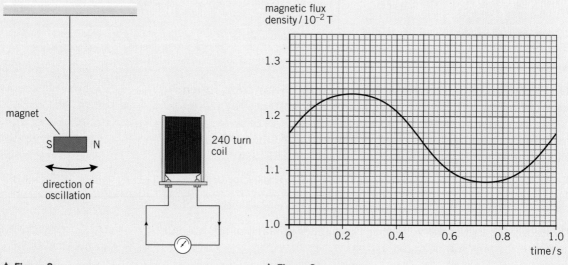

▲ **Figure 2**   ▲ **Figure 3**

Figure 3 shows how the magnetic flux density, B, through the coil varies with time, t, for one complete oscillation of the magnet. The magnetic flux density through the coil can be assumed to be uniform.

(i)   Calculate the maximum emf induced in the coil.

(ii) Sketch a graph to show how the induced emf in the coil varies during the same time interval.

(iii) Explain how the pendulum may be modified to double the frequency of oscillation of the magnet.

(iv) The frequency of oscillation of the magnet is increased without changing the amplitude.
      Explain why this increases the maximum induced emf.

(v)   State **two** other ways of increasing the maximum induced emf.

(*11 marks*)

AQA, 2003

3   **Figure 4** shows a system used by an engineer to determine the rate of revolution of a rotating axle.

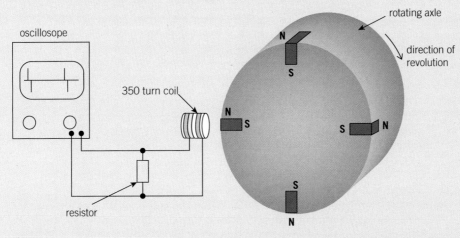

▲ **Figure 4**

Four small bar magnets are embedded in the axle as shown. The N-pole of each magnet is towards the outside of the axle. A voltage is produced between the terminals of a coil placed close to the rotating axle. The voltage produced is monitored using an oscilloscope. The waveform produced is shown in **Figure 5**.

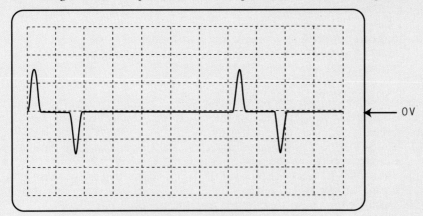

Oscilloscope grid marked in cm
The Y amplifier setting $= 5\,\mathrm{mV\,cm^{-1}}$
The time-base setting $= 10\,\mathrm{ms\,cm^{-1}}$

▲ **Figure 5**

(a) Determine the number of revolutions made by the axle in one minute. (3 marks)
(b) (i) Use Faraday's law to explain how the voltage pulses are produced.
    (ii) The coil has 350 turns. Determine the maximum rate of change of flux through the coil. (6 marks)
(c) Use Lenz's law to explain the production of positive and negative voltage pulses. (3 marks)
(d) Draw on a copy of **Figure 5** the waveform that shows the changes you would expect to see when the rate of revolution of the axle increases. (3 marks)

AQA, 2007

4   (a) Copy and complete the diagram in **Figure 6** to show a current balance, which may be used to measure the magnetic flux density between the poles of the ceramic magnets. Clearly label the directions of the current and the magnetic force acting on the conductor in the field. (3 marks)

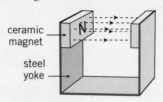

ceramic magnet

steel yoke

▲ Figure 6

(b) (i) The armature of a simple motor consists of a square coil of 20 turns and carries a current of 0.55 A just before it starts to move. The lengths of the sides of the coil are 0.15 m and they are positioned perpendicular to a magnetic field of flux density 40 mT. Calculate the force on each side of the coil.
    (ii) Explain why the current falls below 0.55 A once the coil of the motor is rotating.
    (iii) The resistance of the coil is $0.50\,\Omega$. When the coil is rotating at a constant rate the minimum current in the coil is found to be 0.14 A. Calculate the maximum rate at which the flux is cut by the coil. (8 marks)

AQA, 2003

5   (a) Explain what is meant by the term *magnetic flux linkage*. State its unit. (2 marks)
(b) Explain, in terms of electromagnetic induction, how a transformer may be used to step down voltage. (4 marks)
(c) A minidisc player is provided with a mains adapter. The adapter uses a transformer with a turns ratio of 15 : 1 to step down the mains voltage from 230 V.
    (i) Calculate the output voltage of the transformer.
    (ii) State **two** reasons why the transformer may be less than 100% efficient. (4 marks)

AQA, 2004

6   (a) A transformer, operating from 230 V, supplies a 12 V garden lighting system consisting of 8 lamps. Each lamp is rated at 30 W and they are connected in parallel.
    (i) The primary coil of the transformer has 3000 turns. Calculate the number of turns on the secondary coil.
    (ii) Show that the total resistance of the lamps when they are working at normal brightness is $0.60\,\Omega$.
    (iii) Calculate the power input to the transformer, assuming that the transformer is perfectly efficient. (8 marks)

(b) **Figure 7** shows a brass pendulum bob swinging through the magnetic field above a strong magnet. Its oscillations are observed to be quite heavily damped.

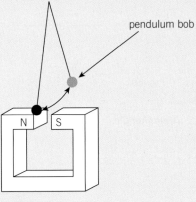

pendulum bob

▲ **Figure 7**

Explain, using the principles of electromagnetic induction, why this pendulum is heavily damped.

(4 marks)
AQA, 2006

7 **(a)** Electrical power is transmitted through cables of total resistance $1.8\,\Omega$, operated with alternating current at an rms voltage of $11\,kV$. The power supplied to the input of the cables is $960\,kW$. Calculate
   (i) the peak value of the current in the cables,
   (ii) the percentage of the input power that is available at the output end of the cables. *(7 marks)*

(b) When public electricity supplies were first introduced, many of the power stations generated direct current. This meant that premises that were a large distance from the power station could not be supplied economically with electricity.
Discuss this by reference to the physical principles involved, stating why ac is now preferred for distributing electrical power. *(5 marks)*

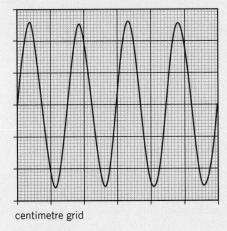

centimetre grid

▲ **Figure 8**

8 An alternating current source is connected to a $470\,\Omega$ resistor to form a complete circuit. **Figure 8** shows the waveform of the alternating pd across the resistor.
   (a) The time base setting of the oscilloscope was $20\,ms\,cm^{-1}$. Calculate the the frequency of the alternating pd.
   (b) The Y-gain of the oscilloscope is $0.10\,V\,cm^{-1}$.
      (i) Calculate the peak value and the rms value of the alternating pd.
      (ii) Calculate the mean power dissipated in the resistor.

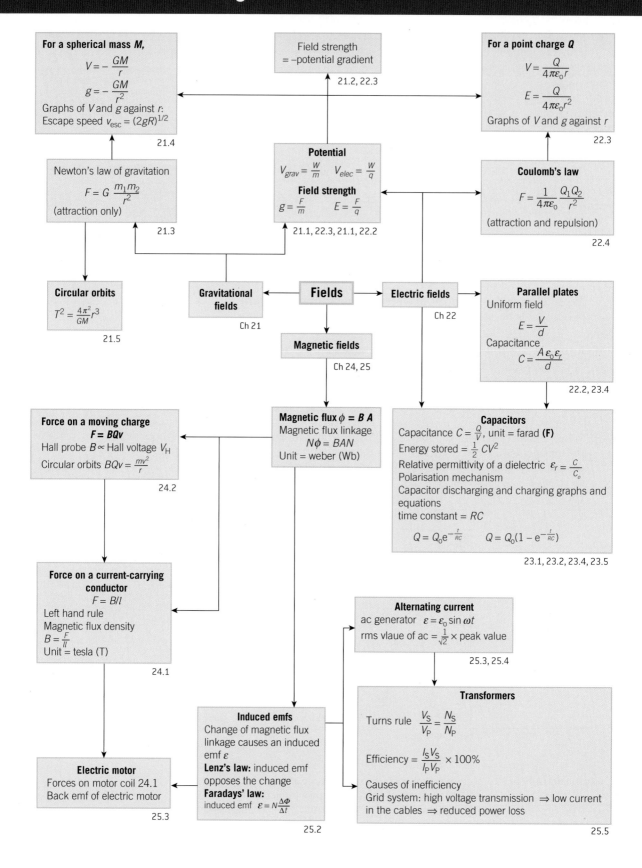

For a spherical mass **M**,
$$V = -\frac{GM}{r}$$
$$g = -\frac{GM}{r^2}$$
Graphs of V and g against r:
Escape speed $v_{esc} = (2gR)^{1/2}$

21.4

Field strength
= −potential gradient

21.2, 22.3

For a point charge **Q**
$$V = \frac{Q}{4\pi\varepsilon_0 r}$$
$$E = \frac{Q}{4\pi\varepsilon_0 r^2}$$
Graphs of V and g against r

22.3

Newton's law of gravitation
$$F = G\frac{m_1 m_2}{r^2}$$
(attraction only)

21.3

Potential
$$V_{grav} = \frac{W}{m} \quad V_{elec} = \frac{W}{q}$$
Field strength
$$g = \frac{F}{m} \quad E = \frac{F}{q}$$

21.1, 22.3, 21.1, 22.2

Coulomb's law
$$F = \frac{1}{4\pi\varepsilon_0}\frac{Q_1 Q_2}{r^2}$$
(attraction and repulsion)

22.4

Circular orbits
$$T^2 = \frac{4\pi^2}{GM}r^3$$

21.5

Gravitational fields

Ch 21

Fields

Electric fields

Ch 22

Parallel plates
Uniform field
$$E = \frac{V}{d}$$
Capacitance
$$C = \frac{A\varepsilon_0\varepsilon_r}{d}$$

22.2, 23.4

Magnetic fields

Ch 24, 25

Force on a moving charge
F = BQv
Hall probe $B \propto$ Hall voltage V_H
Circular orbits $BQv = \frac{mv^2}{r}$

24.2

Magnetic flux $\phi = B\,A$
Magnetic flux linkage
$$N\phi = BAN$$
Unit = weber (Wb)

Capacitors
Capacitance $C = \frac{Q}{V}$, unit = farad **(F)**
Energy stored $= \frac{1}{2}CV^2$
Relative permittivity of a dielectric $\varepsilon_r = \frac{C}{C_0}$
Polarisation mechanism
Capacitor discharging and charging graphs and equations
time constant = RC
$$Q = Q_0 e^{-\frac{t}{RC}} \qquad Q = Q_0(1 - e^{-\frac{t}{RC}})$$

23.1, 23.2, 23.4, 23.5

Force on a current-carrying conductor
$$F = BIl$$
Left hand rule
Magnetic flux density
$$B = \frac{F}{Il}$$
Unit = tesla (T)

24.1

Alternating current
ac generator $\varepsilon = \varepsilon_0 \sin \omega t$
rms vlaue of ac $= \frac{1}{\sqrt{2}} \times$ peak value

25.3, 25.4

Transformers
Turns rule $\dfrac{V_S}{V_P} = \dfrac{N_S}{N_P}$

Efficiency $= \dfrac{I_S V_S}{I_P V_P} \times 100\%$

Causes of inefficiency
Grid system: high voltage transmission $\Rightarrow$ low current in the cables $\Rightarrow$ reduced power loss

25.5

Electric motor
Forces on motor coil 24.1
Back emf of electric motor

25.3

Induced emfs
Change of magnetic flux linkage causes an induced emf ε
Lenz's law: induced emf opposes the change
Faradays' law:
induced emf $\varepsilon = N\dfrac{\Delta\Phi}{\Delta t}$

25.2

Practical skills

In this section you have met the following skills:

- use of an electroscope or an electrometer to detect electric charge
- use of a data logger or a microammeter, and a high-resistance voltmeter or oscilloscope, as appropriate, to measure the pd across a capacitor when it is charged or discharged.
- use of a top pan balance or a current balance to measure the force on a current-carrying conductor in a uniform magnetic field
- use of a Hall probe to measure magnetic flux density
- use of a parallel plate capacitor to measure the relative permittivity of a substance
- use of a data logger to measure an induced emf
- use of an oscilloscope to measure the peak value and the frequency of an alternating pd.

Maths skills

In this section you have met the following skills:

- calculate the strength of an electric or a gravitational field from the force on an appropriate object
- use graphs to relate field strength and potential for gravitational and electric fields
- solve problems involving an inverse square law (e.g., time period of a satellite in circular orbit) or an inverse linear relationship (e.g., electric potential near a point charge)
- carry out calculations on capacitors (e.g., energy stored, relative permittivity, use of $V = V_0 e^{-\frac{t}{RC}}$)
- carry out calculations on magnetic fields (e.g., force on a current-carrying conductor and on a moving charged particle in a magnetic field
- calculate the emf induced in a coil from a graph of magnetic flux or flux linkage, or from an appropriate equation (e.g., $\varepsilon = \varepsilon_0 \sin \omega t$)
- relate the rms value of an alternating current or pd to the peak value
- carry out transformer calculations using appropriate equations
- plot a graph from data provided or found experimentally (e.g., ln V against t for capacitor discharge)
- relate $y = mx + c$ to a linear graph with physics variables to find what the gradient and intercept of the graph represent (e.g., finding the time constant of a capacitor discharge in the above point).

Extension task

Metal detectors work by inducing a current in a piece of metal. Use other books and the internet to find out

- why a metal detector bleeps when it is near a piece of metal
- what materials shield metals from being detected
- what the range of a metal detector is. Present your findings as a presentation suitable for your class.

Practice questions: Fields

1 **(a)** (i) State **one** similarity of electric and gravitational fields.
(ii) State **one** difference between electric and gravitational fields. *(2 marks)*

(b) A satellite of mass 165 kg has the radius of its orbit reduced from 4.24×10^7 m to 8.08×10^6 m.
Calculate the change in potential energy of the satellite and state whether it is an increase or a decrease. *(3 marks)*

(c) The orbital change mentioned in part (b) reduces the period of the orbit from 24 hours to 2 hours. State and explain why each of these orbits is useful for information collection or transfer. *(3 marks)*
AQA 2005

2 **(a)** State the factors that affect the gravitational field strength at the surface of a planet. *(2 marks)*

(b) **Figure 1** shows the variation, called an anomaly, of gravitational field strength at the Earth's surface in a region where there is a large spherical granite rock buried in the Earth's crust.

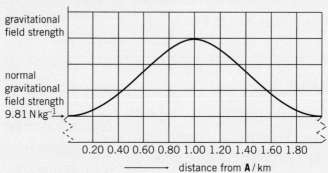

▲ Figure 1

The density of the granite rock is 3700 kg m^{-3} and the mean density of the surrounding material is 2200 kg m^{-3}.
(i) Show that the difference between the mass of the granite rock and the mass of an equivalent volume of the surrounding material is 5.0×10^{10} kg.
(ii) Calculate the difference between the gravitational field strength at B and that at point A on the Earth's surface that is a long way from the granite rock.
(iii) Describe how the graph of gravitational field strength would change if the granite rock were buried deeper in the Earth's crust. *(9 marks)*
AQA 2003

3 The hydrogen atom may be represented as a central proton with an electron moving in a circular orbit around it as shown in **Figure 2**. When the atom is in the ground state, the radius of the electron's orbit is 5.3×10^{-11} m.

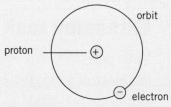

▲ Figure 2

(a) By applying this model to the hydrogen atom in the ground state, calculate:
(i) the force of electrostatic attraction between the electron and the proton,
(ii) the speed of the electron,
(iii) the ratio of the de Broglie wavelength of the electron to the circumference of the orbit. *(6 marks)*

(b) The total energy of the electron in a hydrogen atom may be shown to have discrete values given, in J, by

$$E = \frac{2.2 \times 10^{-18}}{n^2}$$

where $n = 1$ for the ground state, $n = 2$ for the first excited state, and so on.

(i) Calculate the wavelength of the light emitted when the electron returns to the ground state from the first excited state.

(ii) Explain why visible light will not be produced by any transition in which the electron returns to the ground state.

(5 marks)

AQA 2005

4   Two capacitors **A** and **B** are separately charged to a pd of 40.0 V before being discharged through the same resistor of value 10.0 kΩ. The discharge curves are shown in **Figure 3**.

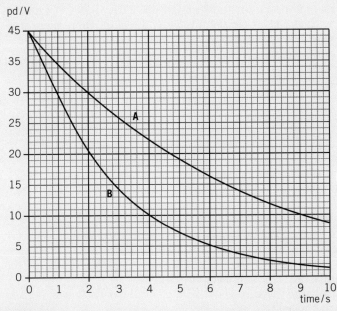

▲ Figure 3

(a) Show that the time for the pd across capacitor **A** to halve is always approximately 4.5 s.

(b) Calculate the capacitance of **A**.

(c) Calculate the energy dissipated in the resistor in the interval between $t = 0$ and $t = 4.0$ s as capacitor **B** discharges.

(8 marks)

AQA 2007

5   A steel wire of diameter 0.24 mm is stretched between two fixed points 0.71 m apart. A U-shaped magnet is placed at the centre of the wire so that the wire passes between its poles, as shown in **Figure 4**.

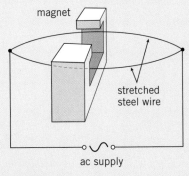

▲ Figure 4

(a) (i) Explain why the wire vibrates when an alternating current is passed through it.

(ii) Explain why the wire vibrates strongly in its fundamental mode when the frequency of the alternating current is 290 Hz.

(iii) Show that the speed of the waves on the wire is 410 m s⁻¹.
(6 marks)

(b) The speed, *c*, of waves on a wire of mass per unit length, μ, is related to the tension, *T*, in the wire by

$$c = \sqrt{\frac{T}{\mu}}$$

(i) The wire in **Figure 4** is at a tension of 60 N. Calculate its mass per unit length.

(ii) Hence calculate the density of the metal.
(5 marks)
AQA 2005

6   **Figure 5** shows a particle **P** with charge $+6.4 \times 10^{-19}$ C about to enter a region where there is a uniform electric field of strength 2.0×10^4 N C⁻¹. **Figure 6** shows the same charged particle about to enter a region where there is a uniform magnetic field of flux density 0.17 T directed into the paper.

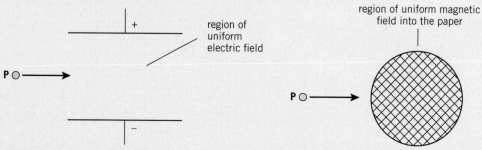

▲ Figure 5   ▲ Figure 6

(a) Sketch the paths taken by the particle when in each field.
(2 marks)

(b) (i) State what is meant by *uniform electric field strength*.

(ii) The separation of the plates in **Figure 5** is 0.045 m. Calculate the potential difference between the plates.
(4 marks)

(c) (i) Calculate the magnitude of the force on the particle when it is in the electric field.

(ii) Calculate the initial velocity of the charged particle for which the magnitude of the force on the particle is the same in each field.
(4 marks)

(d) Explain why the speed of the particle changes in one of the above situations but remains constant in the other.
(6 marks)
AQA 2005

7   (a) A satellite moves in a circular orbit at constant speed. Explain why its speed does not change even though it is acted on by a force.
(3 marks)

(b) At a certain point along the orbit of a satellite in uniform circular motion, the Earth's magnetic flux density has a component of 56 μT towards the centre of the Earth and a component of 17 μT in a direction perpendicular to the plane of the orbit.

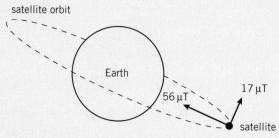

▲ Figure 7

    (i)   Calculate the magnitude of the resultant magnetic flux density at this point.

    (ii) The satellite has an external metal rod pointing towards the centre of the Earth. Calculate the angle between the direction of the resultant magnetic field and the rod.

    (iii) Explain why an emf is induced in the rod in this position.   *(4 marks)*

                                                                     AQA 2002

8 In a television cathode ray tube, electrons are accelerated through a potential difference of 12 kV in a vacuum before striking the screen.

  (a) (i)   Calculate the speed of an electron accelerated through this potential difference.

       (ii) The beam current is 25 mA. Calculate the number of electrons that strike the screen in one second.   *(4 marks)*

  (b) The electron beam is deflected in the television tube by a changing magnetic field produced by currents in coils placed around the tube.

       (i)   Explain how this changing magnetic field can lead to induction effects in other electrical circuits in the television.

       (ii) Explain how this changing magnetic field could lead to faults in these other electrical television circuits.

       (iii) The electron beam is moved from the left-hand side of the screen to the right-hand side by uniformly varying the field from -3.5×10^{-4} T to $+3.5 \times 10^{-4}$ T in a time of 50 μs. Each turn of a 250-turn coil of wire in this changing field has an area of 4.0×10^{-3} m^2

             Calculate the **maximum** emf that can appear in the 250-turn coil.

       (iv) Explain why the answer to part (b)(iii) is a maximum value.   *(8 marks)*

                                                                      AQA 2004

9 **(a)** Eddy currents in the cores of transformers are one cause of inefficiency in transformers.

       (i)   Explain what is meant by an eddy current, how eddy currents are produced in transformer cores and why they lead to inefficient operation of the transformer.

       (ii) Explain how the design of a transformer minimises the inefficiency due to eddy currents.   *(6 marks)*

  (b) **Figure 8** shows a laboratory arrangement for demonstrating the operation of a transformer.

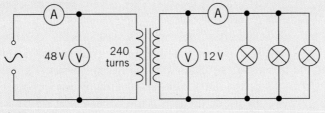

▲ **Figure 8**

  The supply voltage was adjusted until the three 12 V, 36 W lamps operated normally. The voltmeter readings are shown on **Figure 8**. The efficiency of the transformer was 100%.

  Calculate:

    (i)   the reading of the ammeter in the secondary circuit,

    (ii) the reading of the ammeter in the primary circuit,

    (iii) the number of turns on the secondary coil of the transformer.   *(5 marks)*

                                                                      AQA 2005

Section 8
Nuclear physics

Introduction

This section extends your A Level Physics Year 1 studies on the structure of the atom and mechanics. You will develop your understanding of the nucleus by looking at Rutherford's landmark α-scattering experiment, which established the nuclear model of the atom. You will then consider the properties of α, β, and γ radiation, including ionisation, absorption, and range in air, as well as the experimental verification of the inverse-square law for γ radiation. You will discuss the relative hazards to humans of exposure and the safe handling and storage of radioactive sources. You will look in depth at the random nature of radioactive decay. This includes the theory of radioactive decay and half-life calculations such as those involved in radioactive dating and the safe storage of radioactive material. You will study nuclear instability by using equations to represent the changes that take place when α, β^-, β^+, and electron capture occur in relation to stable nuclei plotted on an N–Z graph. You will also consider the evidence from γ radiation studies for nuclear excited states, and study the use of γ radiation from these states. You will then move on to calculate the density of the nucleus from experimental measurements. In the last part of this section, you will consider why energy is released in nuclear fission and fusion, how to calculate the energy released in such changes, how the release of fission energy is controlled in a thermal nuclear reactor, and how safety is ensured in the operation of nuclear reactors and in the storage of nuclear waste.

Working scientifically

In this final part of the course, you will bring together all of the maths skills and physics knowledge that you learnt from previous parts of the course on topics such as radioactive decay and particle physics. You will gain a deeper understanding of the applications of nuclear physics, including the use of radioactive isotopes and the principles and operation of nuclear reactors. By studying radioactivity, you will further develop and broaden your practical skills. This might include measuring radioactivity by using Geiger counters. Through such practical work, you will also develop your awareness of the necessary safety regulations when studying and working with radioactivity. The study of nuclear physics will give you insight into the historical importance of experimental work in how our present knowledge of the structure of the atom was developed.

The maths skills you will develop in this section include further arithmetical, graphical, and algebraic skills involving inverse, inverse-square, and exponential functions, as well as logarithms. By using the maths skills you have learnt from earlier parts of the A Level Physics course, you will develop your understanding of concepts such as half-life, and you will meet new concepts in nuclear physics, such as binding energy. You will develop further awareness beyond dynamics and fields by studying rates of change in radioactive decay. These maths skills will give you further confidence in your abilities that will help you beyond A level.

What you already know

From your A Level Physics Year 1 studies on particle physics, you should know that:

○ Every atom has a positively charged nucleus surrounded by electrons. The nucleus is composed of protons and neutrons, except for the ^1_1H nucleus, which is a single proton.

○ An unstable nucleus becomes stable or less unstable by emitting an α particle or a β^- particle or a β^+ particle or a γ photon or by capturing an inner-shell electron.

○ α emission occurs from nuclei with too many neutrons and protons, β^- emission occurs from nuclei with too many neutrons, and β^+ emission or electron capture occurs from nuclei with too many protons.

○ An α particle consists of two neutrons and two protons, a β^- particle is an electron created and emitted by a nucleus, and a β^+ particle is a positron created and emitted by a nucleus.

○ Ionisation is the process of creating ions (i.e., charged atoms) from uncharged atoms – the radiation from unstable nuclei is ionising radiation and is therefore harmful.

○ Energy can be changed into mass and mass can be changed into energy as given by $E = mc^2$.

From your A Level Physics Year 1 studies on mechanics and materials, you should know that:

○ For an object of mass m moving at speed v, its kinetic energy $= \frac{1}{2}mv^2$ (if $v \ll c$, the speed of light in free space).

○ Momentum is conserved in any interaction where no external forces act.

○ Energy cannot be created or destroyed.

○ Density $= \dfrac{\text{mass}}{\text{volume}}$.

Learning objectives:

→ State how big the nucleus is.

→ Describe how the nucleus was discovered.

→ Explain why it was not discovered earlier.

Specification reference: 3.8.1.1

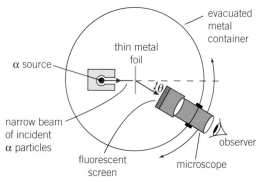

▲ **Figure 1** *Rutherford's α-scattering apparatus*

Ernest Rutherford 1871–1937

Ernest Rutherford arrived in Britain from New Zealand in 1895. By the age of 28, he was a professor. He made important discoveries about radioactivity and was awarded the Nobel prize for Chemistry in 1908 for his investigations into the disintegration of radioactive substances. He worked in the universities of Montreal, Manchester and Cambridge. When at Manchester, he put forward the nuclear model of the atom and proved it experimentally using α-scattering experiments. He was knighted in 1914 and made Lord Rutherford of Nelson in 1931. His co-worker Otto Hahn described him as a 'very jolly man'. In 1915, he expressed the hope that 'no one discovers how to release the intrinsic energy of radium until man has learned to live at peace with his neighbour'. After his death in 1937, his ashes were placed close to Newton's tomb in Westminster Abbey.

The nuclear model of the atom was proposed by Ernest Rutherford in 1911. He knew from the work of J.J. Thomson that every atom contains one or more electrons. Thomson had shown that the electron is a negatively charged particle inside every atom but no-one knew until Rutherford's theory was confirmed experimentally in 1913 how the positive charge in the atom was distributed. Thomson thought the atom could be like a 'currant bun' – with electrons dotted in the atom like currants in a bun. In this model, the positive charge was supposedly spread throughout the atom like the dough of the bun.

Rutherford knew that the atoms of certain elements were unstable and emitted radiation. It had been shown that there were three types of such radiation, referred to as **alpha radiation** (symbol α), **beta radiation** (symbol β), and **gamma radiation** (symbol γ). Rutherford knew that α radiation consisted of fast-moving positively charged particles. He and his co-workers used this type of radiation to probe the atom. He reckoned that a beam of α particles directed at a thin metal foil might be scattered slightly by the atoms of the foil if the positive charge was spread throughout each atom. He was astonished when he discovered that some of the particles bounced back from the foil – in his own words 'as incredible as if you fired a 15 inch naval shell at tissue paper and it came back'.

Let's consider Rutherford's experiment in more detail. Rutherford used a narrow beam of α particles, all of the same kinetic energy, in an evacuated container to probe the structure of the atom. Figure 1 shows an outline of the arrangement he used. A thin metal foil was placed in the path of the beam. α particles scattered by the metal foil were detected by a detector which could be moved round at a constant distance from the point of impact of the beam on the metal foil.

 ### What made Rutherford originate the nuclear model of the atom?

Rutherford had previously investigated the scattering of alpha particles by gas atoms and noted on rare occasions an α particle was scattered through a very large angle when most particles were hardly scattered, if at all. He realised there could be a 'nucleus' of positively charged matter inside the atom which could explain these rare events.

1 The α particles must have the same speed otherwise slow α particles would be deflected more than faster α particles on the same initial path.

2 The container must be evacuated or else the α particles would be stopped by air molecules.

3 The source of the α particles must have a long half-life otherwise later readings would be lower than earlier readings due to radioactive decay of the source nuclei.

> **Q:** Describe the type of force that Rutherford believed would scatter an α particle.

Rutherford used a microscope to observe the pinpoints of light emitted when α particles hit the atoms of a fluorescent screen. He measured the number of α particles reaching the detector per minute for different angles of deflection from zero to almost 180°. His measurements showed that:

1 most α particles passed straight through the foil with little or no deflection; about 1 in 2000 were deflected,

2 a small percentage of α particles (about 1 in 10 000) were deflected through angles of more than 90°.

Imagine throwing tennis balls at a row of vertical posts separated by wide gaps. Most of the balls would pass between the posts and therefore would not be deflected much. However, some would rebound as a result of hitting a post. Rutherford realised the α-scattering measurements could be explained in a similar way by assuming every atom has a 'hard centre' much smaller than the atom. His interpretation of each result was that

* most of the atom's mass is concentrated in a small region, the **nucleus**, at the centre of the atom,

* the nucleus is positively charged because it repels α particles (which carry positive charge) that approach it too closely.

Figure 3 shows the paths of some α particles which pass near a fixed nucleus.

* α particle C collides 'head-on' with the nucleus and rebounds back so its angle of deflection is 180°.

* α particles A, B, and D are deflected through different angles. The closer the initial direction of an α particle is to the 'head-on' direction,

  * the greater its deflection is because the electrostatic force of repulsion between the α particle and the nucleus increases with decreased separation between them,

  * the smaller the least distance of approach of the α particle to the nucleus is.

* α particle E does not approach the nucleus closely enough to be significantly deflected.

Using Coulomb's law of force (i.e., the law of force between charged objects) and Newton's laws of motion, Rutherford used his nuclear model to explain the exact pattern of the results. By testing foils of different metal elements, he also showed that the magnitude of the charge of a nucleus is $+Ze$, where e is the charge of the electron and Z is the **atomic number** of the element.

Estimate of the size of the nucleus

About 1 in 10 000 α particles are deflected by more than 90° when they pass through a metal foil. The foil must be very thin otherwise α particles are scattered more than once.

For such single scattering by a foil that has n layers of atoms, the probability of an α particle being deflected by a given atom is therefore about 1 in 10 000n. This probability depends on the effective area

▲ **Figure 2** *Ernest Rutherford*

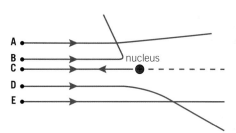

▲ **Figure 3** *α-scattering paths*

Study tip

Nuclear diameter is of the order of 10^{-15} m.

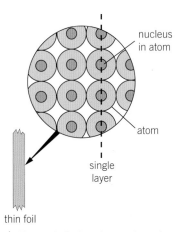

▲ **Figure 4** *Estimating nuclear size*

Notes

In a head-on impact, the α particle stops momentarily at the least distance of approach, d. At this point, the potential energy of the α particle in the electric field of the nucleus is equal to the initial kinetic energy, E_K, of the α particle. So, using the formula in Topic 22.5 for potential energy gives

$$E_K = \frac{Q_\alpha Q_N}{4\pi\varepsilon_0 d},$$ where Q_α is the charge of the α particle $(+2e)$ and Q_N is the charge of the nucleus $(= +Ze)$ and d is the least distance of approach. Prove for yourself that the least distance of approach of an α particle with kinetic energy of 8×10^{-13} J to an aluminium nucleus $(Z = 13)$ is about 7×10^{-15} m which is about 10^5 times smaller than the diameter of a typical atom. A more accurate method of determining the radius of a nucleus is explained in Topic 26.9.

of cross-section of the nucleus to that of the atom. Therefore, for a nucleus of diameter d in an atom of diameter D, the area ratio is equal to $\dfrac{\frac{1}{4}\pi d^2}{\frac{1}{4}\pi D^2}$. So $d^2 = \dfrac{D^2}{10000n}$.

A typical value of $n = 10^4$ gives $d = \dfrac{D}{10000}$.

In other words, the size of a nucleus relative to an atom is about the same as a football to a football stadium!

Summary questions

1   a   In the Rutherford α particle scattering experiment, most of the α particles passed straight through the metal foil. What did Rutherford deduce about the atom from this discovery?

   b   A small fraction of the α particles were deflected through large angles. What did Rutherford deduce about the atom from this discovery?

2   In Rutherford's α particle scattering experiment, why was it essential that:

   a   the apparatus was in an evacuated chamber,

   b   the foil was very thin,

   c   the α particles in the beam all had the same speed,

   d   the beam was narrow?

3   An α particle collided with a nucleus and was deflected by it, as shown in Figure 5.

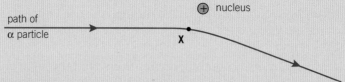

▲ Figure 5

   a   Copy the diagram and show on it

      i   the direction of motion of the α particle,

      ii the direction of the force on the α particle when it was at the position marked X.

   b   Describe how

      i   the kinetic energy of the α particle,

      ii the potential energy of the α particle changed during this interaction.

4   In the α particle scattering experiment, about 1 in 10 000 α particles are deflected by more than 90°.

   a   For a metal foil which has n layers of atoms, explain why the probability of an α particle being deflected by a given atom is therefore about 1 in 10 000 n.

   b   Assuming this probability is equal to the ratio of the cross-sectional area of the nucleus to that of the atom, estimate the diameter of a nucleus for atoms of diameter 0.5 nm in a metal foil of thickness 10 µm.

   c   Calculate the least distance of approach of an α particle to a gold nucleus $(Z = 79)$ if the initial kinetic energy of the α particle is 8.0×10^{-13} J.

   $\varepsilon_0 = 8.85 \times 10^{-12}$ F m^{-1}, $e = 1.6 \times 10^{-19}$ C

The discovery of radioactivity

Marie Curie established the nature of radioactive materials. She showed how radioactive compounds could be separated and identified. She and her husband, Pierre, and Henri Becquerel won the 1903 Nobel prize for physics for their discovery of radioactivity. After Pierre's death in 1906, she continued her painstaking research and was awarded a second Nobel Prize in 1911 – an unprecedented honour.

In 1896, Henri Becquerel was investigating materials that glow when placed in an X-ray beam. He wanted to find out if strong sunlight could make uranium salts glow. He prepared a sample and placed it in a drawer on a wrapped photographic plate, ready to test the salts on the next sunny day. When he developed the film, he was amazed to see the image of a key. He had put the key on the plate in the drawer and then put the uranium salts on top of the key. He realised that uranium salts emit radiation which can penetrate paper and blacken a photographic film. The uranium salts were described as being radioactive. The task of investigating radioactivity was passed on by Becquerel to one of his students, Marie Curie. Within a few years, Marie Curie discovered other elements which are radioactive. One of these elements, radium, was found to be over a million times more radioactive than uranium.

Rutherford's investigations on radioactivity

Rutherford wanted to find out what the radiation emitted by radioactive substances was and what caused it. He found that the radiation:

- ionised air, making it conduct electricity. He made a detector which could measure the radiation from its ionising effect.
- was of two types. One type which he called alpha (α) radiation was easily absorbed. The other type which he called beta (β) radiation was more penetrating. A third type of radiation, called gamma (γ) radiation, even more penetrating than β radiation, was discovered a year later.

Further tests showed that a magnetic field deflects α and β radiation in opposite directions and has no effect on γ radiation. From the deflection direction, it was concluded that α radiation consists of positively charged particles and β radiation consists of negatively charged particles. γ radiation was later shown to consist of high energy **photons**.

Learning objectives:

→ Define α, β, and γ radiation.

→ Explain why it is dangerous.

→ Describe the properties of α, β, and γ radiation.

Specification reference: 3.8.1.2

▲ **Figure 1** *Marie Curie (1867–1934)*

Study tip

Don't confuse the radiation with the source; the source is radioactive not the radiation.

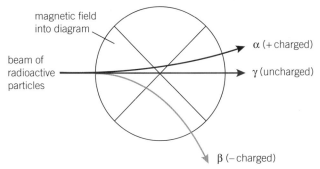

▲ **Figure 2** *Deflection by a magnetic field*

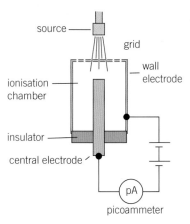

▲ **Figure 3** *Investigating ionisation*

Synoptic link

α particles from a given source have the same range in air as each other, whereas β particles do not. The α particles from a given isotope are always emitted with the same kinetic energy. This is because each α particle and the nucleus that emits it move apart with equal and opposite amounts of momentum. This isn't the case with β particles because a neutrino or antineutrino is emitted as well. In β emission, the nucleus, the β particle and a neutrino or antineutrino share the energy released in variable proportions. See Topic 1.2, Stable and unstable nuclei.

Radioactivity experiments

Ionisation

The ionising effect of each type of radiation can be investigated using an ionisation chamber and a picoammeter, as shown in Figure 3. The chamber contains air at atmospheric pressure. Ions created in the chamber are attracted to the oppositely charged electrode where they are discharged. Electrons pass through the picoammeter as a result of ionisation in the chamber. The current is proportional to the number of ions per second created in the chamber.

- α radiation causes strong ionisation. However, if the source is moved away from the top of the chamber, ionisation ceases beyond a certain distance. This is because α radiation has a range in air of no more than a few centimetres.

- β radiation has a much weaker ionising effect than α radiation. Its range in air varies up to a metre or more. A β particle therefore produces fewer ions per millimetre along its path than an α particle does.

- γ radiation has a much weaker ionising effect than either α or β radiation. This is because photons carry no charge so they have less effect than α or β particles do.

Cloud chamber observations

A cloud chamber contains air saturated with a vapour at a very low temperature. Due to ionisation of the air, an α or a β particle passing through the cloud chamber leaves a visible track of minute condensed vapour droplets. This is because the air space is supersaturated. When an ionising particle passes through the supersaturated vapour, the ions produced trigger the formation of droplets.

- α particles produce straight tracks that radiate from the source and are easily visible. The tracks from a given isotope are all of the same length, indicating that the α particles have the same range.

- β particles produce wispy tracks that are easily deflected as a result of collisions with air molecules. The tracks are not as easy to see as α particle tracks because β particles are less ionising than α particles.

 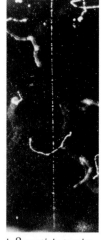

a α *particle tracks*   b β *particle tracks*

▲ **Figure 4** *Cloud chamber photographs*

Absorption tests

Figure 5 shows how a Geiger tube and a counter may be used to investigate absorption by different materials. Each particle of radiation that enters the tube is registered by the counter as a single count.

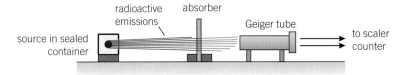

▲ **Figure 5** *Investigating absorption*

The number of counts in a given time is measured and used to work out the **count rate** which is the number of counts divided by the time taken. Before the source is tested, the count rate due to background radioactivity must be measured. This is the count rate without the source present.

- The count rate is then measured with the source at a fixed distance from the tube without any absorber present. The background count rate is then subtracted from the count rate with the source present to give the corrected (i.e., true) count rate from the source.

- The count rate is then measured with the absorber in a fixed position between the source and the tube. The corrected count rates with and without the absorber present can then be compared.

By using absorbers of different thickness of the same material, the effect of the absorber thickness can be investigated. Figure 6 shows a typical set of measurements for the absorption of β radiation by aluminium. Note that the count rate scale is a logarithmic scale (see Topic 30.3).

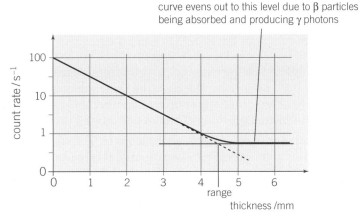

▲ **Figure 6** *Count rate against absorber thickness*

Absorption summary

- α radiation is absorbed completely by paper and thin metal foil.
- β radiation is absorbed completely by about 5 mm of metal.
- γ radiation is absorbed completely by several centimetres of lead.

The Geiger tube

The Geiger tube is a sealed metal tube that contains argon gas at low pressure. The thin mica window at the end of the tube allows α and β particles to enter the tube. γ photons can enter the tube through the tube wall as well. A metal rod down the middle of the tube is at a positive potential as shown in Figure 7. The tube wall is connected to the negative terminal of the power supply and is earthed.

Notes

The *dead time* of the tube, the time taken to regain its non-conducting state after an ionising particle enters it, is typically of the order of 0.2 ms. Another particle that enters the tube in this time will not cause a voltage pulse. Therefore, the count rate should be no greater than about $5000\,\mathrm{s}^{-1}\left(=\dfrac{1}{0.2}\,\mathrm{ms}\right)$.

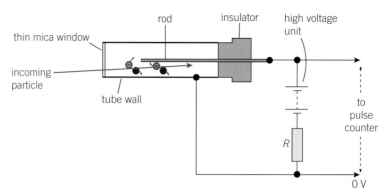

▲ **Figure 7** *A Geiger tube*

When a particle of **ionising radiation** enters the tube, the particle ionises the gas atoms along its track. The negative ions are attracted to the rod and the positive ions to the wall. The ions accelerate and collide with other gas atoms, producing more ions. These ions produce further ions in the same way so that within a very short time, many ions are created and discharged at the electrodes. A pulse of charge passes round the circuit through resistor R, causing a voltage pulse across R which is recorded as a single count by the pulse counter.

Range in air

The arrangement in Figure 5 without the absorbers may be used to investigate the range of each type of radiation in air. The count rate is measured for different distances between the source and the tube, starting with the source close to the tube. The background count rate must also be measured in the absence of the source so the corrected count rate can be calculated for each distance.

- α radiation has a range of only a few centimetres in air. The count rate decreases sharply once the tube is beyond the range of the α particles. This can be seen in Figure 4 as the tracks from the source are the same length indicating that the particles from a given source have the same range and therefore the same initial kinetic energy. The range differs from one source to another indicating that the initial kinetic energy differs from one source to another.

- β radiation has a range in air of up to about a metre. The count rate gradually decreases with increasing distance until it is the same as the background count rate at a distance of about 1 m. The reason for the gradual decrease of count rate as the distance increases is that the β particles from any given source have a range of initial kinetic energies up to a maximum. Faster β particles travel further in air than slower β particles as they have greater initial kinetic energy.

- γ radiation has an unlimited range in air. The count rate gradually decreases with increasing distance because the radiation spreads out in all directions, as shown in Figure 2. The proportion of the γ photons from the source entering the tube decreases according the inverse square law. See Topic 26.3 for more on the properties of α, β and γ radiation.

▼ **Table 1** *Nature and properties of α, β and γ radiation*

	α radiation	β radiation	γ radiation
nature	2 protons + 2 neutrons	β^- = electron (β^+ = positron)	photon of energy of the order of MeV
range in air	fixed range, depends on energy, can be up to 100 mm	range up to about 1 m	follows the inverse square law
deflection in a magnetic field	deflected	opposite direction to α particles, and more easily deflected	not deflected
absorption	stopped by paper or thin metal foil	stopped by approx 5 mm of aluminium	stopped or significantly reduced by several centimetres of lead
ionisation	produces about 10^4 ions per mm in air at standard pressure	produces about 100 ions per mm in air at standard pressure	very weak ionising effect
energy of each particle/photon	constant for a given source	varies up to a maximum for a given source	constant for a given source

Summary questions

1 A beam of radiation from a radioactive substance passes through paper and is then stopped by an aluminium plate of thickness 5 mm.

   a What type of particles are in this beam?

   b Describe a further test you could do to check your answer in **a**.

2 What type of radioactivity was responsible for the image of the key seen by Becquerel in the effect described at the beginning of this topic?

   a Explain why an image of the key was produced on the photographic plate.

   b Which type of radiation from a radioactive source is
   • least ionising,
   • most ionising?

   c When an α-emitting source above an ionisation chamber grid was moved gradually away from the grid, the ionisation current suddenly dropped to zero. Explain why the current suddenly dropped to zero.

3 In an absorption test as shown in Figure 5 using a β-emitting source and a Geiger counter, a count rate of 8.2 counts per second was obtained without the absorber present and a count rate of 3.7 counts per second was obtained with the absorber present. The background count rate was 0.4 counts per second. What percentage of the β particles hitting the absorber

   a pass through it,

   b are stopped by the absorber?

4 In an investigation to find out the type of radiation emitted by a radioactive source, a Geiger tube was placed near the source and its count rate was significantly more than the background count rate. When an aluminium plate was placed between the source and the tube, the count rate was reduced but it was still significantly more than the background count rate.

   a What can be concluded from these observations?

   b When the distance from the source to the tube was doubled with the aluminium plate still present, the corrected count rate decreased to about 25%. What conclusion can be drawn from this observation?

26.3 More about α, β, and γ radiation

Learning objectives:

→ Describe what happens to the nucleus in a radioactive change.

→ Describe how the intensity of γ radiation changes as it spreads out.

→ Explain how to represent the change in a nucleus when it emits α, β, or γ radiation.

Specification reference: 3.8.1.2

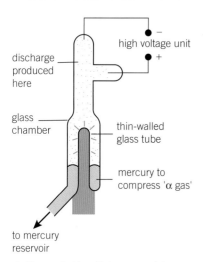

high voltage unit

discharge produced here

glass chamber

thin-walled glass tube

mercury to compress 'α gas'

to mercury reservoir

▲ **Figure 1** *Identifying α particles*

Synoptic link

γ photons are emitted from 'excited' nuclei with energies of the order of MeV (where 1 MeV = 1.6×10^{-13} J), which means their wavelength is of the order of 10^{-11} m or less. See Topic 1.3, Photons.

The nature of α, β, and γ radiation

α radiation consists of positively charged particles. Each α particle is composed of two protons and two neutrons, the same as the nucleus of a helium atom. Rutherford devised an experiment in which α particles were collected as a gas in a glass tube fitted with two electrodes. When a voltage was applied to the electrodes, the gas conducted electricity and emitted light. Using a spectrometer, he proved that the spectrum of light from the tube was the same as from a tube filled with helium gas.

Rutherford made the discovery that neutralised α particles are the same as helium some years before his discovery that every atom contains a nucleus. After he established the nuclear model of the atom, it was realised that the nucleus of the hydrogen atom, the lightest known atom, was a single positively charged particle which became known as the **proton**. Rutherford realised that other nuclei contain protons and he predicted the existence of neutral particles of similar mass, **neutrons**, in the nucleus. For example, the helium nucleus carries twice the charge of the hydrogen nucleus and therefore contains two protons. However, its mass is four times the mass of the hydrogen nucleus so Rutherford predicted that it contained two neutrons as well as two protons. The existence of the neutron was established in 1932 by James Chadwick, one of Rutherford's former students.

β radiation from naturally occurring radioactive substances consists of fast-moving electrons. This was proved by measuring the deflection of a beam of β particles using electric and magnetic fields. The measurements were used to work out the specific charge (which is the charge/mass) of the particles. This was shown to be the same as the specific charge of the electron. An electron is created and emitted from a nucleus with too many neutrons as a result of a neutron suddenly changing into a proton.

A nucleus with too many protons is also unstable and emits a **positron**, the antiparticle of the electron, when a proton changes to a neutron. Such unstable nuclei are not present in naturally occurring radioactive substances. They are created when high-energy protons collide with nuclei. As outlined in Chapter 1, the theory that for every type of particle there is a corresponding antiparticle was put forward by Paul Dirac in 1928. The first antiparticle to be discovered, the positron, was discovered by Carl Anderson four years later.

γ radiation consists of photons with a wavelength of the order of 10^{-11} m or less. This discovery was made by using a crystal to diffract a beam of γ radiation in a similar way to the diffraction of light by a diffraction grating.

The inverse square law for γ radiation

The **intensity** I of the radiation is the radiation energy per second passing normally through unit area.

- For a point source that emits n γ photons per second, each of energy hf, the radiation energy per second from the source = nhf
- At distance r from the source, all the photons emitted from the source pass through a total area of $4\pi r^2$ (the surface area of a sphere of radius r).

So the intensity I of the radiation at this distance

$$= \frac{\text{radiation energy per second}}{\text{total area}} = \frac{nhf}{4\pi r^2}$$

Therefore $I = \dfrac{k}{r^2}$, where the constant $k = \dfrac{nhf}{4\pi}$.

Thus I varies with the inverse square of distance r.

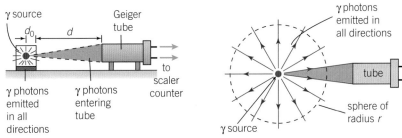

▲ **Figure 2** *Verifying the inverse square law for γ radiation*

The equations for radioactive change

A **nuclide** $_Z^A X$ contains Z protons and $A - Z$ neutrons.

- Its charge = $+Ze$, where e is the charge of the proton
- Its mass in **atomic mass units** $\approx A$

α emission

An α particle consists of 2 neutrons and 2 protons so is represented by the symbol $_2^4\alpha$. As explained in Topic 1.2, the equation below represents the change that takes place when a nuclide $_Z^A X$ emits an α particle to form a new nuclide $_{Z-2}^{A-4}Y$.

$$_Z^A X \rightarrow {_2^4}\alpha + {_{Z-2}^{A-4}}Y$$

β⁻ emission

A negative β particle (i.e., electron) is represented by the symbol $_{-1}^{\ 0}\beta$ (or β^-) as its charge = $-e$ and it is not a neutron or a proton.

The equation below represents the changes that take place when a nucleus $_Z^A X$ emits a β⁻ particle. In effect, a neutron in a neutron-rich nucleus changes into a proton and an (electron) antineutrino $\bar{v}_e$ is emitted at the same time as the β⁻ particle is created.

$$_Z^A X \rightarrow {_{-1}^{\ 0}}\beta + {_{Z+1}^{A}}Y + \bar{v}_e$$

β⁺ emission

A positive β particle (i.e., positron) is represented by the symbol $_{+1}^{\ 0}\beta$ (or β^+) as its charge = $+e$ and it is not a neutron or a proton.

The equation below represents the changes that take place when a nucleus $_Z^A X$ emits a positron. In effect, a proton in a proton-rich nucleus changes into a neutron and an electron neutrino v_e is emitted at the same time as the β⁺ particle is created.

$$_Z^A X \rightarrow {_{+1}^{\ 0}}\beta + {_{Z-1}^{A}}Y + v_e$$

Notes

1. If the intensity at distance r_0 is I_0, then:
   - at distance $2r_0$, the intensity is $\dfrac{I_0}{4}$
   - at distance $3r_0$, the intensity is $\dfrac{I_0}{9}$, etc.

   Radiation workers use remote handling devices (and lead screens) to keep as far as possible from γ sources to reduce their exposure to the radiation.

2. **To verify the inverse square law for a γ source**, use a Geiger counter to measure the count rate C at different measured distances, d, from the tube and the background count rate, C_0, without the source present. The corrected count rate $C - C_0$ is proportional to the intensity of the radiation.

   As shown in Figure 2 the source is at unknown distance, d_0, inside its sealed container. Using the inverse square law for γ radiation gives the corrected count rate

   $$C - C_0 = \frac{\text{constant}}{(d + d_0)^2}$$

   Therefore, a graph of d (on the y-axis) against $\dfrac{1}{(C - C_0)^{\frac{1}{2}}}$ should give a straight line with a negative intercept $-d_0$.

Electron capture

Some proton-rich nuclides can capture an inner-shell electron. This causes a proton in the nucleus to change into a neutron with the emission of an electron neutrino v_e at the same time.

$$^A_Z X + ^0_{-1} e \rightarrow ^A_{Z-1} Y + v_e$$

The inner-shell vacancy is filled by an outer-shell electron, as a result causing an X-ray photon to be emitted by the atom.

γ emission

No change occurs in the number of protons or neutrons of a nucleus when it emits a γ photon.

A γ photon is emitted if a nucleus has excess energy after it has emitted an α or a β⁻ particle.

Study tip

Remember for the inverse square law, doubling the distance reduces the intensity to a quarter and trebling the distance reduces it to a ninth!

Hint

1 atomic mass unit (abbreviated 'u') = $\frac{1}{12}$ × the mass of a $^{12}_6 C$ atom.

The mass of a proton is 1.00728 u and the mass of a neutron is 1.00867 u.

The difference between A and the mass of a nucleus in atomic mass units, although very small for most atoms, is important in nuclear energy calculations which you will meet in Topic 27.1.

Summary questions

1 Copy and complete each of the following equations representing α emission.

a $^{238}_{92} U \rightarrow _{90} Th +$ b $_{90} Th \rightarrow ^{224}_{88} Ra +$

2 Copy and complete each of the following equations representing β⁻ emission.

a $^{64}_{29} Cu \rightarrow _{30} Zn + ^0 \beta +$ b $_{15} P \rightarrow ^{32} S + ^0 \beta +$

3 The bismuth isotope $^{213}_{83} Bi$ decays by emitting a β⁻ particle to form an unstable isotope of polonium (Po) which then decays by emitting an α particle to form an unstable isotope of lead (Pb). This isotope then decays by emitting a β⁻ particle to form a stable isotope of bismuth.

a Write down the symbol for each of the three product nuclides in this sequence.

b Write down the number of protons and the number of neutrons in a nucleus of

i the bismuth isotope $^{213}_{83} Bi$,

ii the stable bismuth isotope.

4 A point source of γ radiation is placed 200 mm from the end of a Geiger tube. The corrected count rate was measured at 12.7 counts per second. Calculate:

a the corrected count rate if the source was moved to a distance of 400 mm from the tube,

b the distance between the source and the tube for a corrected count rate of 20 counts per second.

26.4 The dangers of radioactivity

The hazards of ionising radiation

Ionising radiation is hazardous because it damages living cells. Such radiation includes **X-rays**, protons and neutrons as well as α, β, and γ radiation. Ionising radiation affects living cells because:

- it can destroy cell membranes which causes cells to die, or
- it can damage vital molecules such as DNA directly or indirectly by creating 'free radical' ions which react with vital molecules. Normal cell division is affected and nuclei become damaged. Damaged DNA may cause cells to divide and grow uncontrollably, causing a tumour which may be cancerous. Damaged DNA in a sex cell (i.e., an egg or a sperm) may cause a mutation which may be passed on to future generations.

As a result of exposure to ionising radiation, living cells die or grow uncontrollably or mutate, affecting the health of the affected person (somatic effects) and possibly affecting future generations (genetic effects). High doses of ionising radiation kill living cells. Cell mutation and cancerous growth occurs at low doses as well as at high doses. There is no evidence of the existence of a threshold level of ionising radiation below which living cells would not be damaged.

Radiation monitoring

Anyone using equipment that produces ionising radiation must wear a **film badge** to monitor his or her exposure to ionising radiation. The badge contains a strip of photographic film in a light-proof wrapper. Different areas of the film are covered by absorbers of different materials and different thicknesses. When the film is developed, the amount of exposure to each form of ionising radiation can be estimated from the blackening of the film. If the badge is overexposed, the wearer is not allowed to continue working with the equipment.

Learning objectives:
→ Explain why ionising radiation is harmful.
→ State the factors that determine whether α, β, or γ are the most dangerous.
→ Discuss how exposure to ionising radiation can be reduced.

Specification reference: 3.8.1.2

▲ **Figure 1** *A radioactive warning sign*

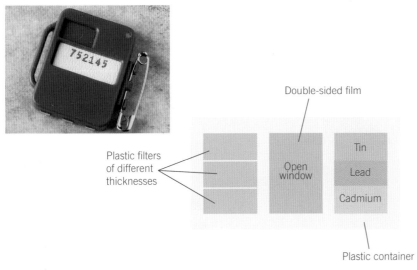

▲ **Figure 2** **a** *a film badge* **b** *inside a film badge*

Double-sided film

Plastic filters of different thicknesses

Open window

Tin

Lead

Cadmium

Plastic container

The biological effect of ionising radiation depends on the dose received and the type of radiation. The dose is measured in terms of the energy absorbed per unit mass of matter from the radiation. The same dose of different types of ionising radiation has different effects. For example, α radiation produces far more ions per millimetre than γ radiation in the same substance so it is far more damaging. However, α radiation from a source outside the body cannot penetrate the skin's outer layer of dead cells so is much less damaging than if the source were inside the body.

Radiation dose limits

For any dose of ionising radiation, its dose equivalent, measured in sieverts (Sv), is the dose due to 250 kV X-rays that would have the same effect. For example, 1 millisievert of α radiation has the same biological effect as 10 millisieverts of 250 kV X-rays.

Maximum permissible exposure limits are recommended safety limits for the annual dose equivalent which a radiation worker should not exceed. The recommended limit is 15 mSv per year, although the average dose due to occupation is much lower at 2 mSv per year. This is based on the death rates of survivors of the atomic bombs dropped on Hiroshima and Nagasaki which is estimated at 3 deaths per 100 000 survivors for each millisievert of radiation. Thus the risk to a radiation worker exposed to 2 mSv per year for 5 years would be 3 in 10 000.

Scientists think there is no lower limit below which ionising radiation is harmless. They recommend that exposure to ionising radiation should be as low as reasonably practical or achievable (abbreviated as ALARP in the UK or ALARA in the USA). It would be unethical to carry out tests that expose people or animals to ionising radiation to find out how they react to it. So the evidence for 'no lower limit' comes from the study of groups of people such as Chernobyl survivors who have been exposed to ionising radiation.

Q: Look at Figure 3. State two sources of background radiation that affect everyone all the time.

Study tip

ALARA stands for 'as low as reasonably achievable'. Risks are always reduced by increasing the distance from sources and shortening the time exposure.

Background radiation

We are all subject to **background radiation** which occurs naturally due to cosmic radiation and from radioactive materials in rocks, soil and in the air. Background radiation does vary with location due to local geological features. For example, radon gas which is radioactive can accumulate in poorly ventilated areas of buildings in certain locations. Figure 3 shows the sources of background radiation in the UK.

Safe use of radioactive materials

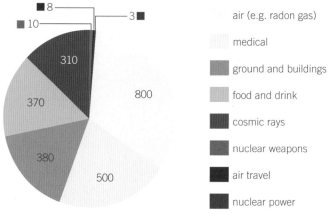

▲ **Figure 3** *Sources of background radiation in the UK in terms of the average radiation dose per person per year in μSv*

Because radioactive materials produce ionising radiation, they must be stored and used with care. In addition, disposal of a radioactive substance must be carried out in accordance with specific regulations. Only approved institutions are allowed to use radioactive materials. Approval is subject to regular checks and approved institutions are categorised according to purpose.

1 **Storage of radioactive materials** should be in lead-lined containers. Most radioactive sources produce γ radiation as well as α or β radiation so the lead lining of a container must be thick enough to reduce the γ radiation from the sources in the container to about the background level. In addition, regulations require that the containers are under 'lock and key' and a record of the sources is kept.

2 **When using radioactive materials**, established rules and regulations must be followed. No source should be allowed to come into contact with the skin.

- Solid sources should be transferred using handling tools such as tongs or a glove-box or using robots. The handling tools ensure the material is as far from the user as practicable so the intensity of the γ radiation from the source at the user is as low as possible and the user is beyond the range of α or β radiation from the source.

- Liquid and gas sources and solids in powder form should be in sealed containers. This is to ensure radioactive gas cannot be breathed in and radioactive liquid cannot be splashed on the skin or drunk.

- Radioactive sources should not be used for longer than is necessary. The longer a person is exposed to ionising radiation, the greater is the dose of radiation received.

Synoptic link

The intensity of a γ beam that passes through an absorber decreases exponentially with the thickness of the absorber. If a certain thickness of a material cuts the intensity of a γ beam to half, twice the thickness will cut the intensity to a quarter, and so on. See Topic 26.5, Radioactive decay.

Hint

A typical radioactive source used in a school might produce of the order of 10^5 radioactive particles per second, each typically of energy of the order of MeV ($1\,\text{MeV} = 1.6 \times 10^{-13}\,\text{J}$). Show for yourself that the energy transfer per second from such a source is of the order of $10^{-8}\,\text{J s}^{-1}$. In 15 minutes or so, the source would transfer $10^{-5}\,\text{J}$ to its surroundings. If this amount of energy were to be absorbed by about 10 kg of living tissue, the dose would be about $10^{-6}\,\text{Sv}$ ($= 1\,\mu\text{Sv}$) which is not insignificant.

Summary questions

1 What is meant by ionisation? Explain why a source of α-radiation is not as dangerous as a source of β radiation provided the sources are outside the body.

2 **a** Discuss the reasons why ionising radiation is hazardous to a person exposed to the radiation.

  b **i** What is the purpose of a film badge worn by a radiation worker?

  ii With the aid of a diagram, describe what is in a film badge and how the film badge is tested.

3 **a** Explain why a radioactive source should be

  i kept in a lead-lined storage box when not in use,

  ii transferred using a pair of tongs with long handles.

  b Discuss the precautions you would take when carrying out an experiment using a source of γ radiation.

4 **a** State **four** sources of ionising radiation which we are all exposed to.

  b State **two** sources of ionising radiation which are likely to affect people in certain occupations more than the general public.

  c Radon gas is a source of α radiation that can seep into buildings from the ground. Explain why the presence of this gas in the air in a building is a serious health hazard.

26.5 Radioactive decay

Learning objectives:

→ State what is meant by the activity of a radioactive isotope.

→ Define the half-life of a radioactive isotope.

→ Discuss whether anything affects radioactive decay.

Specification reference: 3.8.1.3

Synoptic link

Although in theory a radioactive decay curve never falls to zero, in practice it eventually falls to a level which is indistinguishable from background radiation. See Topic 26.4, The dangers of radioactivity.

Study tip

Know how to use a calculator to raise 0.5 to a power.

Note

Reminder about molar mass and the Avogadro constant.

For an element with a mass number A,

- its molar mass M is its mass number in grams,
- one mole of the element contains N_A atoms where N_A is the Avogadro constant,
- mass m of the element contains $\left(\dfrac{m}{M}\right)N_A$ atoms.

See Topic 20.2 for more information.

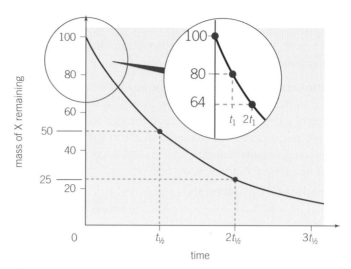

▲ **Figure 1** *A radioactive decay curve*

Half-life

When a nucleus of a radioactive isotope emits an α or a β particle, it becomes a nucleus of a different element because its proton number changes. The number of nuclei of the initial radioactive isotope therefore decreases. The mass of the initial isotope decreases gradually as the number of nuclei of the isotope decreases. Figure 1 shows how the mass decreases with time. The curve is referred to as a **decay curve**. The mass of the isotope decreases with time at a slower and slower rate. Measurements show that the mass decreases exponentially which means that the mass drops by a constant factor (e.g., ×0.8) in equal intervals of time. For example, if the initial mass of the radioactive isotope is 100 g and the mass decreases to a factor of ×0.8 every 1000 s, then

- after 1000 s, the mass remaining = 80 g (= 0.8 × 100 g),
- after 2000 s, the mass remaining = 64 g (= 0.8 × 0.8 × 100 g),
- after 3000 s, the mass remaining = 51 g (= 0.8 × 0.8 × 0.8 × 100 g)

A convenient measure for the rate of decrease is the time taken for a decrease by half. This is the half-life of the process.

The half-life, $T_{\frac{1}{2}}$, of a radioactive isotope is the time taken for the mass of the isotope to decrease to half the initial mass. This is the same as the time taken for the number of nuclei of the isotope to decrease to half the initial number.

Consider a sample of a radioactive isotope X which initially contains 100 g of the isotope.

- After 1 half-life, the mass of X remaining = 0.5 × 100 = 50 g
- After 2 half-lives from the start, the mass of X remaining = $0.5^2 \times 100 = 25$ g
- After 3 half-lives from the start, the mass of X remaining = $0.5^3 \times 100 = 12.5$ g
- After n half-lives from the start, the mass of X remaining = $0.5^n m_0$, where m_0 = the initial mass

The mass of X decreases exponentially. This is because radioactive decay is **a random** process and the number of nuclei that decay in a certain time is in proportion to the number of nuclei of X remaining. To understand this idea, consider a game of dice where there are 1000 dice, each representing a nucleus of X. The throw of a dice is a random process in which each face has an equal chance of being uppermost.

*   1st throw; when the dice are all thrown, you would expect $\frac{1}{6}$ of the dice to show the same figure on the upper surface. Let all the dice that show '1' uppermost represent nuclei that have disintegrated, an expected total of 167 $\left(= \frac{1000}{6}\right)$. If these are removed, then 833 dice remain.

*   2nd throw; the remaining dice are thrown to give $\frac{1}{6}$ of 833 as the expected number of '1's. So 694 dice $\left(= 833 - \frac{833}{6}\right)$ remain.

*   3rd throw; the remaining dice are thrown to give $\frac{1}{6}$ of 694 as the expected number of '1's. So 578 dice $\left(= 694 - \frac{694}{6}\right)$ remain.

*   4th throw; the remaining dice are thrown to give $\frac{1}{6}$ of 578 as the expected number of '1's. So 482 dice $\left(= 578 - \frac{578}{6}\right)$ remain.

The analysis shows that 4 throws are needed to reduce the number of dice to less than half the initial number. Prove for yourself that a further 4 throws would reduce the number of dice to 25% of the initial number. Figure 2 shows how the number of dice remaining decreases with time. The curve has the same shape as Figure 1. The half-life of the process is 3.8 'throws'.

Activity

The activity A of a radioactive isotope is the number of nuclei of the isotope that disintegrate per second. In other words, it is the rate of change of the number of nuclei of the isotope. The unit of activity is the **becquerel (Bq),** where 1 Bq = 1 disintegration per second.

The activity of a radioactive isotope is proportional to the mass of the isotope. Because the mass of a radioactive isotope decreases with time due to radioactive decay, the activity decreases with time. Figure 3 shows an experiment in which the activity of a radioactive isotope of protactinium $^{234}_{91}$Pa is measured and recorded using a Geiger tube and a counter. This isotope is a β-emitter produced by the decay of the radioactive isotope of thorium $^{234}_{90}$Th. In this experiment, an organic solvent in a sealed bottle is used to separate protactinium from thorium to enable the activity of the protactinium to be monitored.

Before the experiment is carried out, the background count rate is measured without the bottle present. The bottle is then shaken to mix the aqueous and solvent layers and then placed near the end of the Geiger tube. The layers are allowed to separate as shown in Figure 3. The protactinium is collected by the solvent and the thorium by the aqueous layer. The Geiger tube detects β particles emitted by the decay of the protactinium nuclei in the solvent layer.

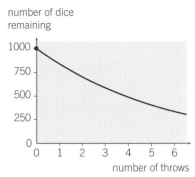
number of dice remaining

▲ **Figure 2** *Exponential decrease*

Synoptic link

To see how to model the dice model in a computer spreadsheet, see Topic 30.6, Graphical and computational modelling.

Study tip

Don't confuse the activity *A* with the mass number. This context will tell you which quantity *A* represents.

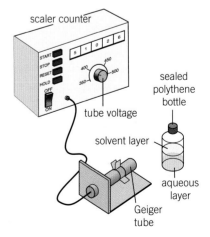

▲ **Figure 3** *Measuring the activity of protactinium*

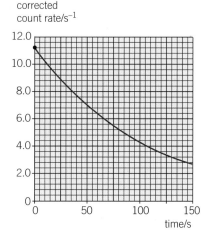

corrected count rate/s^{-1}

▲ **Figure 4** *A radioactive decay curve*

The counter is used to measure the number of counts every 10 s. The count rate is the number of counts in each ten second interval divided by 10 s. The background count rate is subtracted to give the corrected count rate. Since the activity is proportional to the corrected count rate, a graph of the corrected count rate against time, as in Figure 4, shows how the activity of the protactinium decreases with time.

Activity and power

For a radioactive source of activity A that emits particles (or photons) of the same energy E, the energy per second released by radioactive decay in the source by the radiation is the product of its activity and the energy of each particle. In other words, the power of the source = AE.

The energy transfer per second from a radioactive source = AE

If the source is in a sealed container and emits only α particles which are all absorbed by the container, the container gains thermal energy from the absorbed radiation equal to the energy transferred from the source. For example, for a source that has an activity of 30 MBq and emits particles of energy 2.5 MeV, the energy transfer per second from the source = 30×10^6 Bq $\times$ 2.5 MeV = 7.5×10^7 MeV s^{-1} = 1.2×10^{-5} J s^{-1}.

Summary questions

$N_A = 6.02 \times 10^{23}$ mol^{-1}, 1 MeV = 1.6×10^{-13} J

1 Figure 5 shows how the mass of a certain radioactive isotope decreases with time. Use the graph to work out

   a the half-life of this isotope,

   b the mass of the isotope remaining after 120 s.

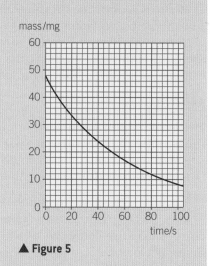

▲ **Figure 5**

2 A freshly prepared sample of a radioactive isotope X contains 1.8×10^{15} atoms of the isotope. The half-life of the isotope is 8.0 hours. Calculate:

   a the number of atoms of this isotope remaining after

      i 8 h,   ii 24 h.

   b the number of atoms of X that would have decayed after

      i 8 h,   ii 24 h.

   c the energy transfer from the sample in 24 h if the isotope emits α particles of energy 5 MeV.

3 $^{131}_{53}$I is a radioactive isotope of iodine which has a half-life of 8.0 days. A sample of this isotope has an initial activity of 38 kBq. Calculate the activity of this sample

   a 8.0 days later,   b 32 days later.

4 $^{137}_{55}$Cs is a radioactive isotope of caesium which has a half-life of 35 years. A sample of this isotope has a mass of 1.0×10^{-3} kg.

   a Calculate the number of atoms in 1.0×10^{-3} kg of this isotope.

   b Calculate the number of atoms of the isotope remaining in the sample after 70 years.

26.6 The theory of radioactive decay

The random nature of radioactive decay

An unstable nucleus becomes stable by emitting an α or a β particle or a γ photon. This is an unpredictable event. Every nucleus of a radioactive isotope has an equal probability of undergoing radioactive decay in any given time interval. Therefore, for a large number of nuclei of a radioactive isotope, the number of nuclei that disintegrate in a certain time interval depends only on the total number of nuclei present.

Consider a sample of a radioactive isotope X that initially contains N_0 nuclei of the isotope. Let N represent the number of nuclei of X remaining at time t after the start. Suppose in time Δt, the number of nuclei that disintegrate is ΔN.

Because radioactive disintegration is a random process, ΔN is proportional to:

1   N, the number of nuclei of X remaining at time t,

2   the duration of the time interval Δt.

Therefore $\Delta N = -\lambda N \Delta t$, where λ is a constant referred to as the **decay constant**. The minus sign is necessary because ΔN is a decrease.

So the rate of disintegration $\dfrac{\Delta N}{\Delta t} = -\lambda N$.

For a given radioactive isotope, its activity, A, is the rate of disintegration $\dfrac{\Delta N}{\Delta t}$.

Therefore, the activity A of N atoms of a radioactive isotope is given by

$$A = \lambda N$$

The solution of the equation $\dfrac{\Delta N}{\Delta t} = -\lambda N$ is $N = N_0 e^{-\lambda t}$, where e^x is the exponential function. See below.

Figure 1 shows that a graph of N against t gives a decay curve. The number of nuclei N decreases exponentially with time. In other words,

• in one half-life, the remaining number of nuclei $N_1 = 0.5 N_0$
• in two half-lives, the remaining number of nuclei $N_2 = 0.25 N_0$
• in n half-lives, the remaining number of nuclei $N = 0.5^n N_0$

The graph of the number of nuclei N against time t as represented by the equation $N = N_0 e^{-\lambda t}$ is shown in Figure 1. It is a curve with exactly the same shape as Figure 1 in Topic 26.5.

The mass, m, of a radioactive isotope decreases from initial mass m_0 in accordance with the equation $m = m_0 e^{-\lambda t}$ because the mass m is proportional to the number of nuclei, N, of the isotope.

The activity A of a sample of N nuclei of an isotope decays in accordance with the equation

$$A = A_0 e^{-\lambda t}$$

This is because the activity A = the number of disintegrations per second = λN. So $A = \lambda N_0 e^{-\lambda t} = A_0 e^{-\lambda t}$ where $A_0 = \lambda N_0$.

Learning objectives:
→ Discuss whether a radioactive source can decay completely.
→ Define exponential decrease.
→ Explain why a radioactive decay is a random process.

Specification reference: 3.8.1.3

> **Study tip**
>
> The unit of decay constant is s^{-1}. A large value of λ means fast decay and short half-life.

> **Synoptic link**
>
> To learn how to model the rate of decay in a computer spreadsheet, see Topic 30.6, Graphical and computational modelling.

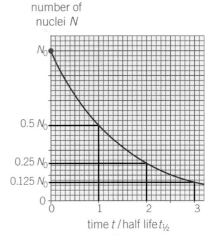

▲ **Figure 1** $N = N_0 e^{-\lambda t}$

Notes

You do *not* need to know the following points. But they might help you improve your understanding of this topic.

1   The exponential function appears in any situation where the rate of change of a quantity is in proportion to the quantity itself. This is because the rate of change of each term in the function sequence is equal to the previous term in the sequence.

2   The exponential function,
$$e^x = 1 + x + \frac{x^2}{2!} + \frac{x^3}{3!} + \dots$$
Differentiating e^x with respect to x gives e^x (i.e., $\frac{d(e^x)}{dx} = e^x$) because differentiating each term in the expression for e^x gives the previous term.
The exponential function is indicated on a calculator as 'e^x' or 'inv ln'. See Topic 30.3.

3   The natural logarithm function, $\ln x$, is the inverse exponential function. In other words, if $y = e^x$, then $\ln y = x$. Therefore, $N = N_0 e^{-\lambda t}$ may be written $\ln N = \ln N_0 - \lambda t$.
The graph of $\ln N$ against t is therefore a straight line with:
- a gradient $= -\lambda$, and
- a y-intercept $= \ln N_0$

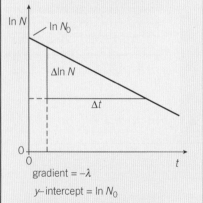

▲ **Figure 2** *ln N against t*

4   The exponential decrease formula is also used in the theory of capacitor discharge. See Topic 23.3.

The corrected count rate C due to a sample of a radioactive isotope at a fixed distance from a Geiger tube is proportional to the activity of the source. Therefore, the count rate decreases with time in accordance with the equation $C = C_0 e^{-\lambda t}$, where C_0 is the count rate at time $t = 0$.

The above equations for the number of nuclei N, the activity A, and the count rate C are all of the same general form, namely $x = x_0 e^{-\lambda t}$, where x represents N or A or C and x_0 represents the initial value.

Worked example

A sample of a radioactive material initially contains 1.2×10^{20} atoms of the isotope. The decay constant for the isotope is $3.6 \times 10^{-3}\,\text{s}^{-1}$. Calculate:

a   the number of atoms of the isotope remaining after $1000\,\text{s}$,
b   the activity of the sample after $1000\,\text{s}$.

Solution

a $N_0 = 1.2 \times 10^{20}$, $\lambda = 3.6 \times 10^{-3}\,\text{s}^{-1}$, $t = 1000\,\text{s}$,
$\lambda t = 3.6 \times 10^{-3} \times 1000 = 3.6$
$\therefore N = N_0 e^{-\lambda t} = 1.2 \times 10^{20}\,e^{-3.6} = 1.2 \times 10^{20} \times 2.7 \times 10^{-2}$
$= 3.2 \times 10^{18}$

b   Activity $A = \lambda N = 3.6 \times 10^{-3} \times 3.2 \times 10^{18} = 1.2 \times 10^{16}\,\text{Bq}$

The decay constant

The decay constant λ is the probability of an individual nucleus decaying per second. If there are $10\,000$ nuclei present and 300 decay in $20\,\text{s}$, the decay constant is $0.0015\,\text{s}^{-1}\left(= \frac{\left(\frac{300}{10\,000}\right)}{20}\,\text{s}^{-1}\right)$.

In general, if the change of the number of nuclei ΔN in time Δt is given by $\Delta N = -\lambda N \Delta t$, then the probability of decay $= \frac{\Delta N}{N} = \lambda \Delta t$ (the minus sign is not needed here as reference is made to decay).

So the probability per unit time $= \frac{\Delta N}{N}/\Delta t = \lambda$.

As explained in Topic 26.5, the half-life, $T_{1/2}$ of a radioactive isotope is the time taken for half the initial number of nuclei to decay. The longer the half-life, the smaller the decay constant because the probability of decay per second is smaller.

The half-life $T_{1/2}$ is related to the decay constant λ according to the equation

$$T_{1/2} = \frac{\ln 2}{\lambda}$$

As $\ln 2 = 0.693$, this equation may be written as $T_{1/2} = \frac{0.693}{\lambda}$

Proof of $T_{1/2} = \dfrac{\ln 2}{\lambda}$

You do *not* need to know this proof. But it might help you improve your understanding of this topic. And you *do* need to know how to rearrange an exponential equation using natural logs.

Let the number of nuclei $N = N_0$ at time $t = 0$

Therefore at time $t = T_{1/2}$, $N = 0.5 N_0$

Inserting $t = T_{1/2}$, $N = 0.5 N_0$ into $N = N_0 e^{-\lambda t}$ gives $0.5 N_0 = N_0 e^{-\lambda T_{1/2}}$

Cancelling N_0 therefore gives $0.5 = e^{-\lambda T_{1/2}}$

Taking the natural logarithm (ln) of each side gives $\ln 0.5 = -\lambda T_{1/2}$

Because $\ln 0.5 = -\ln 2$, then $\ln 2 = \lambda T_{1/2}$

Rearranging this equation gives $T_{1/2} = \dfrac{\ln 2}{\lambda}$

Hint

To calculate N at time t, given values of N_0 and $T_{1/2}$,

- **either** calculate λ using $\lambda = \dfrac{\ln 2}{T_{1/2}}$ then use the equation $N = N_0 e^{-\lambda t}$,

- **or** calculate the number of half-lives, n, using $n = \dfrac{t}{T_{1/2}}$ then use $N = 0.5^n N_0$.

Synoptic link

Radioactive waste from a nuclear reactor contains a range of unstable isotopes with different half-lives. The waste products must be stored for many years until their activity is no more than background. See Topic 27.4, The thermal nuclear reactor.

Summary questions

$N_A = 6.02 \times 10^{23}\,\text{mol}^{-1}$, $1\,\text{MeV} = 1.6 \times 10^{-13}\,\text{J}$

1 $^{131}_{53}\text{I}$ is a radioactive isotope of iodine which has a half-life of 8.0 days. A fresh sample of this isotope contains 4.2×10^{16} atoms of this isotope. Calculate:

   a   the decay constant of this isotope,

   b   the number of atoms of this isotope remaining after 24 h.

2 A radioactive isotope has a half-life of 35 years. A fresh sample of this isotope has an activity of 25 kBq. Calculate:

   a   the decay constant in s^{-1},

   b   the activity of the sample after 10 years.

3 a   Calculate the number of atoms present in 1.0 kg of $^{226}_{88}\text{Ra}$.

   b   The isotope $^{226}_{88}\text{Ra}$ has a half-life of 1620 years. For an initial mass of 1.0 kg of this isotope, calculate:

   i   the mass of this isotope remaining after 1000 years,

   ii   how many atoms of the isotope will remain after 1000 years.

4 A fresh sample of a radioactive isotope has an initial activity of 40 kBq. After 48 h, its activity has decreased to 32 kBq. Calculate:

   a   the decay constant of this isotope,

   b   its half-life.

Radioactive isotopes are used for many purposes. The choice of an isotope for a particular purpose depends on its half-life and on the type of radiation it emits. For some uses, the choice also depends on how the isotope is obtained and on whether or not it produces a stable decay product. In addition, the toxicity and biochemical suitability of the pharmaceuticals to which it is attached need to be considered in medical and related applications. The following examples are intended to provide a wider awareness of important uses of radioactive substances and to set contexts in which knowledge and understanding of radioactivity is developed further.

Radioactive dating

Carbon dating

Living plants and trees contain a small percentage of the radioactive isotope of carbon, $^{14}_{6}C$, which is formed in the atmosphere as a result of cosmic rays knocking out neutrons from nuclei. These neutrons then collide with nitrogen nuclei to form carbon-14 nuclei.

$$^{1}_{0}n + ^{14}_{7}N \rightarrow ^{14}_{6}C + ^{1}_{1}P$$

Carbon dioxide from the atmosphere is taken up by living plants as a result of photosynthesis. So a small percentage of the carbon content of any plant is carbon-14. This isotope has a half-life of 5570 years so there is negligible decay during the lifetime of a plant. Once a tree has died, no further carbon is taken in so the proportion of carbon-14 in the dead tree decreases as the carbon-14 nuclei decay. Because activity is proportional to the number of atoms still to decay, measuring the activity of the dead sample enables its age to be calculated, provided the activity of the same mass of living wood is known.

Worked example

A certain sample of dead wood is found to have an activity of 0.280 Bq. An equal mass of living wood is found to have an activity of 1.30 Bq. Calculate the age of the sample. Give your answer in years to 3 significant figures.

The half-life of carbon-14 is 5570 years.

Solution

The half-life, $T_{1/2}$, in seconds $= 5570 \times 365 \times 24 \times 3600\,s = 1.76 \times 10^{11}\,s$

So, the decay constant of carbon-14 $\lambda = \dfrac{0.693}{T_{1/2}} = \dfrac{0.693}{1.76 \times 10^{11}}$

$$= 3.95 \times 10^{-12}\,s^{-1}$$

Using activity $A = A_0 e^{-\lambda t}$ where $A = 0.280\,Bq$ and $A_0 = 1.30\,Bq$ gives

$0.280 = 1.30\,e^{-\lambda t}$ so $e^{-\lambda t} = \left(\dfrac{0.280}{1.30}\right) = 0.215$

$\therefore \lambda t = 1.535$

$t = \dfrac{1.535}{\lambda} = \dfrac{1.535}{3.95 \times 10^{-12}}s = 3.88 \times 10^{11}\,s = 12\,300$ years

Hint

A useful check is to estimate the number of half-lives needed for the activity to decrease from 1.30 Bq to 0.28 Bq. You should find that just over 2 half-lives are needed, corresponding to about 11 000 years.

Argon dating

Ancient rocks contain trapped argon gas as a result of the decay of the radioactive isotope of potassium, $^{40}_{19}\text{K}$, into the argon isotope $^{40}_{18}\text{Ar}$. This happens when its nucleus captures an inner shell electron. As a result, a proton in the nucleus changes into a neutron and a neutrino is emitted.

The equation for the change is $^{40}_{19}\text{K} + {}^{0}_{-1}\text{e} \rightarrow {}^{40}_{18}\text{Ar} + \nu_e$

The potassium isotope $^{40}_{19}\text{K}$ also decays by β^- emission to form the calcium isotope $^{40}_{20}\text{Ca}$. This process is 8 times more probable than electron capture.

$$^{40}_{19}\text{K} \rightarrow {}^{0}_{-1}\beta + {}^{40}_{20}\text{Ca} + \bar{\nu}_e$$

The effective half-life of the decay of $^{40}_{19}\text{K}$ is 1250 million years. The age of the rock (i.e., the time from when it solidified) can be calculated by measuring the proportion of argon-40 to potassium-40.

For every N potassium-40 atoms now present, if there is 1 argon-40 atom present, there must have originally been $N + 9$ potassium atoms. (i.e., 1 that decayed into argon-40 + 8 that decayed into calcium-40 + N remaining). The radioactive decay equation $N = N_0 e^{-\lambda t}$ can then be used to find the age of the sample.

For example, suppose for every 4 potassium-40 atoms now present, a certain rock now has 1 argon-40 atom. Therefore, $N = 4$ and $N_0 = 13$. Substituting these values into the equation $N = N_0 e^{-\lambda t}$ gives $4 = 13e^{-\lambda t}$. Therefore $e^{-\lambda t} = \dfrac{4}{13} = 0.308$ which gives $t = \dfrac{-\ln 0.308}{\lambda}$.

Substituting $\dfrac{0.693}{T_{1/2}}$ for λ into this equation gives $t = \left(\dfrac{-\ln 0.308}{0.693}\right) T_{1/2} = 1.70\,T_{1/2}$. The age of the sample is therefore 2120 million years.

Radioactive tracers

A radioactive tracer is used to follow the path of a substance through a system. Table 1, overleaf, gives some examples. In general, the radioactive isotope(s) in the tracer should:

- have a half-life which is stable enough for the necessary measurements to be made and short enough to decay quickly after use,
- emit β radiation or γ radiation so it can be detected outside the flow path.

Industrial uses of radioactivity

The examples below are just three of a wide range of applications of radioactivity in industry and technology.

Engine wear

The rate of wear of a piston ring in an engine can be measured by fitting a ring that is radioactive. As the ring slides along the piston compartment, radioactive atoms transfer from the ring to the engine oil. By measuring the radioactivity of the oil, the mass of radioactive metal transferred from the ring can be determined and the rate of wear calculated. A metal ring can be made radioactive by exposing it to neutron radiation in a nuclear reactor. Each nucleus that absorbs a neutron becomes unstable and disintegrates by β^- emission.

Hint

A useful check is to estimate the number of half-lives needed for n to decrease from 13 to 4. You should find that between 1 and 2 half-lives are needed, corresponding to an age of between 1250 and 2500 million years.

Synoptic link

The metastable isotope of technetium $^{99}_{43}\text{Tc}^m$ is widely used in medical diagnosis because it is a γ emitter with a half-life of 6 hours and it can be prepared 'on site'. See Topic 26.8, More about decay modes.

▲ Figure 1 Using tracers

▲ Figure 2 Measuring engine wear

▼ **Table 1** *Examples of radioactive tracers*

application	method	tracer
detecting underground pipe leaks	radioactive tracer injected into the flow. A detector on the surface above the pipeline is used to detect leakage	injected fluid contains a β-emitter or a γ⁻ emitter (depending on factors such as depth, soil density) as α-radiation would be absorbed by the pipes
modelling oil reservoirs mathematically to improve oil recovery	water containing a radioactive tracer is injected into an oil reservoir at high pressure, forcing some of the oil out. Detectors at the production wells monitor breakthrough of the radioactive isotope	'tritiated' water $^3_1$HO, a β-emitter with a half-life of 12 years
investigating the uptake of fertilisers by plants	plant watered with a solution containing a fertiliser. By measuring the radioactivity of the leaves, the amount of fertiliser reaching them can be determined	fertiliser contains phosphorus, $^{32}_{15}$P, a β-emitter with a half-life of 14 days
monitoring the uptake of iodine by the thyroid gland	patient is given a solution containing sodium iodide which contains a small quantity of radioactive iodine, $^{133}_{53}$I. The activity of the patient's thyroid and the activity of an identical sample prepared at the same time is measured 24 h later	solution of sodium iodide contains iodine $^{131}_{53}$I, a β-emitter with a half-life of 8 days

Thickness monitoring

Metal foil is manufactured by using rollers to squeeze plate metal on a continuous production line. A detector measures the amount of radiation passing through the foil. If the foil is too thick, the detector reading drops. A signal from the detector is fed back to the control system to make the rollers move closer together and so make the foil thinner. The source used is a β⁻ emitter with a long half-life. α radiation would be absorbed completely by the foil and γ radiation would pass straight through without absorption.

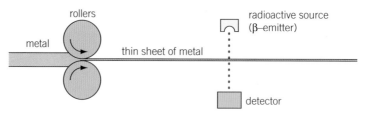

if the radiation reaching the detector changes the detector makes the rollers move further apart or closer together

▲ **Figure 3** *The manufacture of metal foil*

Power sources for remote devices

Satellites, weather sensors, and other remote devices can be powered using a radioactive isotope in a thermally insulated sealed container which absorbs all the radiation emitted by the isotope. A thermocouple attached to the container produces electricity as a result of the container becoming warm through absorbing radiation.

Study tip

The choice of a radioactive isotope for a particular application is determined by:

- the half-life and the type of radiation needed,
- the toxicity and biochemical suitability of the pharmaceuticals to which it is attached,
- whether or not a stable product is needed.

For mass m of the isotope, its activity $A = \lambda N$, where N is the number of radioactive atoms present in mass m. If each disintegration of a nucleus releases energy E, the energy transfer per second from the source = $\lambda N E$. The source needs to have a reasonably long half-life, so it does not need to be replaced frequently, but a very long half-life may require too much mass to generate the necessary power.

Summary questions

$N_A = 6.02 \times 10^{23}\ mol^{-1}$, $1\ MeV = 1.6 \times 10^{-13}\ J$

1 a Explain why living wood is slightly radioactive.

b A sample of ancient wood of mass 0.50 g is found to have an activity of 0.11 Bq. A sample of living wood of the same mass has an activity of 0.13 Bq. Calculate the age of the sample of wood.

The half-life of radioactive carbon $^{14}_{6}C$ is 5570 years.

2 The radioactive isotope of iodine, $^{131}_{53}I$, is used for medical diagnosis of the kidneys. The isotope has a half-life of 8 days. A sample of the isotope is to be given to a patient in a glass of water. The passage of the isotope through each kidney is then monitored using two detectors outside the body. The isotope is required to have an activity of 800 kBq at the time it is given to the patient.

a Calculate:

i the activity of the sample 24 hours after it was given to the patient,

ii the activity of the sample when it was prepared 24 hours before it was given to the patient,

iii the mass of $^{131}_{53}I$ in the sample when it was prepared.

b The reading from the detector near one of the patient's kidneys rises then falls. The reading from the other detector which is near the other kidney rises and does not fall. Discuss the conclusions that can be drawn from these observations.

3 a In the manufacture of metal foil, describe how the thickness of the foil is monitored using a radioactive source and a detector.

b Explain why the source needs to:

i be a β-emitter, not an α-emitter or a γ-emitter,

ii have a long half-life.

4 A cardiac pacemaker is a device used to ensure that a faulty heart beats at a suitable rate. The required electrical energy in one type of pacemaker is obtained from the energy released by a radioactive isotope. The radiation is absorbed inside the pacemaker. As a result, the absorbing material gains thermal energy and heats a thermocouple attached to the absorbing material. The voltage from the thermocouple provides the source of electrical energy for the pacemaker.

a i Discuss whether the radioactive source should be an α-emitter or a β-emitter or a γ-emitter.

ii The radioactive source needs to have a reasonably long half-life, otherwise it would need to be replaced frequently. Discuss the disadvantages of using a radioactive source with a very long half-life.

b The energy source for a remote weather station is the radioactive isotope of strontium, $^{90}_{38}Sr$ which has a half-life of 28 years. It emits β particles of energy 0.40 MeV. For a mass of 10 g of this isotope, calculate:

i its activity,

ii the energy released per second.

26.8 More about decay modes

Learning objectives:

→ Discuss what you can tell about radioactive isotopes from an N–Z chart.

→ Explain why naturally occurring isotopes don't emit β^+ radiation.

→ Describe what happens to an unstable nucleus that emits γ radiation.

Specification reference: 3.8.1.4

Study tip

Neutron–proton ratio is high for β^- and low for β^+ emitters.

There are no α-emitters for $Z <$ about 60.

The largest stable nuclide is $^{209}_{83}\text{Bi}$.

Most α-emitting nuclides are larger than $^{209}_{83}\text{Bi}$ and therefore lie above the top of the stability belt of the N-Z plot.

Those α-emitting nuclides that are smaller than $^{209}_{83}\text{Bi}$ generally lie below the N-Z stability belt.

The N–Z graph

A useful way to survey nuclear stability is to plot a graph of the neutron number N against the proton number Z for all known isotopes, as shown in Figure 1. Each isotope is plotted on the graph according to its values of N and Z. The graph shows that stable nuclei lie along a belt curving upwards with an increasing neutron–proton ratio from the origin to $N = 120$, $Z = 80$ approximately.

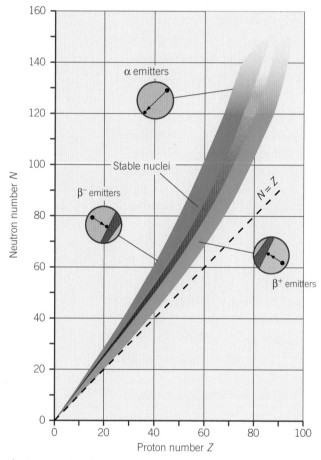

▲ **Figure 1** The N–Z graph

- **For light isotopes** (Z from 0 to no more than 20), the stable nuclei follow the straight line $N = Z$. Such nuclei have equal numbers of protons and neutrons.

- **As Z increases beyond about 20**, stable nuclei have more neutrons than protons. The neutron/proton ratio increases. The extra neutrons help to bind the nucleons together without introducing repulsive electrostatic forces as more protons would do.

- α **emitters** occur beyond about Z = 60, most of them with more than 80 protons and 120 neutrons. These nuclei have more neutrons than protons but they are too large to be stable. This is because the **strong nuclear force** between the nucleons is unable to overcome the electrostatic force of repulsion between the protons.

- β⁻ emitters occur to the left of the stability belt where the isotopes are neutron-rich compared to stable isotopes. As explained previously, neutron-rich isotopes become stable or less unstable by 'converting' a neutron into a proton and emitting a β⁻ particle (and an electron antineutrino) at the same time.

- β⁺ emitters occur to the right of the stability belt where the isotopes are proton-rich compared to stable isotopes. As explained previously, proton-rich isotopes become stable or less unstable by 'converting' a proton into a neutron and emitting a β⁺ particle (and an electron neutrino) at the same time. Electron capture also takes place in this region.

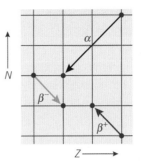

▲ **Figure 2** *N–Z changes*

The change that takes place when an unstable nucleus becomes stable or less unstable can be represented on the *N–Z* graph as shown in general in Figure 1 and in detail in Figure 2.

- A nucleus that emits an α particle loses two protons and two neutrons so it moves diagonally downwards to the left across two grid squares.

- A nucleus that emits a β⁻ particle loses a neutron and gains a proton so it moves diagonally downwards to the right across one grid square.

- A nucleus that emits a β⁺ particle (or captures an electron) loses a proton and gains a neutron so it moves diagonally upwards to the left across one grid square.

Radioactive series

Many radioactive isotopes decay to form another isotope which might itself be unstable. When an unstable nucleus emits an α particle, its position on the *N–Z* plot moves downwards parallel to the *N* = *Z* line to a new position with a greater neutron–proton ratio. If the 'daughter' nucleus is also unstable, it will decay to form a nucleus of a different isotope (which may itself be stable or unstable) by either emitting:

- a further α particle, or:
- a β⁻ particle if its position is to the left of the stability belt, or:
- a β⁺ particle or by undergoing electron capture if, in both cases, its position is to the right of the stability belt.

Thus an unstable nucleus, before it becomes stable, may undergo a series of isotopic changes in which each change involves an emission of an α particle or a β particle. Naturally occurring radioactive isotopes decay through a series of such changes with one or more of the changes having a very long half-life – hence the reason why such isotopes have not decayed completely.

For example, the uranium isotope $^{238}_{92}\text{U}$ has a half-life of 4500 million years and decays through a series of changes outlined below.

$$^{238}_{92}\text{U} \xrightarrow{\ \alpha\ } {}^{234}_{90}\text{Th} \xrightarrow{\ \alpha\ } {}^{230}_{88}\text{Ra} \xrightarrow{\ \beta^-\ } {}^{230}_{89}\text{Ac} \xrightarrow{\ \beta^-\ } {}^{230}_{90}\text{Th} \xrightarrow{\ \alpha\ } {}^{226}_{88}\text{Ra} \rightarrow \dots \rightarrow {}^{206}_{82}\text{Pb}$$

Any radioactive series may be represented on the *N–Z* graph by a sequence of 'decay arrows' as shown in Figure 3, which represents the first five changes above. Notice there are no β⁺ emissions in the sequence because such an emission after an α emission would

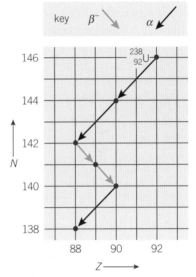

▲ **Figure 3** *Part of a radioactive decay series*

Synoptic link

β⁻ emitters can also be manufactured by bombarding stable isotopes with neutrons. β⁺ emitters can only be produced by bombarding stable isotopes with protons. The protons need to have sufficient kinetic energy to overcome coulomb repulsion from the nucleus. See Topic 26.1, The discovery of the nucleus.

Notes

1   In Figure 4 the presence of two excited states is indicated by the fact that the two smaller γ energies add up to the largest γ energy.

2   In this example, the parent nucleus can also decay by β⁻ emission to the excited state of $^{27}_{13}$Al at 0.83 MeV then by emission of a γ photon of energy 0.83 MeV to the ground state. The β⁻ emission for this change is shown on Figure 4 by the dashed arrow.

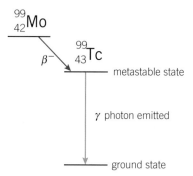

▲ **Figure 5** *The metastable state of technetium $^{99}_{43}$Tc*

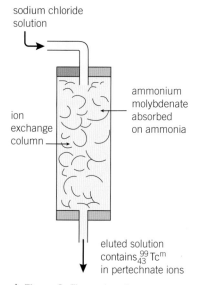

▲ **Figure 6** *The technetium generator*

make the nucleus more neutron-rich, so moving it away from the stability belt. A β⁻ emission after an α emission makes the nucleus less neutron-rich causing it to be more stable and nearer the stability belt.

Nuclear energy levels

After an unstable nucleus emits an α or a β particle or undergoes electron capture, it might emit a γ photon. Emission of a γ photon does not change the number of protons or the number of neutrons in the nucleus but it does allow the nucleus to lose energy. This happens if the 'daughter' nucleus is formed in an **excited state** after it emits an α or a β particle or undergoes electron capture. The excited state is usually short-lived and the nucleus moves to its lowest energy state, its **ground state**, either directly or via one or more lower-energy excited states. We can represent such changes by means of an energy level diagram, as shown by the example in Figure 4 in which:

*   a magnesium $^{27}_{12}$Mg nucleus decays by β⁻ emission (shown as a blue arrow) to form an aluminium $^{27}_{13}$Al nucleus in an excited state 1.02 MeV above the ground state at zero energy,

*   the aluminium nucleus de-excites by emitting (as shown by the green arrows) either a 1.02 MeV γ photon (1), or a 0.19 MeV γ photon (2) followed by a 0.83 MeV γ photon (3).

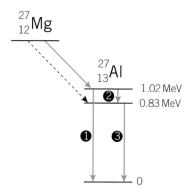

▲ **Figure 4** *Nuclear energy states*

The technetium generator

The technetium generator is used in hospitals to produce a source which emits γ radiation only. Some radioactive isotopes such as the technetium isotope $^{99}_{43}$Tc form in an excited state after an α emission or a β emission and stay in the excited state long enough to be separated from the parent isotope. Such a long-lived excited state is said to be a **metastable state**. Nuclei of the technetium isotope $^{99}_{43}$Tc form in a metastable state (indicated by the symbol $^{99}_{43}$Tc^m) after β⁻ emission from nuclei of the molybdenum isotope $^{99}_{42}$Mo which has a half-life of 67 h. $^{99}_{43}$Tc^m has a half-life of 6 h and decays to the ground state by γ emission.

Technetium $^{99}_{43}$Tc in the ground state is a β⁻-emitter with a half-life of 500 000 years and it forms a stable product. Therefore, a sample of $^{99}_{43}$Tc^m with no molybdenum present effectively emits only γ photons. Such samples of technetium $^{99}_{43}$Tc^m are used in medical diagnosis applications, as outlined below.

The technetium generator consists of an ion exchange column containing ammonium molybdenate exposed to neutron radiation several days earlier to make a significant number of the molybdenum nuclei unstable. When a solution of sodium chloride is passed through the column, some of the chlorine ions exchange with pertechnate ions but not with molybdenate ions so the solution that emerges contains $^{99}_{43}Tc^m$ nuclei.

Examples of diagnostic uses of $^{99}_{43}Tc^m$

- **Monitoring blood flow** through the brain using external detectors after a small quantity of sodium pertechnate solution is administered intravenously.

- **The γ camera** is designed to 'image' internal organs and bones by detecting γ radiation from sites in the body where a γ-emitting isotope such as $^{99}_{43}Tc^m$ nuclei is located. For example, bone deposits can be located using a phosphate tracer labelled with $^{99}_{43}Tc^m$. The γ camera itself consists of detectors called photomultiplier tubes in a lead shield behind a lead collimator grid which ensures each tube only detects γ photons emitted from nuclei located at a well-defined spot directly in front of the tube.

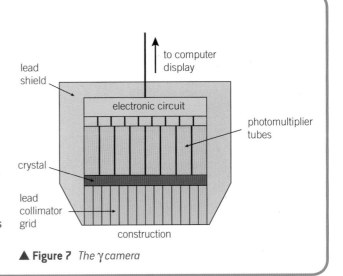

▲ **Figure 7** The γ camera

Summary questions

1 a Sketch an N–Z graph to show how N varies with Z for the known stable isotopes.

   b Show on your diagram possible locations of an isotope that is

   i an α-emitter,   ii a β⁻-emitter,   iii a β⁺-emitter.

2 A nucleus of the polonium isotope $^{216}_{84}Po$ decays to form a stable nucleus X by emitting in succession an α particle, a β⁻ particle, a further β⁻ particle, then another α particle.

   a Determine the number of protons and the number of neutrons in X.

   b Show the changes on the grid of an N–Z chart.

3 a Explain what is meant by electron capture.

   b State one similarity and one difference between electron capture and positron emission.

4 The germanium isotope $^{77}_{32}Ge$ has a metastable state which decays to the ground state by emission of a 0.16 MeV γ photon. The isotope decays by β⁻ emission to form the arsenic isotope $^{77}_{33}As$ in an excited state 0.48 MeV above the ground state of $^{77}_{33}As$.

   a Copy and complete the energy level diagram to show these changes.

   b The excited state of $^{77}_{33}As$ also decays to the ground state via an excited state which is 0.27 MeV above the ground state. Calculate the energies of the γ photons emitted in this decay.

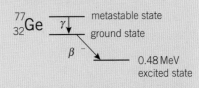

26.9 Nuclear radius

Learning objectives:

→ Discuss whether more massive nuclei are wider.

→ Describe how the radius of a nucleus depends on its mass number A.

→ Describe how dense the nucleus is.

Specification reference: 3.8.1.5

Notes

You do *not* need to know points 2 and 3 below. But they might help you to improve your understanding of this topic.

1   The wavelength of high-energy electrons is calculated using the equation $\lambda = \dfrac{hc}{E}$, where E is the energy of the electrons. The speed of a high-energy electron is very close to c, the speed of light in a vacuum. Therefore, $\lambda = \dfrac{h}{mv} = \dfrac{h}{mc} = \dfrac{hc}{E}$, as $E = mc^2$. In practice, electrons need to be accelerated through pds greater than about 100 million volts to be diffracted significantly by the nucleus.

2   The angle θ_{min} depends on the radius R of the nucleus in accordance with the equation $R \sin \theta_{min} = 0.61\lambda$, where λ is the de Broglie wavelength of the electrons. The equation is derived by applying wave theory to plane waves passing at normal incidence through a circular gap.

3   Prove for yourself that values of $E = 420\,\text{MeV}$ and $\theta_{min} = 44°$ obtained for oxygen nuclei give a de Broglie wavelength of 3.0 fm and a nuclear radius for the oxygen nucleus of 2.6 fm. See Reference data at the end of the book for values of h and c. Note that 1 MeV = 1.6×10^{-13} J.

High-energy electron diffraction

In Topic 26.1, we estimated the diameter of the nucleus to be about a femtometre (1 fm = 10^{-15} m) using two methods. In this topic we will look at a much more accurate method to measure the diameter of different nuclides using high-energy electrons. When a beam of high-energy electrons is directed at a thin solid sample of an element, the incident electrons are diffracted by the nuclei of the atoms in the foil. The beam is produced by accelerating electrons through a potential difference of the order of a hundred million volts. The electrons are diffracted by the nuclei because the **de Broglie wavelength** of such high-energy electrons is of the order of 10^{-15} m which is about the same as the diameter of the nucleus. A detector is used to measure the number of electrons per second diffracted through different angles.

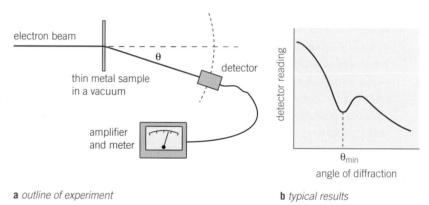

a *outline of experiment*

b *typical results*

▲ **Figure 1** *High-energy electron diffraction*

The measurements show that as the angle θ of the detector to the 'zero order' beam is increased, the number of electrons per second (i.e., the intensity of the beam) diffracted into the detector decreases then increases slightly then decreases again.

- Scattering of the beam electrons by the nuclei occurs due to their charge. This is the same as α-scattering by nuclei except the electrons are attracted not repelled by the nuclei. This effect causes the intensity to decrease as angle θ increases.

- Diffraction of the beam electrons by each nucleus causes intensity maxima and minima to be superimposed on the effect above. This happens provided the de Broglie wavelength of the electrons in the beam is no greater than the dimensions of the nucleus. These superimposed intensity variations are, on a much smaller scale, similar to the concentric bright and dark fringes seen when a parallel beam of monochromatic light is directed at a circular gap or obstacle. The angle of the first minimum from the centre, θ_{min}, is measured and used to calculate the diameter of the nucleus, provided the wavelength of the incident electrons is known. See the notes in the margin.

Dependence of nuclear radius on nucleon number

Using samples of different elements, the radius R of different nuclides can be measured. By plotting a suitable graph as explained below, it can be shown that R depends on mass number A according to $R = r_0 A^{1/3}$, where the constant $r_0 = 1.05\,\text{fm}$.

- A graph of $\ln R$ against $\ln A$ gives a straight line with a gradient of $\frac{1}{3}$ and a y-intercept equal to $\ln r_0$. This is because $\ln R = \ln A^{1/3} + \ln r_0$ $= \frac{1}{3}\ln A + \ln r_0$. See Topic 30.3. Plotting this graph as shown in Figure 2a therefore confirms the power of A in the equation is $\frac{1}{3}$ and it also gives a value for r_0 as the y-intercept is equal to $\ln r_0$.

- A graph of R against $A^{1/3}$ gives a straight line through the origin with a gradient equal to r_0 as shown in Figure 2b. Plotting this graph gives an accurate value of r_0.

- A graph of R^3 against A gives a straight line through the origin with a gradient equal to r_0^3. Plotting this graph would also give an accurate value of r_0.

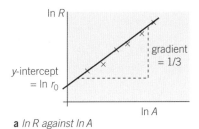

a In R against In A

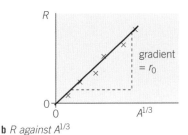

b R against $A^{1/3}$

▲ **Figure 2** Nuclear radius graphs

Nuclear density

Assuming the nucleus is spherical,

its volume $V = \frac{4}{3}\pi R^3 = \frac{4}{3}\pi(r_0 A^{1/3})^3 = \frac{4}{3}\pi r_0^3 A$

This means that the nuclear volume V is proportional to the mass of the nucleus. In other words, the density of the nucleus is constant, independent of its radius, and is the same throughout a nucleus. From this, we can conclude that nucleons are separated by the same distance regardless of the size of the nucleus and are therefore evenly separated inside the nucleus.

We can calculate the density of a nucleus using the volume formula above ($V = \frac{4}{3}\pi r_0^3 A$) and the knowledge that its mass $m = A\text{u}$ where $1\,\text{u} = 1$ atomic mass unit $= 1.661 \times 10^{-27}\,\text{kg}$.

So the density of a nucleus $= \dfrac{A\text{u}}{\frac{4}{3}\pi r_0^3 A} = \dfrac{1\text{u}}{\frac{4}{3}\pi r_0^3} = \dfrac{1.661 \times 10^{-27}}{\frac{4}{3}\pi(1.05 \times 10^{-15})^3}$
$$= 3.4 \times 10^{17}\,\text{kg}\,\text{m}^{-3}$$

- A cubic millimetre of nuclear matter would have a mass of about 340 million kilograms, about the same as the total body mass of about 4 million adults.

- A neutron star is almost as dense as the nucleus of an atom. For example, a neutron star of diameter $25\,\text{km}$ and a mass of about $4 \times 10^{30}\,\text{kg}$ (about 2 solar masses) has a density of $6 \times 10^{16}\,\text{kg}\,\text{m}^{-3}$.

Synoptic link

Log–log graphs are used to find the power n in a relationship of the form $y = kx^n$. If a graph of $\log y$ against $\log x$ is not straight, the relationship is not of this form. See Topic 30.3, Logarithms.

Summary questions

$r_0 = 1.05\,\text{fm}$

1 **a** Explain why a beam of high-energy electrons directed at a target is diffracted by the nuclei in the target.

   b Sketch a graph to show how the intensity of the electrons varies with the angle of diffraction.

   c The radius of a nucleus can be determined from high-energy electron diffraction experiments. Explain why it is important that the electrons in the beam have the same kinetic energy.

2 The volume of a $^{28}_{14}\text{Si}$ nucleus is $1.4 \times 10^{-43}\,\text{m}^3$. Use this value to calculate:

   a the radius of the nucleus,

   b the radius of a $^{120}_{50}\text{Sn}$ nucleus.

3 **a** State the relationship between the radius R of a nucleus and its mass number A.

   b Calculate the radius and the volume of a $^{238}_{92}\text{U}$ nucleus.

4 **a** Compare, without calculations, the scattering by nuclei of α particles and high-energy electrons.

   b Calculate the radius and density of a $^{14}_{7}\text{N}$ nucleus.

1   The neutron–proton model of the nucleus was first put forward by Rutherford to explain the general composition of the nucleus. The existence of the neutron was not proved experimentally until some years later.
   (a) Give **two** reasons why Rutherford's neutron–proton model was considered more than an untested hypothesis when it was first put forward.   (3 marks)
   (b) The α particles from any α-emitting isotope have the same initial kinetic energy and a well-defined range in air at atmospheric pressure. The table below shows the range R in air and the initial kinetic energy E of α particles from several α-emitting isotopes.

R / mm	39	48	53	57	66	78
E / MeV	5.3	6.0	6.5	6.8	7.4	8.3

   Plot a suitable graph to find out if the relationship between R and E is of the form $R = kE^n$, where k and n are constants, and determine a value for n. Explain your choice of graph.   (10 marks)

2   **Figure 1** shows the apparatus in which α particles are directed at a metal foil in order to investigate the structure of the atom.

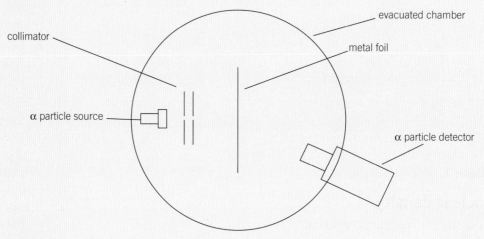

▲ Figure 1

   (a) (i)   Give **two** reasons why the metal foil should be thin.
        (ii) Explain why the incident beam of α particles should be narrow.   (3 marks)
   (b) Describe and explain **one** feature of the distribution of the scattered α particles that suggests the nucleus contains most of the mass of an atom.   (2 marks)
   (c) **Figure 2** shows three α particles with the same constant velocity incident on an atom in the metal foil. They all approach the nucleus close enough to be deflected by at least 10°.

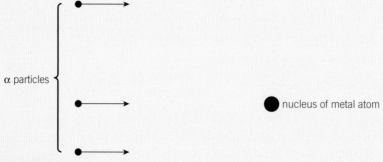

▲ Figure 2

   Draw the paths followed by the three α particles whose initial directions are shown by the arrows.   (3 marks)

AQA, 2007

3   A radioactive nucleus decays with the emission of an alpha particle and a
    gamma-ray photon.
    (a) Describe the changes that occur in the proton number and the nucleon
        number of the nucleus.   (2 marks)
    (b) Comment on the relative penetrating powers of the two types of
        ionising radiation.   (1 mark)
    (c) Gamma rays from a point source are travelling towards a detector. The
        distance from the source to the detector is changed from 1.0 m to 3.0 m.
        Calculate:
        $$\frac{\text{intensity of radiation at 3.0 m}}{\text{intensity of radiation at 1.0 m}}$$ (2 marks)

                                                                                  AQA, 2006

4   (a) A radioactive source gives an initial count rate of 110 counts per second.
        After 10 minutes the count rate is 84 counts per second.
        background radiation = 3 counts per second
        (i)   Give **three** origins of the radiation that contributes to this
              background radiation.
        (ii) Calculate the decay constant of the radioactive source in s^{-1}.
        (iii) Calculate the number of radioactive nuclei in the initial sample
              assuming that the detector counts all the radiation emitted from
              the source.   (7 marks)
    (b) Discuss the dangers of exposing the human body to a source of α
        radiation. In particular compare the dangers when the α source is
        held outside, but in contact with the body, with those when the
        source is placed inside the body.   (3 marks)

                                                                                  AQA, 2004

5   The radioactive isotope of sodium $_{11}^{22}\text{Na}$ has a half-life of 2.6 years.
    A particular sample of this isotope has an initial activity of 5.5×10^5 Bq.
    (a) Explain what is meant by the *random nature* of radioactive decay.   (2 marks)
    (b) Sketch a graph of the activity of the sample of sodium over a period
        of 6 years.   (2 marks)
    (c) Calculate:
        (i)   the decay constant, in s^{-1}, of $_{11}^{22}\text{Na}$,
              1 year = 3.15×10^7 s
        (ii) the number of atoms of $_{11}^{22}\text{Na}$ in the sample initially,
        (iii) the time taken, in s, for the activity of the sample to fall from
              1.0×10^5 Bq to 0.75×10^5 Bq.   (6 marks)

                                                                                  AQA, 2003

6   Iodine-123 is a radioisotope used medically as a tracer to monitor thyroid
    and kidney functions. The decay of an iodine-123 nucleus produces a gamma
    ray which, when emitted from inside the body of a patient, can be detected
    externally.
    (a) Why are gamma rays the most suitable type of nuclear radiation for
        this application?   (2 marks)
    (b) In a laboratory experiment on a sample of iodine-123 the following
        data were collected.

time/h	0	4	8	12	16	20	24	28	32
count rate/counts s^{-1}	512	401	338	279	217	191	143	119	91

        Why was it unnecessary to correct these values for background radiation? (2 marks)
    (c) Draw a graph of count rate against time.   (2 marks)
    (d) Use your graph to find an accurate value for the half-life of iodine-123.
        Show clearly the method you use.   (3 marks)
    (e) Give **two** reasons why radioisotopes with short half-lives are particularly
        suitable for use as a medical tracer.   (2 marks)

                                                                                  AQA, 2004

7   (a)   Sodium-21 $\left(^{21}_{11}\text{Na}\right)$ decays to neon-21 $\left(^{21}_{10}\text{Ne}\right)$. A nucleus of neon-21
        is stable.
        (i)   State the names of the particles emitted when a sodium-21
              nucleus decays.
        (ii) How many neutrons are there in a nucleus of neon-21?   (3 marks)
    (b)   **Figure 3** shows how the activity A of a freshly prepared sample of
          sodium-21 varies as it decays. **Figure 4** shows how N, the number of
          sodium-21 nuclei, varies with time t during the same time interval.

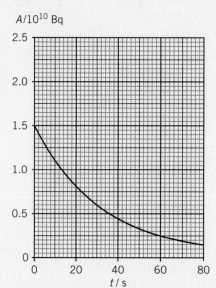

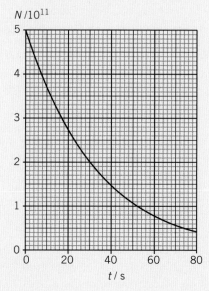

▲ Figure 3   ▲ Figure 4

        (i)   Use the graphs to find the number of active sodium nuclei and
              the corresponding activity one half-life after $t = 0$. Hence find the
              probability of decay of a sodium-21 nucleus.
        (ii) The total energy produced when a sodium-21 nucleus decays is
              5.7×10^{-13} J. Calculate the number of radioactive atoms in a sample
              that is producing 2.6 mJ of energy each second.   (6 marks)
                                                                            AQA, 2003

8   (a)   Calculate the radius of the $^{238}_{92}\text{U}$ nucleus.
          $r_0 = 1.3 \times 10^{-15}$ m   (2 marks)
    (b)   At a distance of 30 mm from a point source of γ rays the corrected count
          rate is C.
          Calculate the distance from the source at which the corrected count rate
          is 0.10 C, assuming that there is no absorption.   (2 marks)
    (c)   The activity of a source of β particles falls to 85% of its initial value in
          52 s. Calculate the decay constant of the source.   (3 marks)
    (d)   Explain why the isotope of technetium, $^{99}\text{Tc}_{m}$, is often chosen as a suitable
          source of radiation for use in medical diagnosis.   (3 marks)
                                                                            AQA, 2006

9 The high-energy electron diffraction apparatus represented in **Figure 5** can be used to determine nuclear radii. The intensity of the electron beam received by the detector is measured at various diffraction angles, *θ*.

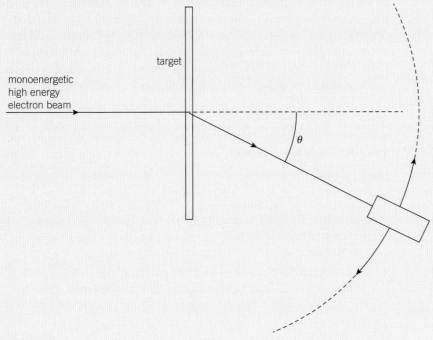

▲ Figure 5

(a) Sketch a graph to show how, in such an electron diffraction experiment, the electron intensity varies with the angle of diffraction, *θ*. *(2 marks)*

(b) (i) Use the data in the table to plot a straight line graph that confirms the relationship

$$R = r_0 A^{1/3}$$

element	radius of nucleus, $R/10^{-15}$ m	nucleon number, A
lead	6.66	208
tin	5.49	120
iron	4.35	56
silicon	3.43	28
carbon	2.66	12

(ii) Estimate the value of r_0 from the graph. *(5 marks)*

(c) Discuss the merits of using high-energy electrons to determine nuclear radii rather than using α particles. *(3 marks)*

AQA, 2005

Learning objectives:

→ Explain $E = mc^2$.

→ Describe what happens to the mass of an object when it gains or loses energy.

→ Calculate the energy released in a nuclear reaction.

Specification reference: 3.8.1.6

In 1905, Einstein published his theory of special relativity. He showed that moving clocks run slower than stationary clocks, fast-moving objects appear shorter than when stationary, the mass of a moving object changes with its speed and no material object can travel as fast as light. He also showed that the mass of an object increases (or decreases) when it gains (or loses) energy, E, in accordance with the equation

$$E = mc^2$$

where m is the change of its mass and c is the speed of light in free space which is $3.0 \times 10^8 \, \text{m s}^{-1}$.

For example,

- a sealed torch that radiates 10 W of light for 10 h (= 36 000 s) would lose 0.36 MJ of energy (= 10 W × 36 000 s). Its mass would therefore decrease by 4.0×10^{-12} kg (= 0.36 MJ/c^2), an insignificant amount compared with the mass of the torch,

- a car of 1000 kg mass that speeds up from a standstill to 30 m s^{-1} would gain 450 kJ of kinetic energy so its mass when moving at 30 m s^{-1} would be 5.0×10^{-12} kg (= 450 kJ/c^2) more than when it is at rest,

- an unstable nucleus that releases a 5 MeV γ photon would lose 8.0×10^{-13} J of energy. Its mass would therefore decrease by 8.9×10^{-30} kg (= 8.0×10^{-13} J/c^2) which is not an insignificant amount compared with the mass of a nucleus.

The equation applies to all energy changes of any object. These three examples show that such changes are important in nuclear reactions but are not usually significant otherwise. A century after Einstein published his theory, the reason why the mass of an object changes when energy is transferred to or from it is still not clearly understood. However, as explained in the AS/Year 1 course, we know that for every type of particle, there is a corresponding antiparticle with the same mass and opposite charge (if charged) and we know that

- when a particle and its corresponding antiparticle meet, they **annihilate** each other and 2 gamma (γ) photons are produced, each of energy mc^2 where m is the mass of the particle or antiparticle.

- a single γ photon of energy in excess of $2 \, mc^2$ can produce a particle and an antiparticle, each of mass m, in a process known as **pair production**.

Energy changes in reactions

Reactions on a nuclear or sub-nuclear scale do involve significant changes of mass. For example, in radioactive decay, if we know the exact rest mass of each particle involved, we can calculate the energy released Q from the difference Δm in the total mass before and after the reaction. In general, for a spontaneous reaction in which no energy is supplied,

the energy released $Q = \Delta mc^2$

In any change where energy is released, such as radioactive decay, the total mass after the change is always less than the total mass before the change. This is because, in the change, some of the mass is converted to energy which is released.

1 In α **decay**, the nucleus recoils when the α particle is emitted so the energy released is shared between the α particle and the nucleus. Applying conservation of momentum to the recoil, you should be able to show that the energy released is shared between the α particle and the nucleus in inverse proportion to their masses.

2 In β **decay**, the energy released is shared in variable proportions between the β particle, the nucleus and the neutrino or antineutrino released in the decay. When the β particle has maximum kinetic energy, the neutrino or antineutrino has negligible kinetic energy in comparison. The maximum kinetic energy of the β particle is very slightly less than the energy released in the decay because of recoil of the nucleus.

3 In **electron capture**, the nucleus emits a neutrino which carries away the energy released in the decay. The atom also emits an X-ray photon when the inner-shell vacancy due to the electron capture is filled.

Worked example

The polonium isotope $^{210}_{84}\text{Po}$ emits α particles and decays to form the stable isotope of lead $^{206}_{82}\text{Pb}$. Write down an equation to represent this process and calculate the energy released when a $^{210}_{84}\text{Po}$ nucleus emits an α particle.

mass of $^{210}_{84}\text{Po}$ nucleus = 209.936 67 u

mass of $^{206}_{82}\text{Pb}$ nucleus = 205.929 36 u

mass of α particle = 4.001 50 u

1 u is equivalent to 931.3 MeV

Solution

$^{206}_{82}\text{Pb} \rightarrow {}^4_2\alpha + {}^{206}_{82}\text{Pb}$ (+ energy released Q)

mass difference = total initial mass − total final mass

= 209.936 67 − (205.929 36 + 4.001 50)

= 5.81×10^{-3} u

energy released Q = mass difference in u × 931.3 = 5.41 MeV

More about the strong nuclear force

As explained in your AS/A level year 1 course, the fact that most nuclei are stable tells us there must be an attractive force, the strong nuclear force, between any two protons or neutrons in the nucleus.

• The strength of the strong nuclear force can be estimated by working out the force of repulsion between two protons at a separation of 1 fm (= 10^{-15} m), the approximate size of the nucleus. The strong nuclear force must be about the same magnitude as this force of repulsion. Prove for yourself, using Coulomb's law of force, that the force of

Notes

1 To calculate the energy corresponding to a mass difference of 1 atomic mass unit (1 u = 1.661 × 10^{-27} kg), using $E = mc^2$ gives
E = 1.661 × 10^{-27} kg × $(3.0 \times 10^8 \text{ m s}^{-1})^2$
= 1.49 × 10^{-10} J
= 931.3 MeV.

2 When calculating Q in beta decay, assume the mass of the neutrino is negligible.

3 If the mass of each atom is given instead of the mass of its nucleus, calculate the mass of each nucleus by subtracting the mass of the electrons ($= Zm_e$) in the atom from the mass of each atom.

Study tip

If you are given the masses of the nuclei and particles involved in atomic mass units (u), calculate the difference between the total initial mass and the total final mass in u then multiply by 931.3 to give the energy released in MeV.

Synoptic link

Coulomb's law of force was covered in Topic 22.4, Coulomb's law.

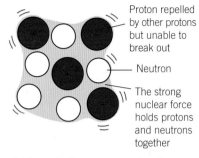

Proton repelled by other protons but unable to break out

Neutron

The strong nuclear force holds protons and neutrons together

▲ **Figure 1** *The strong nuclear force*

repulsion between two protons at a separation of 10^{-15} m is of the order of 200 N. So the strong nuclear force is at least 200 N.

- The range of the strong nuclear force is no more than about 3 to 4×10^{-15} m. The diameter of a nucleus can be measured from high-energy electron scattering experiments (see Topic 26.9). The results show that nucleons are evenly spaced at about 10^{-15} m in the nucleus and therefore the strong nuclear force acts only between nearest neighbour nucleons.

- The energy needed to pull a nucleon out of the nucleus is of the order of millions of electron volts (MeV). This can be deduced because the strong nuclear force is at least about 200 N and it acts over a distance of about 2 to 3×10^{-15} m. The work done by the strong nuclear force over this distance is therefore about 7×10^{-13} J ($= 200$ N $\times 3.5 \times 10^{-15}$ m) which is about 4 MeV, as 1 MeV $= 1.6 \times 10^{-13}$ J.

- The strong nuclear force between two nucleons must become repulsive at separations of about 0.5 fm or less, otherwise nucleons would pull each other closer and closer together and the nucleus would be much smaller than it is.

Summary questions

rest mass of an electron = 9.11×10^{-31} kg

1 u $= 931.3$ MeV

$g = 9.81$ m s^{-2}

1 Calculate the mass increase of:

   a a 10 kg object when it is raised through a height of 2.0 m,

   b an electron when it is accelerated from rest through a pd of

      i 5000 V,   **ii** 5 MV

2 The bismuth isotope $^{212}_{83}$Bi emits α particles and decays to form the stable isotope of thallium $^{208}_{81}$Tl .

   a Write down an equation to represent this process and calculate the energy released.

      mass of $^{212}_{83}$Bi nucleus = 211.945 62 u

      mass of $^{208}_{81}$Tl nucleus = 207.937 46 u

      mass of α particle = 4.001 50 u

   b Explain without calculation why the thallium nucleus in the above decay gains a small proportion of the energy released.

3 The strontium isotope $^{90}_{38}$Sr emits β^- particles and decays to form the stable isotope of yttrium $^{90}_{39}$Y.

   a Write down an equation to represent this process and calculate the energy released.

      mass of $^{90}_{38}$Sr nucleus = 89.886 40 u

      mass of $^{90}_{39}$Y nucleus = 89.885 25 u

      mass of β^- particle = 0.000 55 u

   b Explain without calculation why the kinetic energy of the β^- particle released when the strontium nucleus decays varies from zero up to a maximum.

4 The copper isotope $^{64}_{29}$Cu decays through electron capture to form the stable isotope of nickel $^{64}_{28}$Ni .

   a Write down an equation to represent this process and calculate the energy released.

      mass of $^{64}_{29}$Cu nucleus = 63.913 81 u

      mass of $^{64}_{28}$Ni nucleus = 63.912 56 u

      mass of electron = 0.000 55 u

   b State the name of the particle that takes away the energy released by the nucleus in electron capture.

27.2 Binding energy

Suppose all the nucleons in a nucleus were separated from one another, removing each one from the nucleus in turn. Work must be done to overcome the strong nuclear force and separate each nucleon from the others. The potential energy of each nucleon is therefore increased when it is removed from the nucleus.

The binding energy of the nucleus is the work that must be done to separate a nucleus into its constituent neutrons and protons.

When a nucleus forms from separate neutrons and protons, energy is released as the strong nuclear force does work pulling the nucleons together. The energy released is equal to the binding energy of the nucleus. Because energy is released when a nucleus forms from separate neutrons and protons, the mass of a nucleus is less than the mass of the separated nucleons.

The mass defect Δm of a nucleus is defined as the difference between the mass of the separated nucleons and the mass of the nucleus.

- Calculation of the mass defect of a nucleus of known mass: a nucleus of an isotope $^A_Z X$ is composed of Z protons and $(A - Z)$ neutrons. Therefore, for a nucleus $^A_Z X$ of mass M_{NUC},

  its mass defect $\Delta m = Zm_p + (A - Z)m_n - M_{NUC}$

  where m_p and m_N represent the masses of the proton and the neutron respectively.

- **Calculation of the binding energy of a nucleus:** the mass defect Δm is due to energy released when the nucleus formed from separate neutrons and protons. The energy released in this process is equal to the binding energy of the nucleus. Therefore,

  the binding energy of a nucleus = $\Delta m c^2$

Worked example

The mass of a nucleus of the bismuth isotope $^{212}_{83} Bi$ is 211.800 12 u. Calculate the binding energy of this nucleus in MeV.

mass of a proton, $m_p = 1.007\ 28$ u
mass of a neutron, $m_n = 1.008\ 67$ u.

1 u is equivalent to 931.3 MeV

Solution

Mass defect $\Delta m = 83\ m + (212 - 83)\ m_n - M_{NUC} = 1.922\ 55$ u

Therefore, binding energy = 1.922 55 u × 931.3 MeV/u = 1790 MeV

Learning objectives:

→ Define binding energy.

→ State which nuclei are the most stable.

→ Explain why energy is released when a $^{235}_{92} U$ nucleus undergoes fission.

Specification reference: 3.8.1.6

Synoptic link

Remember 1 MeV = 1.6×10^{-13} J (and 1 eV = 1.6×10^{-19} J). See Topic 1.4, Particles and antiparticles.

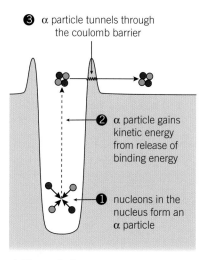

❸ α particle tunnels through the coulomb barrier

❷ α particle gains kinetic energy from release of binding energy

❶ nucleons in the nucleus form an α particle

▲ **Figure 1** *Quantum tunnelling from the nucleus*

Notes

1 The mass of an atom of an isotope $^{A}_{Z}X$ is measured using a mass spectrometer. The mass of a nucleus can then be calculated by subtracting the mass of Z electrons from the atomic mass.

2 The atomic mass unit, $1\,u = 1.661 \times 10^{-27}\,kg$, is defined as $\frac{1}{12}$ th of the mass of an atom of the carbon isotope $^{12}_{6}C$.

3 The energy corresponding to a mass of $1\,u$ is $1.661 \times 10^{-27} \times (3.00 \times 10^{8})^{2}\,J = 931.5\,MeV$. If you are given the mass of the nucleus in kilograms, you can convert this to atomic mass units and use the method above to calculate the mass defect. The values of the mass of the proton and the neutron are given in atomic mass units in the data sheet.

α particle tunnelling

If two protons and two neutrons inside a sufficiently large nucleus bind together as a 'cluster', they may be emitted from the nucleus as an α particle. This is because the binding energy of an α particle is very large at about 7 MeV per nucleon compared with other neutron and proton clusters that may form. The α particle therefore gains sufficient kinetic energy (equal to the binding energy of the cluster) to give it a small probability of 'quantum tunnelling' from the nucleus.

Figure 1 shows how the potential energy of an α particle varies with its distance from outside the nucleus to inside. The 'coulomb' barrier is due to the electrostatic force on the α particle. The 'well' is due to the strong nuclear force. The gain of kinetic energy of the α particle when it forms in the nucleus is sufficient for it to reach the coulomb barrier but not for it to surmount the barrier directly. However, the wave nature of the α particle gives it a small probability of tunnelling through the barrier.

Q: Explain why the probability of a decay increases the higher the kinetic energy of the α particle.

Nuclear stability

The binding energy of each nuclide is different. The **binding energy per nucleon** of a nucleus is the average work done per nucleon to remove all the **nucleons** (protons and neutrons) from a nucleus; it is therefore a measure of the stability of a nucleus. For example, the binding energy per nucleon of the $^{212}_{83}Bi$ nucleus is 8.4 MeV per nucleon (= 1790 MeV/212 nucleons).

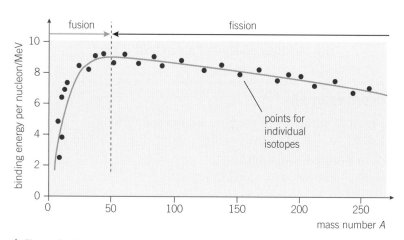

▲ **Figure 2** *Binding energy per nucleon for all known nuclides*

If the binding energies per nucleon of two different nuclides are compared, the nucleus with more binding energy per nucleon is the more stable of the two nuclei. Figure 2 shows a graph of the binding energy per nucleon against mass number A for all the known nuclides. This graph is a curve which has a maximum value of 8.7 MeV per nucleon between $A = 50$ and $A = 60$. Nuclei with mass numbers in this range are the most stable nuclei. As explained below, energy is released in:

- **nuclear fission**, the process in which a large unstable nucleus splits into two fragments which are more stable than the original nucleus. The binding energy per nucleon increases in this process, as shown in Figure 2.
- **nuclear fusion**, the process of making small nuclei fuse together to form a larger nucleus. The product nucleus has more binding energy per nucleon than the smaller nuclei. So the binding energy per nucleon also increases in this process, provided the nucleon number of the product nucleus is no greater than about 50.

Note:
The change of binding energy per nucleon is about 0.5 MeV in a fission reaction and can be more than 10 times as much in a fusion reaction.

Summary questions

mass of a proton, $m_p = 1.007\,28$ u;
mass of a neutron, $m_n = 1.008\,67$ u

1 1 u is equivalent to 931.5 MeV.

  a Explain what is meant by the binding energy of a nucleus.

  b Sketch a curve to show how the binding energy per nucleon of a nucleus varies with its mass number A, showing the approximate scale on each axis.

2 Calculate the binding energy per nucleon, in MeV per nucleon, of

  a a $^{12}_{6}$C nucleus (mass = 12 u by definition),

  b a $^{56}_{26}$Fe nucleus (mass = 55.920 67 u).

3 a Calculate the binding energy per nucleon, in MeV per nucleon, of

  i an α particle,   **ii** a $^{3}_{2}$He nucleus.

  Mass of an α particle = 4.00150 u; mass of a $^{3}_{2}$He nucleus = 3.014 93 u

  b Use the results of your calculations in **a** to explain why an α particle rather than a $^{3}_{2}$He nucleus is emitted by a large unstable nucleus.

4 a Complete the equation below to show the fusion reaction that occurs when two protons fuse together to form a $^{2}_{1}$H nucleus.

$$p + p \rightarrow {}^{2}_{1}H + \beta^{+}$$

  b Calculate the binding energy per nucleon, in MeV, of the $^{2}_{1}$H nucleus.

  mass of $^{2}_{1}$H nucleus = 2.013 55 u

Learning objectives:

→ Describe how much energy is released in a fission or a fusion reaction.

→ Explain why small nuclei can't be split.

→ Explain why large nuclei can't be fused.

Specification reference: 3.8.1.7/8

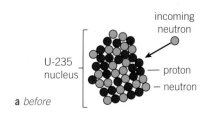

U-235 nucleus

incoming neutron

proton

neutron

a *before*

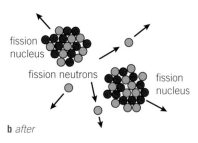

fission nucleus

fission neutrons

fission nucleus

b *after*

▲ **Figure 1** *Induced fission*

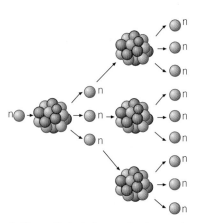

▲ **Figure 2** *A chain reaction in a nuclear reactor*

Induced fission

Fission of a nucleus occurs when a nucleus splits into two approximately equal fragments. This happens when the uranium isotope $^{235}_{92}U$ is bombarded with neutrons, a discovery made by Hahn and Strassmann in 1938. The process is known as **induced fission**. The plutonium isotope, $^{239}_{94}Pu$, is the only other isotope that is fissionable. This isotope is an artificial isotope formed by bombarding nuclei of the uranium isotope $^{235}_{92}U$ with neutrons.

Hahn and Strassman knew that bombarding different elements with neutrons produces radioactive isotopes. Uranium is the heaviest of all the naturally occurring elements; scientists thought that neutron bombardment could turn uranium nuclei into even heavier nuclei. Hahn and Strassmann undertook the difficult work of analysing chemically the products of uranium after neutron bombardment to try to discover any new elements heavier than uranium. Instead, they discovered that many lighter elements such as barium were present after bombardment, even though the uranium was pure before. The conclusion could only be that uranium nuclei were split into two approximately equal fragment nuclei as a result of neutron bombardment.

Further investigations showed that each fission event releases energy and two or three neutrons.

- **Fission neutrons**, the neutrons released in a fission event, are each capable of causing a further fission event as a result of a collision with another $^{235}_{92}U$ nucleus. A **chain reaction** is therefore possible in which fission neutrons produce further fission events which release fission neutrons and cause further fission events and so on. If each fission event releases two neutrons on average, after n 'generations' of fission events, the number of fission neutrons would be 2^n. Prove for yourself that fission of 6×10^{23} $^{235}_{92}U$ nuclei (i.e. 235 g of the isotope) would happen in 79 generations. Each fission event releases about 200 MeV of energy. Because each event takes no more than a fraction of a second, a huge amount of energy is released in a very short time. Using the above figures, complete fission of 235 g of $^{235}_{92}U$ would release about 10^{13} J ($= 6 \times 10^{23} \times$ 200 MeV). This is about a million times more than the energy released as a result of burning a similar mass of fossil fuel.

- **Energy is released** when a fission event occurs because the fragments repel each other (as they are both positively charged) with sufficient force to overcome the strong nuclear force trying to hold them together. The fragment nuclei and the fission neutrons therefore gain kinetic energy. The two fragment nuclei are smaller and therefore more tightly bound than the original $^{235}_{92}U$ nucleus. In other words, they have more binding energy so they are more stable than the original nucleus. The energy released is equal to the change of binding energy. The binding energy of each nucleon increases from about 7.5 MeV to about 8.5 MeV as a result of the fission event. As there are about 240 nucleons in the original nucleus, the energy released in a fission event is of the order of 200 MeV ($= 240 \times$ about 1 MeV).

- Many fission products are possible when a fission event occurs. For example, the following equation shows a fission event in which a $^{235}_{92}U$ nucleus is split into a barium $^{144}_{56}Ba$ nucleus and a krypton $^{90}_{36}Kr$ nucleus and two neutrons are released.

$$^{235}_{92}U + {}^{1}_{0}n \rightarrow {}^{144}_{56}Ba + {}^{90}_{36}Kr + 2{}^{1}_{0}n + \text{energy released, } Q$$

- The energy released, Q, can be calculated using $E = mc^2$ in the form $Q = \Delta m c^2$, where Δm is the difference between the total mass before and after the event.

- In the above equation, the mass difference is

$$\Delta m = M_{U\text{-}238} - M_{Ba\text{-}144} - M_{Kr\text{-}90} - m_{n},$$

where M represents the appropriate nuclear mass and m_n is the mass of the neutron.

Nuclear fusion

Fusion takes place when two nuclei combine to form a bigger nucleus. The binding energy curve (see Topic 27.2, Figure 2) shows that if two light nuclei are combined, the individual nucleons become more tightly bound together. The binding energy per nucleon of the product nucleus is greater than of the initial nuclei. In other words, the nucleons become even more trapped in the nucleus when fusion occurs. As a result, energy is released equal to the increase of binding energy.

Nuclear fusion can only take place if the two nuclei that are to be combined collide at high speed. This is necessary to overcome the electrostatic repulsion between the two nuclei so they can become close enough to interact through the strong nuclear force. Some examples of nuclear fusion reactions are given below.

1   The fusion of two protons produces a nucleus of deuterium (the hydrogen isotope $^{2}_{1}H$), a β^+ particle, a neutrino and 0.4 MeV of energy.

$$^{1}_{1}p + {}^{1}_{1}p \rightarrow {}^{2}_{1}H + {}^{0}_{+1}\beta + v$$

2   The fusion of a proton and a deuterium nucleus $^{2}_{1}H$ produces a nucleus of the helium isotope $^{3}_{2}He$ and 5.5 MeV of energy.

$$^{2}_{1}H + {}^{1}_{1}p \rightarrow {}^{3}_{2}He$$

3   The fusion of two nuclei of helium isotope, $^{3}_{2}He$, produces a nucleus of the helium isotope $^{4}_{2}He$, two protons and 12.9 MeV of energy.

$$^{3}_{2}He + {}^{3}_{2}He \rightarrow {}^{4}_{2}He + 2{}^{1}_{1}p$$

In each case, the energy released in the reaction may be calculated using $E = mc^2$ in the form $Q = \Delta m c^2$, where Δm is the difference between the total mass before and after the event.

Solar energy is produced as a result of fusion reactions inside the Sun. The temperature at the centre of the Sun is thought to be 10^8 K or more. At such temperatures, atoms are stripped of their electrons. Matter in this state is referred to as 'plasma'. The nuclei of the plasma move at very high speeds because of the enormous temperature. When two nuclei collide, they fuse together because they overcome the electrostatic repulsion due to their charge and approach each other closely enough to interact through the strong nuclear force. Protons (i.e., hydrogen nuclei) inside the Sun's core fuse together in stages

Study tip

Energy released in nuclear fission = change of binding energy. Fission of a given nuclide doesn't have a unique outcome.

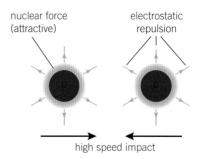

nuclear force (attractive)   electrostatic repulsion

high speed impact

▲ **Figure 3** *Fusion of two protons*

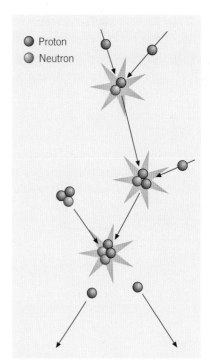

- Proton
- Neutron

▲ **Figure 4** *Fusion reactions inside the Sun*

▲ **Figure 3** *The JET fusion reactor. The plasma is contained in a doughnut-shaped steel container and is heated by passing a very large current through it. A magnetic field is used to confine the plasma so it does not touch the sides of its steel container, otherwise it would lose its energy*

(corresponding to equations 1, 2 and 3 above) to form helium ^4_2He nuclei. For each helium nucleus formed, 25 MeV of energy is released. This corresponds to 6 MeV per proton, considerably more than the energy released per nucleon in a fission event.

Fusion power

Fusion reactors are still at the prototype stage even though scientific teams in several countries have been working on fusion research for more than 50 years. Prototype fusion reactors such as JET, the Joint European Torus, in the United Kingdom have produced large amounts of power but only for short periods of time. JET produces less power than it uses but the less powerful International Thermonuclear Experimental Reactor (ITER) due to start up in 2016 is designed to produce several times more power than it uses.

Energy is released in JET by fusing nuclei of deuterium ^2_1H and tritium ^3_1H to produce nuclei of the helium isotope ^4_2He and neutrons, as below.

$$^2_1\text{H} + {}^3_1\text{H} \rightarrow {}^4_2\text{He} + {}^1_0\text{n} + 17.6\,\text{MeV}$$

The neutrons are absorbed by a 'blanket' of lithium surrounding the reactor vessel. The reaction between the neutrons and the lithium nuclei, as shown below, produces tritium which is then used in the main reaction. Deuterium occurs naturally in water as it forms 0.01% of naturally occurring hydrogen.

$$^6_3\text{Li} + {}^1_0\text{n} \rightarrow {}^4_2\text{He} + {}^3_1\text{H} + 4.8\,\text{MeV}$$

Summary questions

1 u is equivalent to 931 MeV.

1 a Explain why the protons in a nucleus do not leave the nucleus even though they repel each other.

b Explain why the mass of a nucleus is less than the mass of the separated protons and neutrons from which the nucleus is composed.

2 a What is meant by nuclear fission?

b i The incomplete equation below represents a reaction that takes place when a neutron collides with a nucleus of the uranium isotope $^{235}_{92}\text{U}$. Determine the values of a and b in this equation.

$$^{235}_{92}\text{U} + {}^1_0\text{n} \rightarrow {}^{136}_{a}\text{Xe} + {}^b_{36}\text{Kr} + 2\,{}^1_0\text{n} + \text{energy released, } Q$$

ii Calculate the energy, in MeV, released in this fission reaction.

masses: $^{235}_{92}\text{U}$ nucleus 234.993 u,

$^{136}_{a}\text{Xe}$ nucleus 135.877 u,

$^b_{36}\text{Kr}$ nucleus 97.886 u, neutron 1.00867 u

3 a What is meant by nuclear fusion?

b Hydrogen nuclei fuse together to form helium nuclei in the Sun. Two stages in this process are represented by the following equations:

$$^1_1\text{p} + {}^1_1\text{p} \rightarrow {}^2_1\text{H} + {}^0_{+1}\beta$$
$$^2_1\text{H} + {}^1_1\text{p} \rightarrow {}^3_2\text{He}$$

i Describe the reactions that these equations represent.

ii Calculate the energy released in each reaction.

masses: β particle 0.00055 u, proton 1.00728 u,

^2_1H nucleus 2.01355 u, ^3_2He nucleus 3.01493 u,

4 a Explain why light nuclei do not fuse when they collide unless they are moving at a sufficiently high speed.

b Calculate the energy released in the following fusion reaction:

$$^3_2\text{He} + {}^3_2\text{He} \rightarrow {}^4_2\text{He} + 2\,{}^1_1\text{p}$$

masses: proton 1.00728 u,

^3_2He nucleus 3.01493 u,

α particle 4.00150

c Show that about 25 MeV of energy is released when a ^4_2He nucleus is formed from 4 protons.

Inside a nuclear reactor

A thermal nuclear reactor in a nuclear power station contains fuel rods spaced evenly in a steel vessel known as the **reactor core**, as shown in Figure 1. The reactor core also contains **control rods** and a **coolant** (water at high pressure in the pressurised water reactor (PWR) shown in Figure 1) as well as the fuel rods and is connected by means of steel pipes to a **heat exchanger**. A pump is used to force the coolant through the reactor core (where it is heated) and through the heat exchanger where it is used to raise steam to drive the turbines that turn the electricity generators in the power station.

Learning objectives:

→ Explain how a nuclear reactor works.

→ Describe a thermal nuclear reactor.

→ Explain how a nuclear reactor is controlled.

Specification reference: 3.8.1.7/8

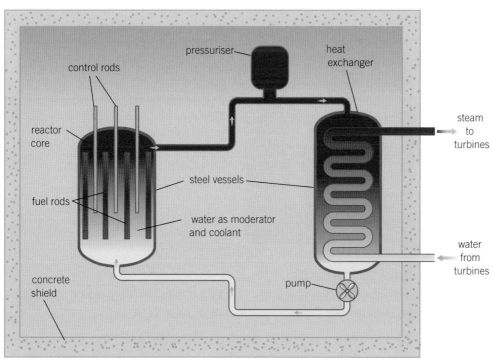

▲ **Figure 1** *Inside a nuclear reactor*

- The fuel rods contain enriched uranium which consists mostly of U-238, the non-fissionable uranium isotope $^{238}_{92}U$, and about 2–3% U-235, the uranium isotope $^{235}_{92}U$ which is fissionable. In comparison, natural uranium contains 99% U-238.

- The function of the control rods is to absorb neutrons. The depth of the control rods in the core is automatically adjusted to keep the number of neutrons in the core constant so that exactly one fission neutron per fission event on average goes on to produce further fission. This condition keeps the rate of release of fission energy constant. If the control rods are pushed in further, they absorb more neutrons so that the number of fission events per second and the rate of release of fission energy is reduced.

- The fission neutrons need to be slowed down significantly to cause further fission of U-235 nuclei otherwise they would be travelling too fast to cause further fission. For this reason, the fuel rods need to be surrounded by a **moderator** so the neutrons are slowed down by

Synoptic link

Neutron loss depends on the surface area; neutron production depends on the mass of material. Below the critical mass, loss/production is too high. The same idea explains why a small object cools faster than a large object. Energy loss depends on the surface area, so temperature loss depends on surface area/mass. See Topic 19.2, Specific heat capacity.

Moderators at work

The atoms of a moderator in a nuclear reactor gain kinetic energy from fission neutrons colliding with them. The transfer of kinetic energy in such a collision is most effective if the mass of the moderator atom is as close as possible to the mass of the neutron. For this reason and taking account of practical considerations such as chemical stability, graphite (which consists of carbon-12 atoms) and water are commonly used as moderators.

Consider an elastic head-on collision between a fission neutron and a carbon-12 nucleus. Let the neutron's speed before and after the collision be u and v respectively, and let V be the recoil velocity of the nucleus. Conservation of momentum gives $u = 12V + v$.

Conservation of kinetic energy gives $u^2 = 12V^2 + v^2$.

Prove for yourself that combining these equations to find v in terms of u gives $v = \dfrac{11u}{13}$. Hence the neutron's kinetic energy after the collision is about $0.72 \times$ its initial kinetic energy (because $\dfrac{v^2}{u^2} = 0.72$).

Q: Estimate how many such collisions would reduce the neutron's kinetic energy from 1 MeV to 1 eV.

repeated collisions with the moderator atoms. The reactor is described as a **thermal nuclear reactor** because the fission neutrons are slowed down to kinetic energies comparable to the kinetic energies of the moderator molecules. In the PWR, the water in the reactor core acts as the moderator as well as acting as a coolant.

- For a chain reaction to occur, the mass of the fissile material (e.g., U-235) must be greater than a minimum mass, referred to as the **critical mass**. This is because some fission neutrons escape from the fissile material without causing fission and some are absorbed by other nuclei without fission. If the mass of fissile material is less than the critical mass needed, too many of the fission neutrons escape because the surface area to mass ratio of the material is too high.

- Different types of thermal reactors are in operation throughout the world. Table 1 shows some of the features of the PWR reactor which operates in many countries including the UK and the Advanced Gas-cooled Reactor (AGR) which operates only in the UK.

▼ **Table 1** *Comparison of thermal reactors*

	AGR	PWR
fuel	uranium oxide in stainless steel cans	uranium oxide in zirconium alloy cans
moderator	graphite	water
coolant	CO_2 gas	water
coolant temperature/K	900	600
typical power output/MW	1300	700

Safety features

A nuclear reactor needs to have a range of safety features to protect its workforce, the wider community and the environment.

1 The reactor core is a thick steel vessel designed to withstand the high pressure and temperature in the core. The thick steel vessel absorbs β radiation and some of the γ radiation and neutrons from the core.

2 The core is in a building with very thick concrete walls which absorb the neutrons and γ radiation that escape from the reactor vessel.

3 Every reactor has an emergency shut-down system designed to insert the control rods fully into the core to stop fission completely.

4 The sealed fuel rods are inserted and removed from the reactor by means of remote handling devices. The rods are much more radioactive after removal than before. This is because the fuel cans

 a before use contain U-235 and U-238 which emit only α radiation and this is absorbed by the fuel cans,

 b after use emit β and γ radiation due to the many neutron-rich fission products that form.

In addition, the spent fuel rods contain the plutonium isotope $^{239}_{94}\text{Pu}$ as a result of the absorption of neutrons by U-238 nuclei. This plutonium isotope is a very active α emitter and if inhaled causes lung cancer.

✚ A nuclear future

To combat climate change, the UK government has legislated to reduce carbon emissions to 50% of 1990 levels by 2027 by introducing measures such as more non-carbon generating capacity. The new nuclear power stations will be built by the private sector. The negotiations to build the first such power station at Hinckley Point in Somerset by 2025 took five years to complete. At present, the UK's nuclear reactors provide about 10 GW of electricity, which is about 20% of total UK electricity production. Most of the present reactors will have been retired by 2023. The new nuclear power stations will provide about 16 GW by 2030, and so will make a significant contribution to the 2027 carbon emission target. However, other measures such as more wind farms are also likely to be needed.

Nuclear power continues to cause concern to many people. The safety features described in this topic are extremely important. Accidents at any nuclear power stations, although rare, alarm people across the world. Two such events are described below.

Chernobyl, 1986

The Chernobyl disaster in Ukraine in 1986 released radioactive materials into the atmosphere and led to the permanent evacuation of all the people in the surrounding area. The disaster was caused when the operators were testing reactor no. 4 to find out if the coolant pumps would keep operating in the event of a loss of power until the emergency diesel generators took over. To carry out the test, the safety systems to shut the reactor down in an emergency were switched off. When the reactor was powered down by pushing the control rods further into the core, the power fell much more than expected, so the movement of the control rods was reversed. This caused an unexpected surge in the rate of fission events, which produced a massive explosion in the reactor core. The fuel rods melted, the reactor cap was blown off, and radioactive fission products were thrown up into the atmosphere. A subsequent inquiry concluded that the main causes of the explosion were design faults in the reactor (e.g., graphite in the cap ignited in the explosion), which are not features of AGR or PWR reactors, and human error (too many control rods being moved at once).

▲ **Figure 2** *Fukushima*

Fukushima, 2011

On 11 March 2011, a tsunami created by a powerful earthquake in Japan killed over 20 000 people and caused a meltdown at the Fukushima nuclear power station. As a result of lessons learned from Chernobyl, over 100 000 people living near the Fukushima power station were evacuated. Public areas and buildings in nearby towns had to be decontaminated and food controls imposed. When the earthquake struck, planned safety procedures were put into effect. All 11 reactors at the site were shut down without stopping their cooling pumps, which prevent the reactors from overheating. However, an hour later, the tsunami hit the site and flooded the generators, causing the reactor pumps to stop. The result was that three older reactors overheated, and hydrogen gas (created by chemical reactions) and radioactive material leaked from broken fuel rods. In addition, the water supply to cooling ponds containing spent fuel rods was damaged, causing the rods in the ponds to overheat until alternative water supplies were provided. Contaminated water leaked into the sea, and traces of plutonium have been found in soil nearby. Such leakage continues to cause problems, and many evacuees are still not allowed to return home.

The new nuclear power stations in the UK are intended to help secure our future electricity supply and to reduce carbon emissions. Hundreds of nuclear reactors are operating satisfactorily in many countries. The risk of another disaster like Chernobyl and Fukushima is small, but the consequences of such a disaster could be severe.

Q: Discuss why Chernobyl is still a dangerous area today.

▲ **Figure 3** *Spent fuel rods in a cooling pond*

Summary questions

1 a What is meant by induced fission?

  b i What is the function of the control rods in a nuclear reactor?

    ii Name a suitable material from which control rods are made.

    iii Describe how the control rods are used to maintain a nuclear reactor so its power output is constant.

2 State the function of the following parts of a thermal nuclear reactor and give an example of a material used for each part:

  a the moderator,

  b the coolant.

3 Explain why the mass of fissile fuel in a nuclear reactor must exceed a critical value in order for fission to be sustained in the reactor.

4 a Explain why the spent fuel rods from a nuclear reactor are more radioactive after removal from the reactor than they were before they were used in the reactor.

  b Explain why radioactive waste must be stored in secure and safe conditions.

Radioactive waste

Radioactive waste is categorised as high-, intermediate-, or low-level waste according to its activity. Most high-level radioactive waste is from nuclear power stations or from specialist users in universities and industry, or from hospitals that use radioactive isotopes for diagnosis or therapy.

Disposal of any form of radioactive waste must be in accordance with legal regulations and by approved disposal companies to ensure that the radioactive waste is stored safely in secure containers until its activity is insignificant. Disposal by dilution, for example diluting radioactive water from nuclear power station cooling systems with large quantities of water and then dispersing it into the sea, is no longer acceptable and has been banned.

- **High-level radioactive waste** such as spent fuel rods from a nuclear power station contains many different radioactive isotopes, including fission fragments as well as unused uranium-235 and uranium-238 and plutonium-239. The spent fuel rods must be removed by remote control and stored underwater in cooling ponds for up to a year because they continue to release heat due to radioactive decay. In Britain, the rods are then transferred in large steel casks to the THORP reprocessing plant at Sellafield in Cumbria where the unused uranium and plutonium is then removed and stored in sealed containers for further possible use. The rest of the material (i.e., the fission products and the fuel cans) is radioactive waste and is stored in sealed containers in deep trenches at Sellafield. Such waste must be stored safely for centuries as it contains long-lived radioactive isotopes which must be prevented from contaminating food and water supplies.

In other countries, high-level radioactive waste is stored in the same way or in underground caverns which are geologically stable. In some countries, the waste is vitrified by mixing it with molten glass and then stored as glass blocks in underground caverns.

The long-term safe storage of high-level radioactive waste remains a major issue in Britain because no one wants such storage in their own locality, nor do people want radioactive waste to be carried through their own locality to storage facilities elsewhere.

- **Intermediate-level waste** such as radioactive materials with low activity and containers of radioactive materials are sealed in drums that are encased in concrete and stored in specially constructed buildings with walls of reinforced concrete.

- **Low-level waste** such as laboratory equipment and protective clothing is sealed in metal drums and buried in large trenches.

1 (a) In a nuclear reactor, some of the $^{238}_{92}$U nuclei absorb neutrons to become uranium $^{239}_{92}$U nuclei. These nuclei decay in two stages to become nuclei of the plutonium isotope $^{239}_{94}$Pu.

   (i) Write down an equation to represent the formation of a $^{239}_{92}$U nucleus from a $^{238}_{92}$U nucleus.

   (ii) What types of particles are emitted when a $^{239}_{92}$U nucleus decays to form a $^{239}_{94}$U nucleus? *(3 marks)*

  (b) The THORP reprocessing plant at Sellafield is used to recover uranium and plutonium from spent fuel rods. In addition to reprocessing nuclear waste from the UK, it also reprocesses waste from nuclear reactors in other countries. Uranium-238 and plutonium-239 can both be used in a type of nuclear reactor called a *fast breeder* reactor, although this type of reactor has not yet been fully appraised in terms of reliability and safety. By using such reactors in the future, the lifetime of the world's reserves of uranium would be extended from a few hundred years to several thousand years. Plutonium can also be used to make nuclear bombs. Discuss the arguments for and against reprocessing nuclear waste. *(5 marks)*

2 (a) In the context of an atomic nucleus,

   (i) state what is meant by *binding energy*, and explain how it arises,

   (ii) state what is meant by *mass difference*,

   (iii) state the relationship between binding energy and mass difference. *(4 marks)*

  (b) Calculate the average binding energy per nucleon, in MeV nucleon^{-1}, of the zinc nucleus $^{64}_{30}$Zn.

   mass of $^{64}_{30}$Zn atom = 63.929 15 u *(5 marks)*

  (c) Why would you expect the zinc nucleus to be very stable? *(1 mark)*
AQA, 2004

3 (a) (i) Describe the physical process of *nuclear fusion*.

   (ii) Describe the physical process of *nuclear fission*.

   (iii) Explain why each of these processes releases energy. *(6 marks)*

  (b) Energy is also released by radioactive decay, such as the decay of radon-220 as represented by the equation

$$^{220}_{86}\text{Rn} \rightarrow {}^{216}_{84}\text{Po} + \alpha$$

Calculate the energy released, in J, by the decay of one nucleus of radon-220.

mass of ^{220}Rn nucleus = 219.964 10 u

mass of ^{216}Po nucleus = 215.955 72 u

mass of α particle = 4.001 50 u *(3 marks)*
AQA, 2007

4 (a) A solar panel of area 2.5 m^2 is fitted to a satellite in orbit above the Earth. The panel produces a current of 2.4 A at a potential difference of 20 V when solar radiation is incident normally on it.

   (i) Calculate the electrical power output of the panel.

   (ii) Solar radiation on the satellite has an intensity of 1.4 kW m^{-2}. Calculate the efficiency of the panel. *(4 marks)*

(b) The back-up power system in the satellite is provided by a radioactive isotope enclosed in a sealed container which absorbs the radiation from the isotope. Energy from the radiation is converted to electrical energy by means of a thermoelectric module.

 (i) The isotope has an activity of 1.1×10^{14} Bq and produces α particles of energy 5.1 MeV.
 Show that the container absorbs energy from the α particles at a rate of 90 J s^{-1}.

 (ii) The isotope has a half-life of 90 years. Calculate the decay constant λ of this isotope.

 (iii) The mass number of the isotope is 239.
 Calculate the mass of isotope needed for an activity of 1.1×10^{14} Bq. *(7 marks)*

 AQA, 2003

5 **Figure 1** shows the general relationship between the nuclear binding energy per nucleon (*B*) and nucleon number (*A*).

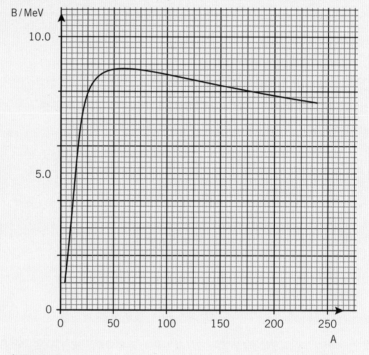

▲ Figure 1

(a) (i) Copy **Figure 1** and mark with the letter S to show the nucleon number and the nuclear binding energy per nucleon for the nuclide with the most stable nuclear structure.

 (ii) Write down the nucleon number and the nuclear binding energy per nucleon for this nuclide.

 (iii) Calculate the total binding energy of this nuclide. *(3 marks)*

(b) A fusion reaction in which two protons combine to form a deuterium nucleus is summarised by the equation:

$$_1^1H + _1^1H \rightarrow _1^2H + _1^0e + \nu + 1.44 \text{ MeV}$$

 (i) What do the symbols $_1^0e$ and ν represent?

 (ii) By considering charge, baryon number and lepton number for each side of the equation, show that this reaction satisfies the conservation laws for these quantities.

 (iii) Subsequently two γ-ray photons are released, each with an energy of 0.51 MeV.
 Calculate the wavelength of these photons. *(18 marks)*

(c) With reference to **Figure 1** explain why the fission of a heavy nucleus is likely to release more energy than when a pair of light nuclei undergo nuclear fusion. You may wish to sketch the general shape of **Figure 1** in order to aid your explanation.

(5 marks)

AQA, 2004

6 (a) With reference to the process of nuclear fusion, explain why energy is released when two small nuclei join together, and why it is difficult to make two nuclei come together. *(3 marks)*

(b) A fusion reaction takes place when two deuterium nuclei join, as represented by

$$^{2}_{1}H + ^{2}_{1}H \rightarrow ^{3}_{2}He + ^{0}_{1}n$$

mass of ^{2}H nucleus = 2.013 55 u
mass of ^{3}He nucleus = 3.014 93 u

Calculate:
(i) the mass difference produced when two deuterium nuclei undergo fusion,
(ii) the energy released, in J, when this reaction takes place. *(3 marks)*

AQA, 2003

7 (a) In the context of the processes that occur in a nuclear power reactor, explain what is meant by:
(i) thermal neutrons,
(ii) induced fission,
(iii) a self-sustaining chain reaction. *(5 marks)*

(b) (i) Describe the process of moderation that takes place in an operational reactor.
(ii) How is the fission rate controlled in a power reactor? *(7 marks)*

8 (a) When a fuel rod has been in use in a nuclear reactor for several years, it produces less output power and presents a greater hazard than it did when first installed. Explain why this is so. *(3 marks)*

(b) Describe how the spent fuel rods are handled and processed after they have been removed from a nuclear reactor. Indicate how the active wastes are dealt with in order to reduce the hazards they could present to future generations. *(5 marks)*

Section 8 Summary

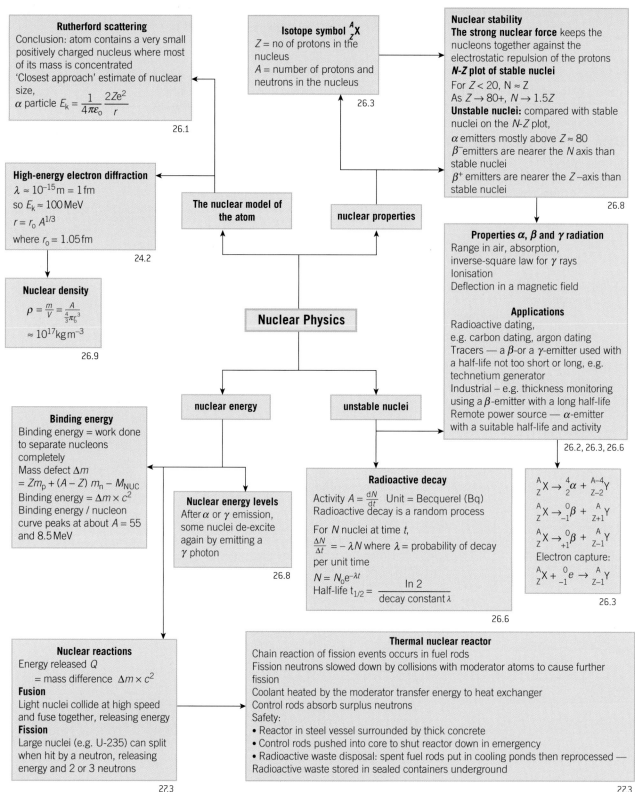

Rutherford scattering
Conclusion: atom contains a very small positively charged nucleus where most of its mass is concentrated
'Closest approach' estimate of nuclear size,
α particle $E_k = \dfrac{1}{4\pi\varepsilon_o}\dfrac{2Ze^2}{r}$

26.1

Isotope symbol $^A_Z X$
Z = no of protons in the nucleus
A = number of protons and neutrons in the nucleus

26.3

Nuclear stability
The strong nuclear force keeps the nucleons together against the electrostatic repulsion of the protons
N-Z plot of stable nuclei
For $Z < 20$, $N \approx Z$
As $Z \to 80+$, $N \to 1.5Z$
Unstable nuclei: compared with stable nuclei on the N-Z plot,
α emitters mostly above $Z \approx 80$
β^- emitters are nearer the N axis than stable nuclei
β^+ emitters are nearer the Z –axis than stable nuclei

26.8

High-energy electron diffraction
$\lambda \approx 10^{-15}\,\mathrm{m} = 1\,\mathrm{fm}$
so $E_k \approx 100\,\mathrm{MeV}$
$r = r_o A^{1/3}$
where $r_o = 1.05\,\mathrm{fm}$

24.2

The nuclear model of the atom

nuclear properties

Properties α, β and γ radiation
Range in air, absorption, inverse-square law for γ rays
Ionisation
Deflection in a magnetic field

Applications
Radioactive dating, e.g. carbon dating, argon dating
Tracers — a β-or a γ-emitter used with a half-life not too short or long, e.g. technetium generator
Industrial – e.g. thickness monitoring using a β-emitter with a long half-life
Remote power source — α-emitter with a suitable half-life and activity

26.2, 26.3, 26.6

Nuclear density
$\rho = \dfrac{m}{V} = \dfrac{A}{\frac{4}{3}\pi r_o^3}$
$\approx 10^{17}\,\mathrm{kg\,m^{-3}}$

26.9

Nuclear Physics

nuclear energy

unstable nuclei

Binding energy
Binding energy = work done to separate nucleons completely
Mass defect Δm
$= Zm_p + (A - Z)\,m_n - M_{NUC}$
Binding energy $= \Delta m \times c^2$
Binding energy / nucleon curve peaks at about $A = 55$ and 8.5 MeV

Nuclear energy levels
After α or γ emission, some nuclei de-excite again by emitting a γ photon

26.8

Radioactive decay
Activity $A = \dfrac{dN}{dt}$ Unit = Becquerel (Bq)
Radioactive decay is a random process
For N nuclei at time t,
$\dfrac{\Delta N}{\Delta t} = -\lambda N$ where λ = probability of decay per unit time
$N = N_o e^{-\lambda t}$
Half-life $t_{1/2} = \dfrac{\ln 2}{\text{decay constant } \lambda}$

26.6

$^A_Z X \to \,^4_2\alpha + \,^{A-4}_{Z-2}Y$
$^A_Z X \to \,^0_{-1}\beta + \,^A_{Z+1}Y$
$^A_Z X \to \,^0_{+1}\beta + \,^A_{Z-1}Y$
Electron capture:
$^A_Z X + \,^0_{-1}e \to \,^A_{Z-1}Y$

26.3

Nuclear reactions
Energy released Q
   = mass difference $\Delta m \times c^2$
Fusion
Light nuclei collide at high speed and fuse together, releasing energy
Fission
Large nuclei (e.g. U-235) can split when hit by a neutron, releasing energy and 2 or 3 neutrons

27.3

Thermal nuclear reactor
Chain reaction of fission events occurs in fuel rods
Fission neutrons slowed down by collisions with moderator atoms to cause further fission
Coolant heated by the moderator transfer energy to heat exchanger
Control rods absorb surplus neutrons
Safety:
• Reactor in steel vessel surrounded by thick concrete
• Control rods pushed into core to shut reactor down in emergency
• Radioactive waste disposal: spent fuel rods put in cooling ponds then reprocessed — Radioactive waste stored in sealed containers underground

27.3

Practical skills

In this section you have met the following skills:

- use of a cloud chamber or a spark counter to observe the range of α particle tracks
- use of a micrometer to measure the thickness of different absorbers
- use of a Geiger tube with a counter to measure
    - background radioactivity
    - the corrected count rate from a radioactive source
- use of a Geiger tube with a counter or a ratemeter to investigate α, β, and γ radiation, including
    - absorption of α, β, or γ radiation by different materials, or different thicknesses of the same material
    - the inverse square for γ radiation.

Maths skills

In this section you have met the following skills:

- use appropriate equations to estimate the size of the nucleus of an atom from the least distance of approach of an α particle
- calculate the mean value of a set of Geiger counter readings and obtain the corrected count rate
- use measured data to plot a straight line graph to verify the inverse square law for γ radiation
- interpret a radioactive decay curve of a radioactive isotope in terms of activity and half-life
- use count rate or activity data to find the decay constant λ of a radioactive isotope, or to plot a log $-$ linear graph, and so determine the half-life of the isotope
- carry out calculations
    - relating the mass m of a radioactive isotope to the number of nuclei N and the activity A of the isotope
    - using the radioactive decay equation $X = X_0 e^{-\lambda t}$ where $X = A, N,$ or m, and solve problems (e.g., radioactive dating)
- calculate the energy of a γ photon from an energy level diagram
- plot a log-log graph to demonstrate the relationship between the radius of a nucleus and its mass number and hence calculate the density of nuclear matter.
- use given data to calculate
    - the mass defect and the binding energy per nucleon of a nucleus
    - the energy released in a nuclear change (e.g., in a fission event)

Extension task

Thorium-232 is a radioactive isotope that can be used to produce nuclear fuel in a nuclear reactor. Use the internet to find out more about thorium-based nuclear reactors and create a presentation for your class. In the presentation, you could include

- how abundant thorium-232 is and how it is used to produce nuclear fuel
- the advantages and disadvantages of using thorium-232 to produce nuclear fuel
- the current extent of the usage of thorium-based nuclear reactors

1   The equation

$$p \rightarrow n + \beta^+ + \nu_e$$

represents the emission of a positron from a proton.

(a)   Energy and momentum are conserved in this emission.
      What other quantities are conserved in this emission?   (*3 marks*)

(b)   Copy and complete the following table using ticks ✓ and crosses ✗.

particle	meson	baryon	lepton
p			
n			
β^+			
ν_e			

(*4 marks*)
AQA, 2005

2   (a)   (i)   Alpha and beta emissions are known as *ionising radiations*. State and explain why
                such radiations can be described as *ionising*.
          (ii) Explain why beta particles have a greater range in air than alpha particles.
                                                                        (*4 marks*)

(b)   **Figure 1** shows the variation with time of the number of radon (^{220}Ra) atoms in a
      radioactive sample.

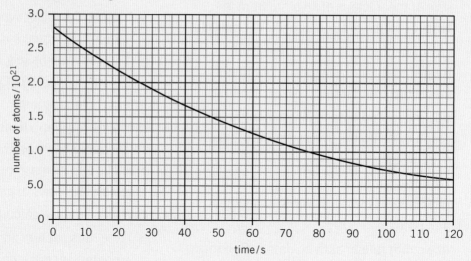

▲ Figure 1

      (i)   Use the graph to show that the half-life of the decay is approximately 53 s. Show
            your reasoning clearly.
      (ii) The decay constant of $^{220}_{88}$Ra is $1.3 \times 10^{-2}\,\text{s}^{-1}$. Use data from the graph to find the
            activity of the sample at a time $t = 72$ s.   (*6 marks*)

(c)   (i)   State **two** origins of background radiation.
      (ii) Suggest why it should be unnecessary to allow for background radiation
            when measuring the activity of the sample described in part (b)(ii).   (*3 marks*)
AQA, 2005

3   A nucleus of plutonium ($^{240}_{94}$Po) decays to form uranium (U) and an alpha particle (α).

(a)   Copy and complete the equation that describes this decay:

$$^{240}_{94}\text{Po} \rightarrow$$

(2 marks)

(b)   **(i)**   Show that about 1 pJ of energy is released when one nucleus decays.

mass of plutonium nucleus = 3.98626×10^{-25} kg

mass of uranium nucleus = 3.91970×10^{-25} kg

mass of alpha particle = 6.64251×10^{-25} kg   *(3 marks)*

(ii)   The plutonium isotope has a half-life of 2.1×10^{-11} s. Show that the decay constant of the plutonium is about 3×10^{-12} s^{-1}.

(iii)   A radioactive source in a school laboratory contains 3.2×10^{21} atoms of plutonium.

Calculate the energy that will be released in one second by the decay of the plutonium described in part (b)(i).

(iv)   Comment on whether the energy release due to the plutonium decay is likely to change by more than 5% during 100 years. Support your answer with a calculation.   *(12 marks)*

AQA, 2004

4   The table shows data for some nuclei.

element	Z	A	nuclear radius r/ 10^{-15} m	binding energy per nucleon/MeV	emission (half-life)
beryllium	4	9	2.5	6.46	stable
sodium	11	23	3.4	8.11	stable
manganese	25	56	4.6	8.74	β^- (2.6 h)

(a)   **(i)**   Show that these data support the rule that

$$r = r_0 A^{1/3}$$

where r_0 is a constant.

(ii)   The mass of a nucleon is about 1.7×10^{-27} kg. Calculate the density of nuclear matter.   *(6 marks)*

(b)   **(i)**   Explain what is meant by the binding energy of a nucleus.

(ii)   Show that the total binding energy of a sodium-23 nucleus is about 3×10^{-11} J.

(iii)   Calculate the mass-equivalent of this binding energy.   *(5 marks)*

(c)   Nuclear structure can be explored by bombarding the nuclei with alpha particles. The de Broglie wavelength of the alpha particle must be similar to the nuclear diameter. Calculate the energy of an alpha particle that could be used to explore the structure of manganese-56.

mass of an alpha particle = 6.8×10^{-27} kg   *(4 marks)*

(d)   **(i)**   State the proton number and nucleon number of the nucleus formed by the decay of manganese-56.

(ii)   The activity of a sample of manganese-56 varies with time according to the equation

$$A = A_0 e^{-\lambda t}$$

What value should be used for λ in calculations involving manganese-56 when t is in seconds?   *(4 marks)*

AQA, 2007

5 **(a)** When an α particle is emitted from a nucleus of the isotope $^{212}_{83}\text{Bi}$, a nucleus of thallium, Tl, is formed. Copy and complete the equation below.

$$^{212}_{83}\text{Bi} \rightarrow \alpha + \text{Tl}$$

(2 marks)

(b) The α particle in part (a) is emitted with 6.1 MeV of kinetic energy.
 (i) The mass of the α particle is 4.0 u. Show that the speed of the α particle immediately after it has been emitted is $1.7 \times 10^7 \,\text{m s}^{-1}$. Ignore relativistic effects.
 (ii) Calculate the speed of recoil of the daughter nucleus immediately after the α particle has been emitted. Assume the parent nucleus is initially at rest.

(6 marks)
AQA, 2002

6 **(a)** A solar panel of area $3.8 \,\text{m}^2$ is fitted to a satellite in orbit above the Earth. The panel produces a current of 2.8 A at a pd of 25 V when solar radiation is incident normally on it.
 (i) Calculate the electrical power output of the panel.
 (ii) Solar radiation on the panel has an intensity of $1.4 \,\text{kW m}^{-2}$. Calculate the efficiency of the panel. *(4 marks)*

(b) The back-up power system of the satellite is provided by a radioactive isotope enclosed in a sealed container which absorbs the radiation from the isotope. Energy from the radiation is converted to electrical energy by means of a thermoelectric module.
 (i) The isotope produces α particles of energy 4.1 MeV and the container absorbs energy from the α particles at a rate of $85 \,\text{J s}^{-1}$. Show that the isotope has an activity of $1.3 \times 10^{14} \,\text{Bq}$.
 (ii) The half-life of the isotope is 200 years. Calculate the decay constant, λ, of this isotope.
$$1 \text{ year} = 3.15 \times 10^7 \,\text{s}.$$
 (iii) The nucleon number of the isotope is 209. Calculate the mass of isotope needed for an activity of $1.3 \times 10^{14} \,\text{Bq}$. *(7 marks)*
AQA, 2006

7 The radioactive isotope $^{40}_{19}\text{K}$ decays by β^- emission to form a stable isotope of calcium (Ca) or by electron capture to form a stable isotope of argon (Ar).
(a) Copy and complete the following equation which represents the β^- decay of $^{40}_{19}\text{K}$.

$$^{40}_{19}\text{K} \rightarrow \beta^- + \text{Ca} + \bar{v}_e$$

(1 mark)

(b) Copy and complete the following equation which represents electron capture by $^{40}_{19}\text{K}$.

$$^{40}_{19}\text{K} + e^- \rightarrow \text{Ar} + v_e$$

(1 mark)

(c) The isotope of argon formed as a result of electron capture by $^{40}_{19}\text{K}$ is found as a trapped gas in ancient rocks. The age of an ancient rock can be determined by measuring the proportion of this isotope of argon to $^{40}_{19}\text{K}$.
An ancient rock is found to contain 1 argon atom for every 4 atoms of $^{40}_{19}\text{K}$.
 (i) The decay of $^{40}_{19}\text{K}$ by β^- emission is 8 times more likely than electron capture. Show that for every argon atom in this rock, there must have originally been 13 atoms of $^{40}_{19}\text{K}$.
 (ii) $^{40}_{19}\text{K}$ has a half-life of 1250 million years. Calculate the age of this rock. *(6 marks)*
AQA, 2004

8 The isotope of uranium, $^{238}_{92}$U, decays into a stable isotope of lead, $^{206}_{82}$Pb, by means of a series of α and β⁻ decays.

(a) In this series of decays, α decay occurs 8 times and β⁻ decay occurs n times. Calculate n. *(1 mark)*

(b) (i) Explain what is meant by the binding energy of a nucleus. *(2 marks)*

(ii) Figure 2 shows the binding energy per nucleon for some stable nuclides.

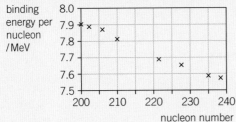

▲ Figure 2

Use Figure 2 to estimate the binding energy, in MeV, of the $^{206}_{82}$Pb nucleus. *(1 mark)*

(c) The half-life of $^{238}_{92}$U is 4.5 × 10⁹ years, which is much larger than all the other half-lives of the decays in the series.

A rock sample when formed originally contained 3.0 × 10²² atoms of $^{238}_{92}$U and no $^{206}_{82}$Pb atoms.

At any given time most of the atoms are either $^{238}_{92}$U or $^{206}_{82}$Pb with a negligible number of atoms in other forms in the decay series.

(i) Sketch graphs on a copy of Figure 3 to show how the numbers of $^{238}_{92}$U atoms and the number of $^{206}_{82}$Pb atoms in the rock sample vary over a period of 1.0 × 10¹⁰ years from its formation. Label your graphs U and Pb.

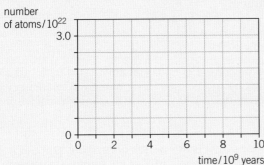

▲ Figure 3

(ii) A certain time, t, after its formation the sample contained twice as many $^{238}_{92}$U atoms as $^{206}_{82}$Pb atoms.

Show that the number of $^{238}_{92}$U atoms in the rock sample at time t was 2.0 × 10²². *(1 mark)*

(iii) Calculate t in years. *(3 marks)*

AQA, 2012

9 (a) Describe the changes made inside a nuclear reactor to reduce its power output and explain the process involved. *(2 marks)*

(b) State the main source of the highly radioactive waste from a nuclear reactor. *(1 mark)*

(c) (i) In a nuclear reactor, neutrons are released with high energies. The first few collisions of a neutron with the moderator transfer sufficient energy to excite nuclei of the moderator. Describe and explain the nature fo the radiation that may be emitted from an excited nucleus of the moderator. *(2 marks)*

(ii) The subsequent collisions of a neutron with the moderator are elastic. Describe what happens to the neutrons as a result of these subsequent collisions with the moderator. *(2 marks)*

AQA, 2013

Further practice questions

Multiple choice questions

1 State which **one** of the following alternatives **A–D** gives the ratio

$$\frac{\text{specific charge of a carbon } {}^{12}_{6}\text{C nucleus}}{\text{specific charge of the proton}}$$

A $\dfrac{1}{3}$ B $\dfrac{1}{2}$ C 1   D 1

2 State how many baryons and mesons there are in an atom of ${}^{9}_{4}$ Be.

	baryons	mesons
A	5	0
B	9	0
C	5	4
D	9	4

3 State which **one** of the following statements is true about β⁺ emission.
A   An up quark changes to a down quark and emits a W⁻ boson.
B   A W⁻ boson decays into an electron and an antineutrino.
C   A down quark changes to an up quark and emits a W⁺ boson.
D   A W⁺ boson decays into a positron and a neutrino

4 A conduction electron at the surface of a metal escapes from the surface after absorbing a photon. The work function of the metal is ϕ and the metal is at zero potential. State which one of the following equations about wavelength λ of the photon is true.

A $\lambda > \dfrac{hc}{\phi}$ B $\lambda > \dfrac{h\phi}{c}$ C $\lambda > \dfrac{\phi}{hc}$ D $\lambda < \dfrac{c}{h\phi}$

5 State which **one** of the following statements about polarisation is true.
A   Sound waves can be polarised.
B   Some electromagnetic waves cannot be polarised.
C   The vibrations of a polarised wave are always in the same plane.
D   Light waves are always polarised.

6 A wire of length L and diameter d is fixed at each end and placed under tension T. At this tension, the frequency of its first harmonic (i.e., its fundamental frequency) is f. State which one of the following alternatives **A–D** gives the density of the material of the wire.

A $\dfrac{T}{\pi d f L}$ B $\dfrac{4T}{\pi d f L}$ C $\dfrac{T}{\pi d^2 f^2 L^2}$ D $\dfrac{4T}{\pi d^2 f^2 L^2}$

7 A parallel beam of monochromatic light is directed at normal incidence at two narrow parallel slits spaced 0.7 mm apart. Interference fringes are formed on a screen which is perpendicular to the direction of the incidence beam at a distance of 0.900 m from the slits. The distance across five fringe spaces is measured at 4.2 mm.
State which **one** of the following alternatives **A–D** gives the wavelength of the light.

A   130 nm   B   450 nm   C   550 nm   D   650 nm

8 A parallel beam of monochromatic light is directed normally at a diffraction grating. The angle of diffraction of the first order beam was found to be 22°.
State which **one** of the following alternatives **A–D** gives the number of transmitted beams including the zero order beam.

A   3   B   5   C   7   D   9

9 Two particles X and Y at the same initial position accelerate uniformly from rest along a straight line. After 1.0 s, X is 0.20 m ahead of Y. State the separation of X and Y after 2.0 s from the start:

A   0.40 m   B   0.80 m   C   1.20 m   D   1.60 m

10 Two trolleys P and Q travelling in opposite directions at the same speed collide and move together after the collision in the direction in which P was originally travelling.
State which **one** of the following statements about the collision is true.

   A   The collision is elastic.
   B   The mass of P is greater than the mass of Q.
   C   The force exerted by P on Q is greater than the force exerted by Q on P.
   D   The change of momentum of P is greater than the change of momentum of Q.

11 A box of mass m slides down a slope at constant velocity as shown in **Figure 1**. The slope is inclined at angle θ to the horizontal. The diagram shows the frictional force F and the normal reaction force N of the surface on the block.
State which **one** of the following statements about forces F and N is true.

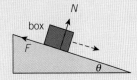

▲ **Figure 1**

   A $F = mg \cos \theta$ **B** $F = N \sin \theta$ **C** $F = N + mg$ **D** $F^2 = (mg)^2 - N^2$

12 Two wires P and Q of the same material have lengths L and $2L$ respectively and different diameters d_p and d_q respectively.
When the same force is applied to each wire, the extension of P is $4 \times$ the extension of Q.
State which **one** of the following alternatives **A–D** gives the ratio $\dfrac{d_p}{d_q}$.

   A $\dfrac{1}{2\sqrt{2}}$ **B** $\dfrac{1}{\sqrt{2}}$ **C** $\sqrt{2}$ **D** $2\sqrt{2}$

13 Four resistors of resistances $1.0\,\Omega$, $2.0\,\Omega$, $3.0\,\Omega$, and $4.0\,\Omega$ are connected together as shown in **Figure 2** and a battery is connected across the $1.0\,\Omega$ resistor.
If I_o is the battery current and I_1 is the current in the $1.0\,\Omega$ resistor, state which **one** of the following alternatives **A–D** gives the ratio $\dfrac{I_o}{I_1}$.

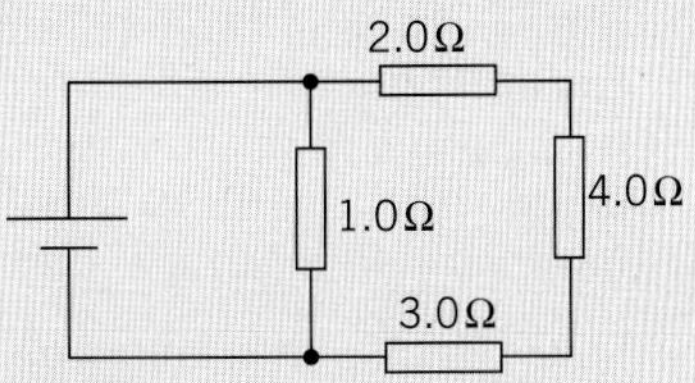

▲ **Figure 2**

   A 0.1   **B** 0.5   **C** 1.5   **D** 10

14 State which **one** of the following alternatives **A–D** in **Figure 2** gives the ratio

$$\frac{\text{the power dissipated in the }1.0\,\Omega\text{ resistor}}{\text{the power dissipated in the }3.0\,\Omega\text{ resistor}}$$

   A 3   **B** 9   **C** 27   **D** $81\,\Omega$

15 Two identical resistors X and Y are connected in series with each other and a 3.0 V cell of negligible internal resistance to form a potential divider. A third identical resistor is then connected in parallel with X.
State which **one** of the following alternatives **A–D** gives the pd across Y.

   A 0.5 V   **B** 1.0 V   **C** 2.0 V   **D** 2.5 V

16 A digital voltmeter connected across the terminals of a battery reads 4.50 V. When a $21.0\,\Omega$ resistor is connected across the battery terminals, the voltmeter reading decreases to 4.20 V.
State which **one** of the following alternatives **A–D** gives the internal resistance of the cell.

   A $1.2\,\Omega$ **B** $1.3\,\Omega$ **C** $1.4\,\Omega$ **D** $1.5\,\Omega$

17 A spherical planet P has a radius R and uniform density ρ. The gravitational potential at the surface of P is V_s.
Another spherical planet Q has a radius $2R$ and uniform density 0.5ρ. State which **one** of the following alternatives **A–D** gives the gravitational potential at the surface of Q.

   A $\dfrac{1}{2}Vs$ **B** Vs **C** $2Vs$ **D** $4Vs$

18 A satellite of mass m is in a circular orbit at height $2R$ above the Earth's surface, where R is the radius of the Earth.

State which **one** of the following alternatives **A–D** gives the ratio

$$\frac{\text{weight of the satellite in its orbit}}{\text{weight of the satellite at the surface of the Earth}}$$

A $\dfrac{1}{9}$ **B** $\dfrac{1}{4}$ **C** $\dfrac{1}{3}$ **D** $\dfrac{1}{2}$

19 Two small spheres carry equal and opposite charges $+Q$ and $-Q$. The force between the spheres is F when the centres of the spheres are at distance d apart.

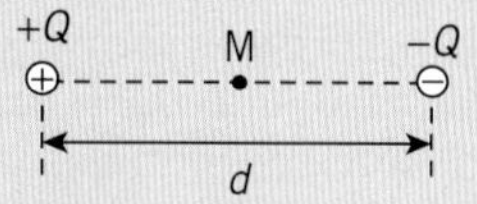

State which **one** of the following alternatives **A–D** gives the magnitude of the electric field strength at the midpoint of their centres.

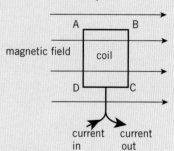

▲ Figure 3

A $\dfrac{F}{Q}$ **B** $\dfrac{2F}{Q}$ **C** $\dfrac{4F}{Q}$ **D** $\dfrac{8F}{Q}$

20 A data logger was used to measure the potential difference across the terminals of a capacitor as it discharges through a $5.0\,\text{M}\Omega$ resistor. The measurements showed that the potential difference across the capacitor decreased from $5.00\,\text{V}$ to $2.75\,\text{V}$ in $300\,\text{s}$.
State which of the following alternatives **A–D** gives the capacitance of the capacitor.

A $100\,\mu\text{F}$ **B** $135\,\mu\text{F}$ **C** $165\,\mu\text{F}$ **D** $300\,\mu\text{F}$

21 **Figure 4** shows a square coil ABCD that has been placed in a uniform horizontal magnetic field such that the magnetic field lines are parallel to the plane of the coil and the sides AD and BC of the coil are vertical. The coil is then connected in series with a battery, a resistor, and an open switch.

State which of the following alternatives **A–D** describes correctly the effect on the coil when the switch is closed.

A Each side experiences a force.

B Only sides AB and CD experience a force.

C Only the sides AD and BC experience a force.

D Opposite sides experience forces in opposite directions.

▲ Figure 4

22 An ac generator turns at a steady frequency of $25\,\text{Hz}$ in a uniform $80\,\text{mT}$ magnetic field. This produces a sinusoidal pd with a peak output of $4.0\,\text{V}$. If the frequency is reduced to $20\,\text{Hz}$ and the magnetic flux density is increased to $100\,\text{mT}$, state the peak output pd.

A $3.2\,\text{V}$ **B** $4.0\,\text{V}$ **C** $5.0\,\text{V}$ **D** $6.3\,\text{V}$

23 A radioactive nucleus $^{A}_{Z}X$ emits alpha, beta, and gamma particles in the following sequence:

$$\text{alpha} \quad \text{beta} \quad \text{gamma}$$

State which of the following alternatives **A–D** represents the nucleus at the end of this sequence.

A $^{A-4}_{Z-2}X$ **B** $^{A-4}_{Z-1}X$ **C** $^{A-2}_{Z-2}X$ **D** $^{A-2}_{Z-1}X$

24 Two radioactive isotopes X and Y have half-lives of 8.0 years and 16.0 years respectively and form stable products. Samples of X and Y are initially pure and contain equal numbers of nuclei.
State which of the following alternatives **A–D** gives the ratio $\dfrac{\text{the activity of X}}{\text{the activity of Y}}$ after 16 years.

A $\dfrac{1}{4}$ **B** $\dfrac{1}{2}$ **C** 1 **D** 2

25 When a $_1^2\text{H}$ nucleus fuses with a $_1^3\text{H}$ nucleus, a $_2^4\text{He}$ is nucleus formed and a neutron is ejected.

Masses/u: neutron 1.00867; $_1^2\text{H}$ 2.01355; $_1^3\text{H}$ 3.01550; $_2^4\text{He}$ 4.00150

$1\,\text{u} = 931.5\,\text{MeV}$

State which of the following alternatives **A–D** gives the binding energy released in this event.

A   6.3 MeV   **B**   17.6 MeV   **C**   21.1 MeV   **D**   960 MeV

In each of Questions **26** to **30**, select from the list, **A–D**, state which relationship correctly describes the connection between y and x.

A y is proportional to x^2.

B y is proportional to $\sqrt{x}$.

C y is proportional to $\dfrac{1}{x}$.

D y is proportional to $\dfrac{1}{x^2}$.

	y	x
26	the speed of an object falling freely from rest	distance fallen by the object
27	maximum kinetic energy of a freely oscillating object	amplitude of the oscillations
28	energy stored by an parallel-plate capacitor connected to a battery	perpendicular distance between the plates
29	intensity of gamma radiation from a point source	distance from the source
30	electric potential energy of two protons	distance between the protons

Longer questions

1   **Figure 5** shows a way to measure the mass of a lorry. The vehicle and its contents are driven onto a platform mounted on a spring. The platform is then made to oscillate vertically, and the mass is found from a measurement of the natural frequency of oscillation.

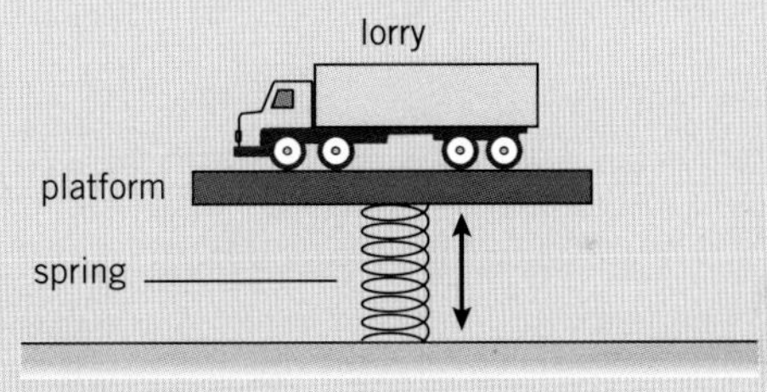

▲ **Figure 5**

(a) (i)   State whether the period of oscillation increases, decreases, or remains unchanged when the amplitude of oscillation of the platform is reduced.

(ii)   The spring constant k of the supporting spring is increased to four times its original value.

State the value of the ratio $\dfrac{\text{new oscillation period}}{\text{old oscillation period}}$

(iii)   The time period of oscillation is T when a lorry is on the platform. The spring constant of the spring is k. Show that the total mass M of lorry and platform is given by $M = \dfrac{kT^2}{4\pi^2}$

(iv)   A lorry and its contents have a total mass of 5300 kg. The spring constant of the supporting spring k is $1.9 \times 10^5\,\text{N m}^{-1}$. The frequency of oscillation of the platform with the lorry resting on it is 0.91 Hz.

Calculate the mass of the platform.   (*7 marks*)

(b)   The driver is required to turn off the vehicle engine whilst the measurement is taking place. The driver of the lorry in part (a)(iv) fails to do this and slowly increases the frequency of vibration of his vehicle from 0.5 Hz to about 4 Hz whilst the measurement is in progress and the platform is free to move. Describe and explain how the amplitude and frequency of the platform vary as this frequency increase occurs. You should use a sketch graph to support your answer.   (*4 marks*)

AQA 2003

2 A dish on a communications satellite is used to transmit a beam of microwaves of wavelength λ. The beam spreads with an angular width $\frac{\lambda}{d}$, in radians, where d is the diameter of the dish.

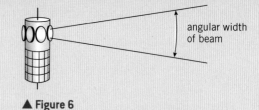

▲ **Figure 6**

(a) (i) Calculate the angular width, in degrees, of a beam of frequency 1200 MHz transmitted using a dish of diameter 1.8 m.

(ii) Show that the beam has a width of 2100 km at a distance of 15 000 km from the satellite. *(4 marks)*

(b) (i) Show that the speed, v, of a satellite in a circular orbit at height h above the Earth is given by

$$v = \sqrt{\frac{GM}{R+h}}$$

where R is the radius of the Earth and M is the mass of the Earth.

(ii) Calculate the speed and the time period of a satellite at a height of 15 000 km in a circular orbit about the Earth.

(iii) The satellite passes directly over a stationary receiver at the North Pole. Show that the beam moves at a speed of 1.3 km s⁻¹ across the Earth's surface and that the receiver can remain in contact with the satellite for no more than 27 minutes each orbit. *(9 marks)*

AQA 2005

3 **Figure 7** shows a motor lifting a small mass. The energy required comes from a charged capacitor.

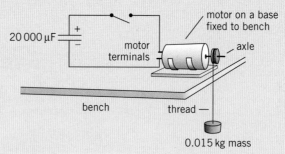

▲ **Figure 7**

The capacitor was charged to a potential difference of 4.5 V and then discharged through the motor.

(a) (i) The motor only operates when the voltage at its terminals is at least 2.5 V. Calculate the energy delivered to the motor when the potential difference across the capacitor falls from 4.5 V to 2.5 V.

(ii) The motor lifted the mass through a distance of 0.35 m. Calculate the efficiency of the transfer of energy from the capacitor to gravitational potential energy of the mass. Give your answer as a percentage.

(iii) Give **two** reasons why the transfer is inefficient. *(7 marks)*

(b) The motor operated for 1.3 s as the capacitor discharged from 4.5 V to 2.5 V. Calculate:

(i) the average useful power developed in lifting the mass,

(ii) the effective resistance of the motor, assuming that it remained constant. *(5 marks)*

AQA 2004

4 (a) (i) Explain why, despite the electrostatic repulsion between protons, the nuclei of most atoms of low nucleon number are stable.

(ii) Suggest why stable nuclei of higher nucleon number have greater numbers of neutrons than protons.

(iii) All nuclei have approximately the same density. State and explain what this suggests about the nature of the strong nuclear force. *(6 marks)*

(b) **(i)** Compare the electrostatic repulsion and the gravitational attraction between a pair of protons the centres of which are separated by 1.2×10^{-15} m.

(ii) Comment on the relative roles of gravitational attraction and electrostatic repulsion in nuclear structure. *(5 marks)*

AQA, 2006

5 **(a)** **Figure 8** shows an arrangement used to investigate the energy stored by a capacitor.

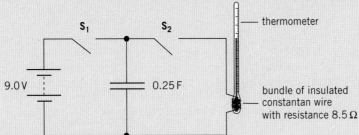

▲ **Figure 8**

The bundle of constantan wire has a resistance of $8.5\,\Omega$. The capacitor is initially charged to a potential difference of 9.0 V by closing S_1.

(i) Calculate the charge stored by the 0.25 F capacitor.

(ii) Calculate the energy stored by the capacitor.

(iii) Switch S_1 is now opened and S_2 is closed so that the capacitor discharges through the constantan wire.
Calculate the time taken for the potential difference across the capacitor to fall to 0.10 V. *(7 marks)*

AQA, 2006

(b) The volume of constantan wire in the bundle in **Figure 8** is 2.2×10^{-7}m^3.
density of constantan = 8900 kg m^{-3}
specific heat capacity of constantan = 420 J kg^{-1}K^{-1}

(i) Assume that all the energy stored by the capacitor is used to raise the temperature of the wire. Use your answer to part (a)(ii) to calculate the expected temperature rise when the capacitor is discharged through the constantan wire.

(ii) Give **two** reasons why, in practice, the final temperature will be lower than that calculated in part (b)(i). *(5 marks)*

AQA, 2006

6 **(a)** A 3.0 kW electric kettle heats 2.4 kg of water from $16\,^\circ$C to $100\,^\circ$C in 320 s.

(i) Calculate the electrical energy supplied to the kettle.

(ii) Calculate the heat energy supplied to the water.
specific heat capacity of water = 4200 J kg^{-1}K^{-1}

(iii) Give **one** reason why not all the electrical energy supplied to the kettle is transferred to the water. *(4 marks)*

(b) The potential difference supplied to the kettle in part (a) is 230 V.

(i) Calculate the resistance of the heating element of the kettle.

(ii) The heating element consists of an insulated conductor of length 0.25 m and diameter 0.65 mm. Calculate the resistivity of the conductor. *(5 marks)*

AQA, 2004

7 **(a)** **(i)** State **three** assumptions concerning the motion of the molecules in an ideal gas.

(ii) For an ideal gas at a temperature of 300 K, show that the mean kinetic energy of a molecule is 6.2×10^{-21} J. *(5 marks)*

(b) **(i)** When no current passes along a metal wire, conduction electrons move about in the wire like molecules in an ideal gas.
Calculate the speed of an electron which has 6.2×10^{-21} J of kinetic energy.

(ii) Describe the motion of conduction electrons in a wire when a pd is applied across the ends of the wire. *(6 marks)*

AQA, 2007

8 **(a)** **(i)** At the surface of a spherical planet of radius R, show that the gravitational potential, V_s, is related to the gravitational field of strength, g_s, by

$$V_s = -g_s R.$$

(ii) The gravitational field strength of the Moon at its surface is $1.6\,N\,kg^{-1}$. Show that the gravitational potential energy of an oxygen molecule at the surface is $-1.4 \times 10^{-19}\,J$.

radius of the Moon = $1700\,km$ molar mass of oxygen = $0.032\,kg\,mol^{-1}$

(5 marks)

(b) Oxygen gas at $400\,K$ is released on the surface of the Moon.

(i) Calculate the mean kinetic energy of an oxygen gas molecule at this temperature.

(ii) The maximum temperature of the surface of the Moon is about $400\,K$. Use the data from part (a)(ii) and the results of your calculations to explain why some of the oxygen gas released at the Moon's surface would escape into space.

(4 marks)

AQA, 2005

Data analysis questions

In an impact investigation by two students, a mass hanger **A** suspended on a thread was displaced from its equilibrium position by a certain distance and released so it collided with a ball **B** suspended on a thread. A horizontal metre ruler fixed in a clamp (not shown) was used to measure the horizontal displacements x_A and x_B of each thread from its equilibrium position at the level of the metre ruler, as shown in the diagram. The vertical distance, d, of the ruler below the upper end of the threads was measured.

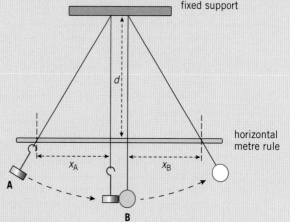

▲ Figure 9

The measurement of x_B was repeated without changing x_A and d for different additional masses added to the mass hanger.

(a) The measurements shown in Table 1 were made in preliminary tests using a total mass m for the mass of the hanger and the additional mass. The students decided to make further measurements between 0.100 and 0.200 kg and above 0.300 kg. Why do you think they made this decision?

(4 marks)

▼ Table 1

m/kg	x_B/mm	
0.100	60	62
0.200	77	75
0.300	80	78

(b) Table 2 shows all their measurements.

▼ Table 2

m/kg	x_B/mm		x_B/mm			$\langle x_B \rangle$/mm	θ/°
0.100	60	62	58	58	60	59.6	8.47
0.120	67	68	68	65	62	66.0	9.37
0.150	68	73	70	68	70	69.8	9.90
0.200	77	75	78	71	76	75.4	10.67
0.300	80	78	79	80	80	79.4	11.22
0.600	86	85	88	85	88		

$d = 400 \, \text{mm}$, $x_A = 60 \, \text{mm}$

The maximum angular displacement θ of the thread from equilibrium can be calculated using the equation $\tan \theta = \langle x_B \rangle / d$, where $\langle x_B \rangle$ is the mean value of x_B. This has been done in Table 2 for all the rows except the last one.

Copy and complete Table 2 by calculating $\langle x_B \rangle$ and θ for $m = 0.600$ kg. (*2 marks*)

(c) **(i)** By considering the energy changes of B after the impact, show its velocity V immediately after the impact is given by $V = \sqrt{(2gh)}$ where h is B's maximum height gain from its equilibrium position. (*2 marks*)

(ii) The height gain h can be calculated using the trigonometry formula $h = l(1 - \cos \theta)$ where l is the distance along the thread from the point of suspension of the ball to its centre. This distance was measured to be 575 mm. For each mass m, Table 3 shows the results of these calculations except for the last row. Copy the table and complete this last row. The last 2 columns are for the next question.

▼ **Table 3**

m/kg	$\theta/°$	h/mm	V/m s^{-1}		
0.100	8.47	6.27	0.351		
0.120	9.37	7.67	0.388		
0.150	9.90	8.56	0.410		
0.200	10.67	9.95	0.442		
0.300	11.22	11.00	0.465		
0.600					

(*2 marks*)

(d) The students found a theoretical analysis of the impact which gave the following equation relating V and m:

$$\frac{1}{V} = \frac{kM}{m} + k$$

where M is the mass of the ball and k is a constant.

(i) Plot a suitable graph to see if this relationship is correct. Show the results of any further calculations you carry out in the last two columns of your own Table 3.

(ii) Using your graph or otherwise, determine values for k and M. (*9 marks*)

(e) **(i)** What conclusions do you draw from the graph?

(ii) Use your results to evaluate your conclusions. (*3 marks*)

Extension question

The theoretical analysis is based on the diagram below in which an object (**A**) of mass m moving at velocity v collides with a stationary object (**B**) of mass M. After the impact, the two objects move apart at velocities v and V in the same direction as A's initial direction.

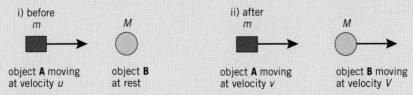

i) before
m M

object **A** moving at velocity u object **B** at rest

ii) after
m M

object **A** moving at velocity v object **B** moving at velocity V

▲ **Figure 10**

The theoretical analysis assumed that the velocity of **B** relative to **A** after the collision $(V - v) = eu$, where e is a constant that depends on the two objects.

(f) Combine the equation above with the equation representing conservation of momentum to derive the theoretical equation given in question 3 where $k = \dfrac{1}{(1 + e)u}$ (*5 marks*)

Section 9, Chapter 28.1
Option A—Astrophysics

Chapters in this section:

Introduction

This option is one of the five optional sections in the AQA A Level Physics specification. Full course notes and further resources can be found online on Kerboodle.

In this option, you will apply fundamental physical principles to astronomical observations and measurements of objects in space. The option starts by considering the optical principles of **telescopes** used by astronomers and how they are designed. In Chapter 1, you will learn how refracting and reflecting telescopes are designed and what is meant by terms such as **magnifying power** and **resolving power**. In addition, you will gain an awareness of how telescope design has developed from the first simple refracting telescopes invented centuries ago, to modern telescopes, such as the Hubble telescope, that have extended our knowledge and understanding of the Universe enormously.

In chapter 2, you will discover how accurate observations and measurements have been used to develop our current astronomical knowledge about **stars** and **galaxies**. You will learn what is meant by the **absolute magnitude** of a star and how the temperature, radius and other properties of stars have been determined. You will discover how astronomers classify stars according to their absolute magnitude and temperature, and how different classes of stars form, evolve, and die.

In the final chapter, you will learn about how astronomers such as Edwin Hubble in the early 20th century deduced from their observations that the distant galaxies are moving away from each other, and why astronomers now believe the **expansion** of the Universe is accelerating. The cause of this acceleration remains unclear at present and astronomers think a mysterious form of energy, known as '**dark energy**', is at work. You will also learn about the search for 'exoplanets'—planets in orbit around stars beyond the Sun, and the use of telescopes designed to detect infra-red radiation.

Working scientifically

Astrophysics is about how astronomical measurements are made and used to develop our knowledge of astronomical objects. In this option, you will develop your awareness of how astronomers use physics to learn about the Universe. For example, how astronomers use measurements of very small angles in working out the distance to a star, or how they use wavelength and intensity measurements of starlight to deduce the properties of stars, such as surface temperature and power output. You will also learn how astronomers have built on such knowledge to classify stars, and how the study of Cepheid variables led to the discovery that stars are concentrated in galaxies vast distances apart. New and better telescopes enabled astronomers to study the light from individual stars in galaxies. Applying the laws of physics to these observations, astronomers have made many discoveries about galaxies, including galactic collisions and supermassive black holes.

What you already know

From your A Level studies on electromagnetic radiation, you should know that:

- ○ for a photon of frequency f, its energy $E = hf$
- ○ an atom emits a photon of light when an electron in the atom moves to a lower energy level
- ○ the electromagnetic spectrum has a continuous range of wavelengths, from less than 10^{-15} m (gamma radiation) to more than 1000 m (radio waves).
- ○ for diffraction to occur when waves pass through a gap, the gap width needs to be comparable with the wavelength of the waves.

From your A Level studies on energy and matter, you should know that:

- ○ the density ρ of a substance is its mass per unit volume
- ○ temperature (K) ≈ temperature (°C) + 273
- ○ the mean kinetic energy of an atom in a gas is proportional to the temperature of the gas in kelvins
- ○ the higher the temperature of an object, the more energy per second it emits as infrared radiation and light
- ○ the spectrum of light
  - from a hot object is a continuous spectrum
  - from a glowing gas is a line emission spectrum
  - from a star is a line absorption spectrum.

Section 9 Summary

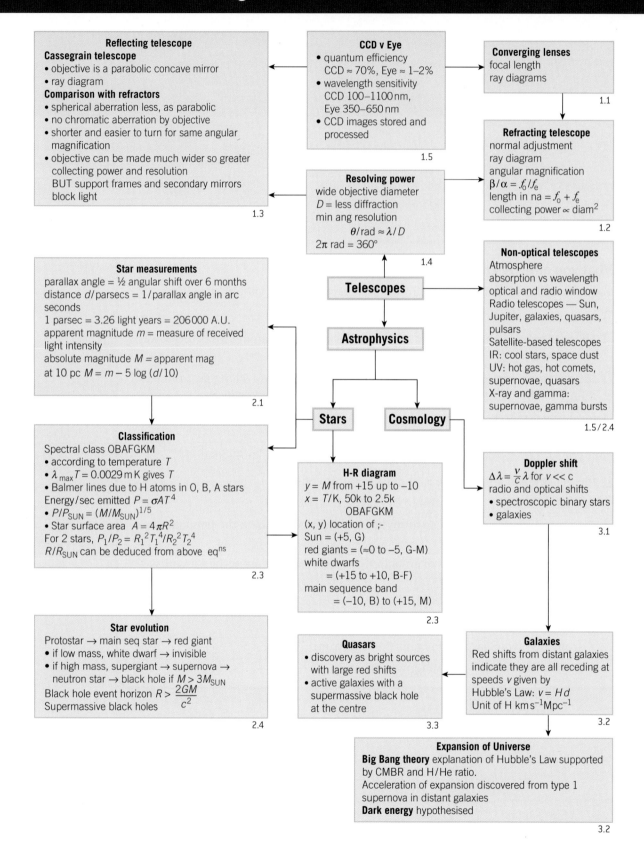

Reflecting telescope
Cassegrain telescope
- objective is a parabolic concave mirror
- ray diagram

Comparison with refractors
- spherical aberration less, as parabolic
- no chromatic aberration by objective
- shorter and easier to turn for same angular magnification
- objective can be made much wider so greater collecting power and resolution
  BUT support frames and secondary mirrors block light

1.3

CCD v Eye
- quantum efficiency CCD ≈ 70%, Eye ≈ 1–2%
- wavelength sensitivity CCD 100–1100 nm, Eye 350–650 nm
- CCD images stored and processed

1.5

Converging lenses
focal length
ray diagrams

1.1

Refracting telescope
normal adjustment
ray diagram
angular magnification
$\beta/\alpha = f_o/f_e$
length in na $= f_o + f_e$
collecting power $\propto$ diam^2

1.2

Resolving power
wide objective diameter
D = less diffraction
min ang resolution
$\theta/\text{rad} \approx \lambda/D$
2π rad $= 360°$

1.4

Non-optical telescopes
Atmosphere
absorption vs wavelength
optical and radio window
Radio telescopes — Sun, Jupiter, galaxies, quasars, pulsars
Satellite-based telescopes
IR: cool stars, space dust
UV: hot gas, hot comets, supernovae, quasars
X-ray and gamma: supernovae, gamma bursts

1.5/2.4

Telescopes

Astrophysics

Stars

Cosmology

Star measurements
parallax angle = ½ angular shift over 6 months
distance d/parsecs = 1/parallax angle in arc seconds
1 parsec = 3.26 light years = 206 000 A.U.
apparent magnitude m = measure of received light intensity
absolute magnitude M = apparent mag at 10 pc $M = m - 5 \log (d/10)$

2.1

Classification
Spectral class OBAFGKM
- according to temperature T
- $\lambda_{max} T = 0.0029$ m K gives T
- Balmer lines due to H atoms in O, B, A stars
Energy/sec emitted $P = \sigma A T^4$
- $P/P_{SUN} = (M/M_{SUN})^{1/5}$
- Star surface area $A = 4\pi R^2$
For 2 stars, $P_1/P_2 = R_1^2 T_1^4/R_2^2 T_2^4$
R/R_{SUN} can be deduced from above eq^{ns}

2.3

H-R diagram
$y = M$ from +15 up to −10
$x = T/K$, 50k to 2.5k
   OBAFGKM
(x, y) location of ;-
Sun = (+5, G)
red giants = (≈0 to −5, G-M)
white dwarfs
   = (+15 to +10, B-F)
main sequence band
   = (−10, B) to (+15, M)

2.3

Doppler shift
$\Delta\lambda = \dfrac{v}{c}\lambda$ for $v \ll c$
radio and optical shifts
- spectroscopic binary stars
- galaxies

3.1

Star evolution
Protostar → main seq star → red giant
- if low mass, white dwarf → invisible
- if high mass, supergiant → supernova → neutron star → black hole if $M > 3M_{SUN}$
Black hole event horizon $R > \dfrac{2GM}{c^2}$
Supermassive black holes

2.4

Quasars
- discovery as bright sources with large red shifts
- active galaxies with a supermassive black hole at the centre

3.3

Galaxies
Red shifts from distant galaxies indicate they are all receding at speeds v given by
Hubble's Law: $v = Hd$
Unit of H km s^{-1}Mpc^{-1}

3.2

Expansion of Universe
Big Bang theory explanation of Hubble's Law supported by CMBR and H/He ratio.
Acceleration of expansion discovered from type 1 supernova in distant galaxies
Dark energy hypothesised

3.2

In this option, you can reinforce skills that you developed in the compulsory topics in your A Level Physics course. By studying this option, you will strengthen your competence in the practical and mathematical skills listed below. In addition, you will develop your data analysis skills by interpreting and analysing astronomical data presented in tables and graphs. The extension task is intended to boost your skills of written communication, and also lets you work within a group by creating presentations on different topics.

Practical skills

In this option you meet the following skills:
- use of distance scales to measure distances (e.g. measurement of focal length)
- safe and correct use of electrical equipment in experiments (e.g. light sources)
- Use of optical equipment safely and correctly in experiments (e.g. estimates of magnifying power).

Maths skills

In this option you meet the following skills:
- use of the radian and its conversion to and from degrees (e.g. angular resolution calculations)
- use of the small angle approximation in calculations of astronomical distances
- calculation of a mean from a set of repeat readings (e.g. measurement of focal length)
- interpret graphs in terms of relevant physics variables (e.g. the Hertzsprung-Russell diagram)
- use of a calculator or a spreadsheet to carry out calculations using data and appropriate formulae, expressing answers to an appropriate number of significant figures and with appropriate units.
- use of base 10 logarithms in magnitude-distance calculations
- use of the inverse square law in relation to intensity variation with distance.

Extension task

Make a chart of the key features of one type of astronomical telescope, including its basic design, how it works, how the image is observed, its limitations and what it may be used for in astronomy. Make a presentation of your findings to present to the class.

Section 9, Chapter 28.2
Option B—Medical physics

Chapters in this section:

Introduction

This option is one of the five optional sections in the AQA A Level Physics specification. Full course notes and further resources can be found online on Kerboodle.

In this option, you will extend your physics knowledge to human health by studying some of the applications of physical principles and techniques used in **diagnosing** and **correcting disorders**. In Chapter 1, you will learn about the optics of the **eye** and how common sight disorders such as short sight and long sight are corrected using suitable lenses. In the next chapter, you will learn how the vibrations caused by sound waves entering the **ear** are detected by the inner ear, and how the human ear responds to sound waves of different frequencies. In addition, you will learn about the **decibel scale**, used to measure the intensity of sound waves, and how it applies to hearing loss.

Chapter 3 provides background information about nerve cells before moving on to study how **electrocardiogram** (ECG) devices work and what causes the ECG signals they detect. In chapters 4 and 5, you will learn how medical physicists use **radiation** to obtain images of organs inside the body. Chapter 4 considers imaging techniques using non-ionising radiation — namely **ultrasound** imaging, **endoscope** imaging, and **magnetic resonance imaging**.

In chapter 5, you will learn in depth about how **X-rays** are used to produce images of organs and structures inside the body. You will also learn about the devices used and measures taken to produce clear X-ray images without unnecessary exposure of ionising radiation to patients and radiographers. In the final chapter, you will extend your knowledge of X-rays and radioactive isotopes to gain an understanding of how high-energy ionising radiation is used to treat disorders.

Working Scientifically

Advances in medical physics often follow from new discoveries and developments. The first medical X-ray images were made soon after the discovery of X-rays in the late 19th century by Wilhelm Rontgen. Improvements in photography enabled doctors to see clear images of broken bones and other disorders in the human body.

Since the discovery of X-rays, medical doctors and medical physicists have always worked together to improve the devices they use and to develop new devices and materials. Improvements in optical materials have led to contact lenses and thinner spectacle lenses used for sight corrections. Advances in electronics have led to smaller devices used to remedy hearing and heart defects. Better imaging techniques have become possible following the discovery of materials — such as piezoelectric materials used in ultrasound imaging, flexible and transparent materials used in endoscopes, and superconducting coils used in magnetic resonance imaging.

Combined with advances in electronics and the use of computers to process and display signals, these and other imaging techniques can give medical doctors vital information necessary to diagnose and treat internal disorders. Where treatment using ionising radiation is necessary, detailed images enable a narrow beam of radiation to be focused on at the site of the disorder.

What you already know

From your A Level studies on light, you should know that:

○ a real image is an image that can be formed on a screen

○ a virtual image is an image formed where light appears to originate from

○ for a light ray

- reflected at a mirror, the angle of incidence = the angle of reflection

- refracted at a plane boundary between two substances, $n_1 \sin \theta_1 = n_2 \sin \theta_2$

- refracted along a plane boundary, $\sin c = \dfrac{n_2}{n_1}$.

From your A Level studies on sound, you should know that:

○ intensity is energy per second per unit area incident on a surface

○ ultrasound is sound of frequency greater than 20 kHz.

From your A Level studies on electromagnetic radiation, you should know that:

○ for a photon of frequency f, its energy $E = hf$

○ an atom emits a photon of light when an electron in the atom moves to a lower energy level

○ the electromagnetic spectrum has a continuous range of wavelengths from less than 10^{-15} m (gamma radiation) to more than 1000 m (radio waves)

○ for diffraction to occur when waves pass through a gap, the gap width needs to be comparable with the wavelength.

From your A level studies on magnetic fields, you should know that:

○ a uniform magnetic field exists in a solenoid when there is a current in the solenoid

○ the unit of magnetic flux density is the tesla (T).

Section 9 Summary

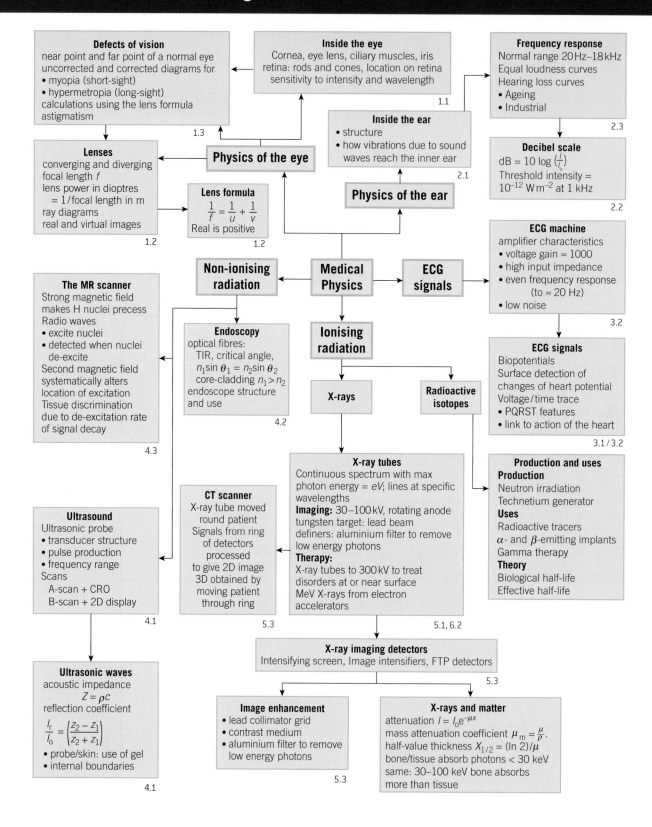

Defects of vision
near point and far point of a normal eye
uncorrected and corrected diagrams for
• myopia (short-sight)
• hypermetropia (long-sight)
calculations using the lens formula
astigmatism

1.3

Inside the eye
Cornea, eye lens, ciliary muscles, iris
retina: rods and cones, location on retina
sensitivity to intensity and wavelength

1.1

Frequency response
Normal range 20 Hz–18 kHz
Equal loudness curves
Hearing loss curves
• Ageing
• Industrial

2.3

Lenses
converging and diverging
focal length f
lens power in dioptres
 = 1/focal length in m
ray diagrams
real and virtual images

1.2

Physics of the eye

Lens formula
$$\frac{1}{f} = \frac{1}{u} + \frac{1}{v}$$
Real is positive

1.2

Inside the ear
• structure
• how vibrations due to sound
 waves reach the inner ear

2.1

Physics of the ear

Decibel scale
dB = 10 log $\left(\frac{I}{I_0}\right)$
Threshold intensity =
10^{-12} W m^{-2} at 1 kHz

2.2

**Non-ionising
radiation**

**Medical
Physics**

**ECG
signals**

ECG machine
amplifier characteristics
• voltage gain ≈ 1000
• high input impedance
• even frequency response
 (to ≈ 20 Hz)
• low noise

3.2

The MR scanner
Strong magnetic field
makes H nuclei precess
Radio waves
• excite nuclei
• detected when nuclei
 de-excite
Second magnetic field
systematically alters
location of excitation
Tissue discrimination
due to de-excitation rate
of signal decay

4.3

Endoscopy
optical fibres:
 TIR, critical angle,
 $n_1\sin\theta_1 = n_2\sin\theta_2$
 core-cladding $n_1 > n_2$
endoscope structure
and use

4.2

**Ionising
radiation**

X-rays

**Radioactive
isotopes**

ECG signals
Biopotentials
Surface detection of
changes of heart potential
Voltage/time trace
• PQRST features
• link to action of the heart

3.1 / 3.2

Ultrasound
Ultrasonic probe
• transducer structure
• pulse production
• frequency range
Scans
 A-scan + CRO
 B-scan + 2D display

4.1

CT scanner
X-ray tube moved
round patient
Signals from ring
of detectors
processed
to give 2D image
3D obtained by
moving patient
through ring

5.3

X-ray tubes
Continuous spectrum with max
photon energy = eV; lines at specific
wavelengths
Imaging: 30–100 kV, rotating anode
tungsten target: lead beam
definers: aluminium filter to remove
low energy photons
Therapy:
X-ray tubes to 300 kV to treat
disorders at or near surface
MeV X-rays from electron
accelerators

5.1, 6.2

Production and uses
Production
Neutron irradiation
Technetium generator
Uses
Radioactive tracers
α- and β-emitting implants
Gamma therapy
Theory
Biological half-life
Effective half-life

Ultrasonic waves
acoustic impedance
 $Z = \rho c$
reflection coefficient
$$\frac{I_r}{I_0} = \left(\frac{z_2 - z_1}{z_2 + z_1}\right)$$
• probe/skin: use of gel
• internal boundaries

4.1

X-ray imaging detectors
Intensifying screen, Image intensifiers, FTP detectors

5.3

Image enhancement
• lead collimator grid
• contrast medium
• aluminium filter to remove
 low energy photons

5.3

X-rays and matter
attenuation $I = I_0 e^{-\mu x}$
mass attenuation coefficient $\mu_m = \frac{\mu}{\rho}$.
half-value thickness $X_{1/2} = (\ln 2)/\mu$
bone/tissue absorb photons < 30 keV
same: 30–100 keV bone absorbs
more than tissue

5.3

In this option, you can reinforce skills that you developed in the compulsory topics in your A Level Physics course. By studying this option, you will strengthen your competence in the practical and mathematical skills listed below. In addition, you will use measurement data and evaluate results in medical and health contexts. The extension task is intended to boost your skills of written communication, and also lets you work within a group by creating presentations on different topics.

Practical skills

In this option you meet the following skills:

- use of distance scales to measure distances (e.g. measurement of focal length)
- use of optical and electrical equipment safely and correctly in experiments (e.g. light sources, total internal reflection)
- use of a signal generator and a loudspeaker to produce sound waves
- use of an oscilloscope to display and measure waveforms (e.g. sound waves).

Maths skills

In this option you meet the following skills:

- use of the radian and its conversion to and from degrees (e.g. angular resolution)
- calculation of a mean from a set of repeat readings (e.g. lens measurements)
- interpret graphs in terms of relevant physics variables (e.g. frequency response of the ear) or relevant changes (e.g. ECG traces)
- use of sines and cosines (e.g. in refraction calculations)
- use of base 10 logarithms (e.g. the decibel scale) and natural logarithms (e.g. in the absorption of X-rays)
- use of a calculator or a spreadsheet to carry out calculations using data and appropriate formulae, expressing answers to an appropriate number of significant figures and with appropriate units.

Extension task

Make a chart of a medical imaging system you have studied. In your chart, include the type of radiation used in the system and an outline of the main features of the system, including radiation hazards and how they are minimised. Make a presentation of your findings for your class.

Section 9, Chapter 28.3
Option C—Engineering physics

Introduction

This option is one of the five optional sections in the AQA A Level Physics specification. Full course notes and further resources can be found online on Kerboodle.

In this option, you will apply fundamental physical principles to two important areas of engineering—rotating machinery and engines. In chapter 1, you will learn about the dynamics of rotating objects, including calculations of **angular acceleration** and **moment of inertia**. In addition, you will learn how the **kinetic energy** of a rotating object depends on its axis of rotation and its moment of inertia about that axis. Also, you will gain an understanding of why moment of inertia is a key consideration in the design of rotating objects such as **flywheels**, gear wheels, washing machines, and many other everyday objects. In the final part of this chapter, you will learn through practical examples what is meant by angular momentum, and what is meant by conservation of angular momentum.

In chapter 2, the focus of the option moves on to an in-depth study of **thermodynamics**, before applying the laws of thermodynamics to energy transfer by **heat engines** and by **heat pumps**. In studying the first law of thermodynamics, you will gain a deeper awareness of the principle of **conservation of energy**, and how it applies to devices designed to do **work** or to transfer energy by **heating**. In addition, you will develop your understanding of **ideal gases**, in particular how the internal energy of an ideal gas changes when it does work or when it is heated. Such understanding is important when you move on to study the thermodynamics of heat engines such as petrol engines and diesel engines. In these studies, you will learn about the second law of thermodynamics and apply it to heat engines to understand why the efficiency of a heat engine is much less than 100%. The final part of this chapter looks at the general principles of heat pumps and refrigerators and what is meant by the coefficient of performance of these devices.

Working Scientifically

In this option, you will gain an appreciation of how physicists and engineers apply physics principles and knowledge of rotational dynamics and thermodynamics to machines, engines, heat pumps, and refrigerators. The principles and knowledge developed here build on mechanics and gases studied earlier in this course. Design engineers use the principles of physics to understand how their designs will work out in practice. For example, they might test how engine efficiency is affected by the temperature of the surroundings.

Physicists often make discoveries that turn out to have unexpected applications in engineering and other areas. For example, the discovery of materials that are superconductors at temperatures below 77 K, the boiling point of liquid nitrogen, led to the very powerful electromagnets used in brain scanners. Physicists and engineers often apply scientific developments in diverse contexts to invent new devices or improve existing ones.

Much of the work in this option builds on and develops your mathematical skills. In addition, in this option you will develop additional skills that engineers and physicists need, such as the ability to explain practical contexts in terms of relevant physical principles, and to interpret the results of calculations in terms of the context from which the data is derived.

What you already know

From your A Level studies on mechanics and circular motion, you should know that:

- ☐ for an object moving in a circular path, its angular speed is its angular displacement per second
- ☐ the moment of a force F about a point = Fd where d is the perpendicular distance from the point to the line of action of F
- ☐ for an object of mass m moving at velocity v, its kinetic energy = $\frac{1}{2} mv^2$ and its momentum = mv
- ☐ the work done by a force $F = Fd$ where d is the distance moved by the force in the direction of the force.
- ☐ power = $\dfrac{\text{work done}}{\text{time taken}}$.

From your A Level studies on thermal physics, you should know that:

- ☐ the internal energy of an object is the sum of the random distribution of the kinetic and potential energies of the particles
- ☐ temperature (kelvins) ≈ temperature (°C) + 273
- ☐ energy transfer due to heating is given by $Q = mc\Delta T$
- ☐ the pressure on a surface = force per unit area acting normally on the surface
- ☐ for an ideal gas $pV = nRT$
- ☐ the efficiency of a machine = $\dfrac{\text{useful energy transferred}}{\text{energy supplied}} \times 100\%$.

Section 9 Summary

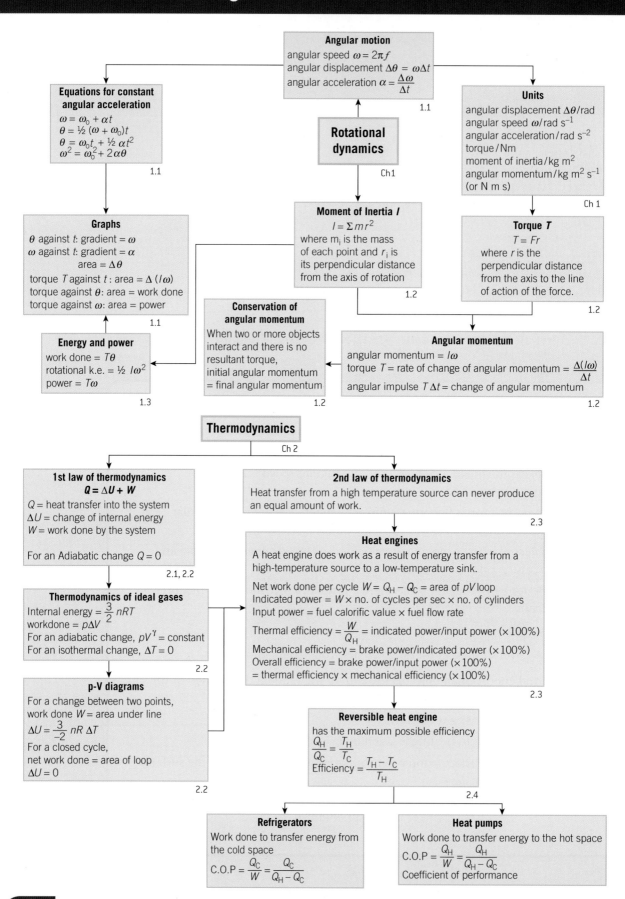

Angular motion
angular speed $\omega = 2\pi f$
angular displacement $\Delta\theta = \omega\Delta t$
angular acceleration $\alpha = \dfrac{\Delta\omega}{\Delta t}$
1.1

Rotational dynamics
Ch1

Equations for constant angular acceleration
$\omega = \omega_0 + \alpha t$
$\theta = \frac{1}{2}(\omega + \omega_0)t$
$\theta = \omega_0 t + \frac{1}{2}\alpha t^2$
$\omega^2 = \omega_0^2 + 2\alpha\theta$
1.1

Units
angular displacement $\Delta\theta$/rad
angular speed ω/rad s^{-1}
angular acceleration/rad s^{-2}
torque/Nm
moment of inertia/kg m^2
angular momentum/kg m^2 s^{-1}
(or N m s)
Ch 1

Graphs
θ against t: gradient = ω
ω against t: gradient = α
 area = $\Delta\theta$
torque T against t : area = $\Delta (I\omega)$
torque against θ: area = work done
torque against ω: area = power
1.1

Moment of Inertia I
$I = \Sigma mr^2$
where m_i is the mass of each point and r_i is its perpendicular distance from the axis of rotation
1.2

Torque T
$T = Fr$
where r is the perpendicular distance from the axis to the line of action of the force.
1.2

Energy and power
work done = $T\theta$
rotational k.e. = $\frac{1}{2} I\omega^2$
power = $T\omega$
1.3

Conservation of angular momentum
When two or more objects interact and there is no resultant torque,
initial angular momentum = final angular momentum
1.2

Angular momentum
angular momentum = $I\omega$
torque T = rate of change of angular momentum = $\dfrac{\Delta(I\omega)}{\Delta t}$
angular impulse $T\Delta t$ = change of angular momentum
1.2

Thermodynamics
Ch 2

1st law of thermodynamics
$Q = \Delta U + W$
Q = heat transfer into the system
ΔU = change of internal energy
W = work done by the system

For an Adiabatic change $Q = 0$
2.1, 2.2

2nd law of thermodynamics
Heat transfer from a high temperature source can never produce an equal amount of work.
2.3

Heat engines
A heat engine does work as a result of energy transfer from a high-temperature source to a low-temperature sink.

Net work done per cycle $W = Q_H - Q_C$ = area of pV loop
Indicated power = $W \times$ no. of cycles per sec $\times$ no. of cylinders
Input power = fuel calorific value $\times$ fuel flow rate

Thermal efficiency = $\dfrac{W}{Q_H}$ = indicated power/input power ($\times 100\%$)

Mechanical efficiency = brake power/indicated power ($\times 100\%$)
Overall efficiency = brake power/input power ($\times 100\%$)
= thermal efficiency $\times$ mechanical efficiency ($\times 100\%$)
2.3

Thermodynamics of ideal gases
Internal energy = $\dfrac{3}{2} nRT$
workdone = $p\Delta V$
For an adiabatic change, pV^γ = constant
For an isothermal change, $\Delta T = 0$
2.2

p-V diagrams
For a change between two points,
work done W = area under line
$\Delta U = \dfrac{3}{-2} nR\,\Delta T$
For a closed cycle,
net work done = area of loop
$\Delta U = 0$
2.2

Reversible heat engine
has the maximum possible efficiency
$\dfrac{Q_H}{Q_C} = \dfrac{T_H}{T_C}$
Efficiency = $\dfrac{T_H - T_C}{T_H}$
2.4

Refrigerators
Work done to transfer energy from the cold space
C.O.P = $\dfrac{Q_C}{W} = \dfrac{Q_C}{Q_H - Q_C}$

Heat pumps
Work done to transfer energy to the hot space
C.O.P = $\dfrac{Q_H}{W} = \dfrac{Q_H}{Q_H - Q_C}$
Coefficient of performance

In this option, you can reinforce skills that you developed in the compulsory topics in your A Level Physics course. By studying this option, you will strengthen your competence in the practical and mathematical skills listed below. In addition, you will develop a deep insight into the application of mathematical knowledge gained in this option to a range of engineering contexts associated with rotational dynamics and thermodynamics. The extension task is intended to boost your skills of written communication, and also lets you work within a group by creating presentations on different topics.

Practical skills

In this option you meet the following skills:

- use of a stopwatch or stopclock to measure the period of rotation of a rotating object
- use of metre rules and vernier callipers to measure distances (e.g. diameters of shafts and wheels)
- use of thermometers or temperature sensors connected safely and accurately to a data logger, in demonstrations and experiments involving heating or cooling (e.g. work done on a gas)
- use of equipment safely and correctly (e.g. in the measurement of the moment of inertia of a flywheel)
- use of a manometer and a pressure gauge to measure the pressure of a gas (e.g. in an adiabatic change).

Maths skills

In this option you meet the following skills:

- use of the radian and its conversion to and from degrees (e.g. angular displacement calculations)
- calculation of a mean from a set of repeat readings (e.g. time period of a rotating object)
- interpret graphs in terms of relevant physics variables (e.g. pressure-volume changes for an ideal gas)
- interpret diagrams such as heat engines in terms of work done and energy transfers
- use a calculator or a spreadsheet to carry out calculations using data and appropriate formulae, expressing answers to an appropriate number of significant figures and with appropriate units
- use appropriate formulae and experimental data from experiments (e.g. moment of inertia of a flywheel).

Extension task

Use the internet to find out more about an application of a device you have studied in this option. For example, you could find out more about flywheels used for energy storage. Or you could find out about the principles of operation of a heat pump or heat engine, such as a diesel engine or a petrol engine. Make a presentation of your findings to give to your class.

Section 9, Chapter 28.4
Option D—Turning points in physics

Introduction

In this option you will study some of the key developments in physics in order to gain a deeper understanding of important concepts, and a greater awareness of the role of experiments in the development of the subject.

In the first chapter, you will learn how the **electron** was discovered and how its properties, such as charge and mass, were determined. The experiments studied in this part of the course will build on your knowledge of the motion of charged particles in electric and magnetic fields. You will learn how physicists measured the **specific charge** of the electron and the value of e, and how they determined that charge only exists in integral multiples of e.

The second chapter develops **wave-particle duality** over several centuries, starting with conflicting theories of light from Newton and Huygens. It explains how experiments, further theories, and conceptual shifts led to the **photon theory of light**. The dual wave-particle nature of matter particles is then extended into de Broglie's hypothesis and a study of how electron microscopes work.

In the final chapter, you will learn about Einstein's theory of **special relativity** and its implications in relation to our understanding of the link between space and time, and the link between mass and energy. You will learn about time dilation, length contraction, and relativistic mass, as well as experimental evidence for the theory of special relativity.

In this option you will learn about the importance of these discoveries in the development of technology. The electronics industry was born from the discovery of the electron. **Quantum theory** underpins modern optoelectronic devices such as the CCD camera and photodiodes as well as electron microscopes. The equation $E = mc^2$ from the theory of special relativity provides the key to understanding nuclear energy and how nuclear reactors work. By studying this option, you will gain a deeper understanding of physics and a real insight into how physics underpins technology.

Working scientifically

In this option, each chapter tells a different story about the link between experiments and theory in physics. In the first chapter, the important experimental discoveries made about electrons and other charged particles led to different models of the atom and on to the establishment of the nuclear model of the atom.

The second chapter shows how experiments were vital in confirming, rejecting, or developing important theories. Young's double slits experiment and Fizeau's measurement of the speed of light established the wave theory of light, which was later called into question by the discoveries on photoelectricity, and then replaced by Einstein's revolutionary photon theory of light. Einstein's idea that a photon is a wave packet of energy $E = hf$ led de Broglie to put forward the hypothesis, subsequently verified by experiment, that matter particles have a wave-like nature and a wavelength $= \dfrac{h}{mv}$.

In the third chapter, an important physical concept, the idea of absolute motion, was challenged by experimenters in a different way when the Michelson-Morley experiment was unable to detect absolute motion. In establishing his theory of special relativity, Einstein assumed that all motion is relative and he developed equations that predicted effects such as relativistic mass and time dilation, since confirmed by experiment.

What you already know

From your previous A Level studies, you should know that:

- [ ] the force on an electron in an electric field of strength $E = eE$
- [ ] the force on an electron moving at speed v in a magnetic field of flux density B is given by $F = Bev$
- [ ] the direction of the force on an electron moving in a magnetic field is at right angles to the motion
- [ ] the kinetic energy of an electron accelerated from rest through a pd $V = eV$
- [ ] $\dfrac{\sin i}{\sin r}$ = constant for a light ray refracted at a boundary
- [ ] Young's fringes are due interference of light waves from the two slits, which causes reinforcement at the bright fringes and cancellation at the dark fringes
- [ ] the electromagnetic spectrum has a continuous range of wavelengths, from less than 10^{-15} m (gamma radiation) to more than 1000 m (radio waves)
- [ ] wave speed = frequency × wavelength
- [ ] for a photon of frequency f, its energy $E = hf$
- [ ] the maximum kinetic energy of an emitted photoelectron $= hf - \phi$
- [ ] the de Broglie wavelength of a matter particle $= \dfrac{h}{mv}$.

Section 9 Summary

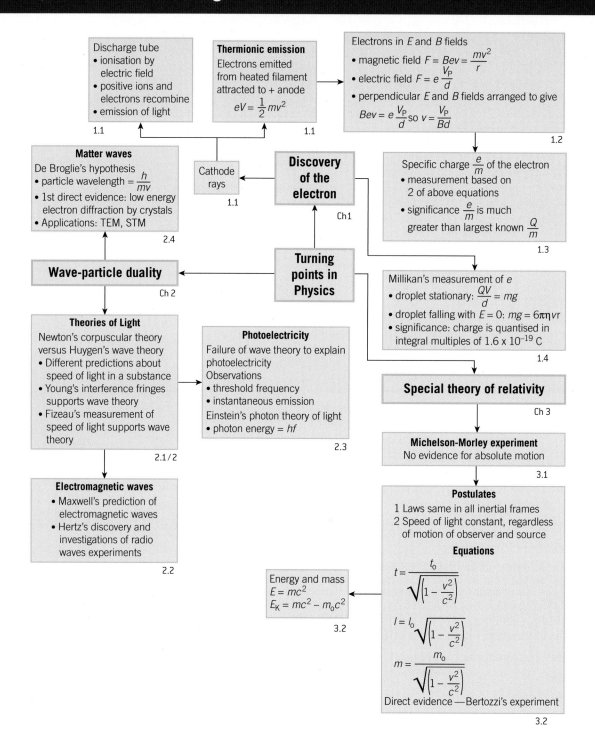

Discharge tube
- ionisation by electric field
- positive ions and electrons recombine
- emission of light

1.1

Thermionic emission
Electrons emitted from heated filament attracted to + anode
$$eV = \frac{1}{2}mv^2$$

1.1

Electrons in E and B fields
- magnetic field $F = Bev = \dfrac{mv^2}{r}$
- electric field $F = e\dfrac{V_P}{d}$
- perpendicular E and B fields arranged to give
$$Bev = e\frac{V_P}{d} \text{ so } v = \frac{V_P}{Bd}$$

1.2

Matter waves
De Broglie's hypothesis
- particle wavelength $= \dfrac{h}{mv}$
- 1st direct evidence: low energy electron diffraction by crystals
- Applications: TEM, STM

2.4

Cathode rays

1.1

Discovery of the electron

Ch1

Specific charge $\dfrac{e}{m}$ of the electron
- measurement based on 2 of above equations
- significance $\dfrac{e}{m}$ is much greater than largest known $\dfrac{Q}{m}$

1.3

Wave-particle duality

Ch 2

Turning points in Physics

Millikan's measurement of e
- droplet stationary: $\dfrac{QV}{d} = mg$
- droplet falling with $E = 0$: $mg = 6\pi\eta vr$
- significance: charge is quantised in integral multiples of 1.6×10^{-19} C

1.4

Theories of Light
Newton's corpuscular theory versus Huygen's wave theory
- Different predictions about speed of light in a substance
- Young's interference fringes supports wave theory
- Fizeau's measurement of speed of light supports wave theory

2.1 / 2

Photoelectricity
Failure of wave theory to explain photoelectricity
Observations
- threshold frequency
- instantaneous emission
Einstein's photon theory of light
- photon energy $= hf$

2.3

Special theory of relativity

Ch 3

Michelson–Morley experiment
No evidence for absolute motion

3.1

Electromagnetic waves
- Maxwell's prediction of electromagnetic waves
- Hertz's discovery and investigations of radio waves experiments

2.2

Postulates
1 Laws same in all inertial frames
2 Speed of light constant, regardless of motion of observer and source

Equations

$$t = \frac{t_0}{\sqrt{\left(1 - \dfrac{v^2}{c^2}\right)}}$$

$$l = l_0\sqrt{\left(1 - \frac{v^2}{c^2}\right)}$$

$$m = \frac{m_0}{\sqrt{\left(1 - \dfrac{v^2}{c^2}\right)}}$$

Direct evidence —Bertozzi's experiment

3.2

Energy and mass
$$E = mc^2$$
$$E_K = mc^2 - m_0c^2$$

3.2

In this option, you can reinforce skills that you developed in the compulsory topics in your A Level Physics course. By studying this option, you will strengthen your competence in the practical and mathematical skills listed below. In addition, you will develop a deep awareness of the way in which experimental methods and mathematical skills are used in the determination of fundamental constants. The extension task is intended to boost your skills of written communication, and also lets you work within a group by creating presentations on different topics.

Practical skills

In this option you meet the following skills:

- use of a stopwatch or stopclock to measure the time of descent of a falling oil droplet
- use of distance scales to measure distances (e.g. electron deflection demonstrations)
- use of electrical equipment safely and correctly in experiments (e.g. in investigations of photoelectricity)
- use of optical equipment safely and correctly in experiments (e.g. refraction and interference of light).

Maths skills

In this option you meet the following skills:

- use of a calculator or a spreadsheet to carry out calculations using data and appropriate formulae, expressing answers to an appropriate number of significant figures and with appropriate units
- use of appropriate formulae and experimental data from experiments (e.g. to determine the specific charge of the electron and fundamental constants including e, h, and c)
- interpret graphs in terms of relevant physics variables (e.g. relativistic mass against speed)
- plot straight line graphs using experimental data given an equation of the form $y = mx + c$ or $y = kx^n$, where n is known, and relate the graph to the physics variables in order to find the gradient and intercept, and understand what the gradient and intercept represent in physics (e.g. in the determination of h from photoelectricty investigations).

Extension task

In this option, you have found out how some of the fundamental constants have been determined, namely c, e, and h. Choose one of these constants and make a presentation summarising the key points in the determination of your chosen constant, and discuss whether or not the three constants could have been determined in a different time sequence.

Section 9, Chapter 28.5
Option E—Electronics

Introduction

In this option, you will learn about **analogue** and **digital** circuits and converters, and how signals are produced, processed, transmitted, and detected. You will also learn about some important electronic components and circuits and what they are used for. The circuits you will study are used with other circuits to make the building blocks of **electronic systems**.

These electronic systems may use analogue and digital circuits at different stages, and the links may differ in each part of a system. The voltage at any point in a digital circuit can only ever be at one of two voltage levels referred to as 'high' and low'. In comparison, the voltage at any point in an analogue circuit can be at any value in the voltage range of the power supply. Examples of analogue circuits include sensor circuits used to monitor physical variables, such as temperature, and amplifier circuits used with microphones and loudspeakers to amplify sounds.

The 5 chapters in this option are organised to introduce you to electronics by considering important electronic devices in chapter 1, before moving to consider analogue and digital circuits and signals in chapter 2. In chapters 3 and 4, you will learn in detail how analogue signals such as radio signals can be filtered and **processed**, and how digital signals can be processed using **logic circuits** and counters. The final chapter brings together the knowledge and understanding gained in the previous chapters by considering the general principles of **data communication**, including how signals carry information and how the different characteristics of each link are matched to the signal characteristics.

Working scientifically

The study of electronics has been at the forefront of our lives ever since radio waves were discovered in 1887. The electronic devices we now rely on have drastically changed the way people communicate, just as the invention of radio transmitters and receivers did many years ago. The electronic chip or integrated circuit was invented over 50 years ago when physicists found out how to make more than one transistor on a single piece of semiconducting material. Since then, successive generations of integrated circuits have been developed, with more and more transistors per chip, to such an extent that the latest everyday computers each have far more computing power than the world's most advanced computers of 50 years ago. At the same time as chip capacity has increased, optical electronic devices such as photodiodes and CCD camera have been developed. These devices became possible as a result of the discovery of new materials, followed by recognition of their potential and intensive development work.

The development of electronics is a story about the potential of scientific discoveries being recognised and exploited to improve our lives. In addition, the story is also about the use by scientists of specialised electronic devices and computers to advance their understanding of complex systems or to make further discoveries. From brain scanners to telescopes that see into the furthest depths of the Universe, electronics have served scientists well.

What you already know

From your previous A Level studies, you should know that:

- ○ resistors are used to limit the current through electronic components such as diodes
- ○ the total pd across two components in series is equal to the sum of the pds across the individual components
- ○ thermistors and LDRs in potential dividers are used to detect changes of temperature and light intensity
- ○ electrons moving across a magnetic field are pushed to one side by the field
- ○ no electric current passes through a fully charged capacitor
- ○ an induced emf opposes the change that causes it
- ○ copper wire and cables can be used to send electrical signals
- ○ light and infrared radiation are used to send signals along optical fibres
- ○ radio waves and microwaves are used to send signals through the atmosphere and through space.

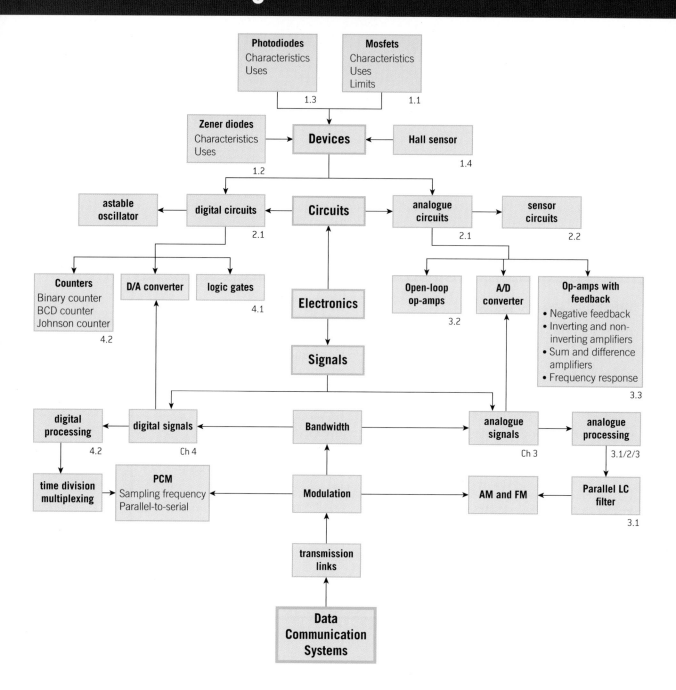

In this option, you can reinforce skills that you developed in the compulsory topics in your A Level Physics course. By studying this option, you will strengthen your competence in the practical and mathematical skills listed below. In addition, you will take an in-depth look at electronic circuits designed or adapted for specific purposes. The extension task is intended to boost your skills of written communication, and also lets you work within a group by creating presentations on different topics.

Practical skills

In this option you meet the following skills:

- use of a stopwatch or stopclock to check a 'seconds' counter
- use of a thermometer to calibrate a temperature sensor
- use of a light sensor and a data recorder to monitor changes in light intensity
- use of an oscilloscope to measure the frequency and amplitude of a waveform
- use of electrical equipment safely and correctly in electronics experiments (e.g. measurement of the frequency response of an operational amplifier).

Maths skills

In this option you meet the following skills:

- interpreting graphs in terms of relevant physics variables (e.g. characteristic curves of a MOSFET)
- convert decimal numbers to binary numbers and vice versa
- interpret and use Boolean algebra in connection with logic systems
- use a calculator or a spreadsheet to carry out calculations using data and appropriate formulae, expressing answers to an appropriate number of significant figures and with appropriate units
- use diagrams to explain communication concepts such as modulation and bandwidth.

Extension task

Choose an electronic device you have studied and make a presentation for your class to summarise its characteristics. Describe an application of its use and how it is used in this application.

Section 10
Skills in A Level Year 2 Physics

Moving on from A Level Year 1

Practical work is just as integral a feature of your A Level Physics Year 2 course as it was in Year 1. Practical activities help you to develop your understanding of important concepts and applications. They also help you to learn how scientists work in practice and to find out how important discoveries have been made and continue to be made. Your Year 2 practical skills embrace the practical skills you have developed in your course in Year 1. You will develop them further in the context of the more demanding knowledge and understanding of the Year 2 part of the course. For example, in your studies on radioactivity, you will learn how to use a Geiger counter when you study how to measure the effect of absorbers on various forms of ionising radiation. In Year 1, you learnt new analytical skills such as using measured data to plot a straight-line graph to confirm a theoretical relationship (e.g., load against extension graph for a spring). In Year 2, you will develop these skills further by using them in the context of more complex theoretical relationships (e.g., measuring capacitor discharge and plotting $\ln V$ against time t to confirm that the decay is exponential).

Assessment of practical skills

- **Indirect** practical skills will be tested *at the end of* your course. To be able to answer written questions on these, you should have successfully completed the set of required practical experiments during your course. The list of practical activities required in A Level Physics Year 2 is shown in Table 1.

- **Direct** practical skills are assessed directly *during* your course. These skills are the practical, technical, and manipulative skills that are essential at A level, such as following instructions, working safely, making measurements, and recording results.

▼ Table 1

Activity	Topic
A Level Physics Year 2 practical experiments	
Investigation of the oscillations of a mass-spring system	18.4
Investigation of the oscillations of a simple pendulum	18.4
Investigation of Boyle's law	20.1
Investigation of Charles' law	20.2
Investigating the charge and discharge of a capacitor	21.3
Investigating the force on a current-carrying wire in a magnetic field	24.1
Using an oscilloscope and a search coil to investigate magnetic flux density	25.2
Investigation of the inverse-square law for gamma radiation	26.3

In any practical assessment at Year 1 or at Year 2 A Level, you are assessed on your ability to:

- plan an investigation,
- carry out practical work,
- analyse data from practical experiments and investigations,
- evaluate the results of practical experiments and investigations.

Planning

- In assessment terms, this might be part of the practical task or activities that you have to carry out or it might be part of the written test in which you have to write about how you would improve an investigation or how you would carry out a different investigation in the same topic area. At Year 2, the topics are more complex than at Year 1, so you can expect assessment questions on planning to be more demanding than at Year 1. For example,

- at Year 1, after carrying out a practical task on how you would investigate the rebound height of a ball released from a certain height, you might be asked to say how you would investigate successive rebound heights to find out if there is a pattern.

- at Year 2, your practical task might involve timing the oscillations of a mass–spring system for different masses. You might then be asked in the written paper to describe how you would investigate the effect of damping on the oscillations of a mass–spring system. Thus the context of the Year 2 'planning' question is more complex than at Year 1 as you have to describe how you would apply damping to a mass–spring system, how you would vary the degree of damping and how you measure the effect of the damping.

Carrying out practical work

At Year 1 and at Year 2, you are required to set up apparatus, use it safely, and possibly rearrange it as part of the procedures you have to follow. You also have to make reliable and accurate measurements and record them, selecting and using the appropriate measuring instruments in the process. The difference between what you have done at Year 1 level and what you have to do at Year 2 lies once again in the more advanced topics in the Year 2 course and in the level of complexity of some of the instruments and procedures you have to use. For example,

- At Year 1, you might be asked to investigate the electrical characteristics of a circuit in a sealed box by setting up a circuit, given the circuit diagram, and making a set of measurements of current and pd in the forward and reverse directions through the box. In addition, you would be asked to record your measurements in a table and, in this case, to plot a graph of your results.

- At Year 2, you might be asked to set up a circuit to charge a capacitor and discharge it through a resistor and then make measurements of the capacitor pd at measured times as it charges or discharges. You would have to decide on the number of measurements to be made, the time interval between successive measurements and the exact procedure of how to make the measurements. The timing element in this investigation adds a further level of complexity beyond Year 1 A Level.

Analysing data

At Year 1 and at Year 2, you generally need to process the data (e.g., calculation of mean values) and plot a graph. In many investigations, the graph has to be related to an equation to check a theoretical relationship or to measure a physical property.

- At Year 1, the theoretical equations you meet in the Year 1 part of the course are generally linear equations (e.g., $s = vt$ for a distance–time graph of an object falling in a viscous fluid at terminal speed) or sometimes simple non-linear equations (e.g. $s = \frac{1}{2}gt^2$ for a distance–time graph of an object falling in air).

- At Year 2, there may be more variables than at Year 1, and the processing of the data might be more complicated, for example the calculation of log values. The graph plotting exercise requires the same skills as at Year 1, but the analysis might be more complicated. For example, you would probably be expected to plot a log graph in a capacitor discharge experiment and use it to determine the time constant RC of the discharge circuit or you might be expected to plot a log–log graph as explained in 30.3 (see Figure 2 in 30.3 and the related text) to establish if two variables x and y relate to each other according to an equation of the form $y = kx^n$.

Evaluating your results

At Year 1 and Year 2 this involves discussing the strength of your conclusions. From your work at Year 1, you should now know how to:

- distinguish between systematic errors (including zero errors) and random errors,
- understand in respect of measurements what is meant by accuracy, uncertainty, sensitivity, linearity, reliability, precision, and validity,
- estimate experimental uncertainties for each measured quantity and use them as outlined on the next page to estimate the overall percentage uncertainty of a result determined from the measured quantities.

If you are unsure about the meaning of any of the above terms, look them up in the glossary of practical terms at the end of this book.

Using experimental uncertainties

As explained in Topic 14.4

- if two measurements are added or subtracted, the uncertainty of the result is the sum of the uncertainties of the two measurements,
- if a quantity in a calculation is raised to a power n, the percentage uncertainty is increased n times.

In addition, if two or more quantities are multiplied or divided by each other in a calculation, the overall percentage uncertainty in the result is the sum of the uncertainties of each quantity. For example, if the % uncertainty in a resistance R is 5% and in a capacitance C is 4%, the % uncertainty in RC is 9%.

Clearly, the more demanding nature of the Year 2 topics compared with Year 1 makes the evaluation of a practical investigation more demanding, but the same general features as outlined above still apply. For example,

- at Year 1, you might be asked to investigate the motion of an object sliding down a slope by measuring the time taken by the object to slide different measured distances down the slope and plotting a distance–time^2 graph. In your evaluation, you might be asked to use a certain measurement (e.g., the smallest) to estimate the percentage uncertainty in the measurement of the dependent and independent variables so you could compare them and discuss how to improve the investigation.

- at Year 2, you might be asked to investigate the oscillations of a mass–spring system and to use the measurements to plot a graph of time period^2 against mass and show the uncertainty in each measurement on the graph, as outlined in Topic 14.4. This would enable you to draw best-fit straight lines with a maximum and minimum gradient to enable you to determine the uncertainty in the gradient as well as its mean value. In this case, there is clearly more to discuss in terms of the closeness of the best-fit lines to the data than at Year 1, where only one best-fit line is drawn.

> **Study tip**
>
> These examples serve to illustrate the point that at Year 2, your practical skills build on your Year 1 skills so you need to continue to practise all the practical skills you met at Year 1. You need to be aware as you study each Year 2 topic that the topics are generally harder than at Year 1 and that consequently the practical work is more demanding, as outlined here.

Further practical activities

Below is a list of optional practical activities that will promote your appreciation and understanding of A Level Physics.

Activity	Topic
Investigating the variation of centripetal force with frequency	17.2
Investigating how displacement varies with time for an oscillating object	18.1
Investigating damped oscillations	18.5
Observing forced oscillations and resonance	18.6
Measuring the specific heat capacity of a liquid	19.2
Measuring the specific latent heat of fusion of water	19.3
Observing the path of a beam of electrons in a uniform electric field	22.2
Measuring the capacitance of a capacitor	21.1
Investigating the parallel-plate capacitor	21.4
Measuring the relative permittivity of a dielectric	21.4
Observing the path of a beam of electrons in a uniform magnetic field	24.2
Using an oscilloscope to measure ac waveforms	25.4
Using a cloud chamber to observe alpha and beta radiation	26.2
Investigating the characteristics of a Geiger tube	26.2
Measuring background radioactivity using a Geiger counter	26.2
Investigating the absorption of beta radiation by aluminium	26.2

At Year 1 A Level Physics, you should have learnt how to:

- use vernier callipers and micrometers to measure small distances,
- use appropriate analogue apparatus to measure angles, length and distance, volume, force, temperature, and pressure,
- use appropriate digital instruments to measure mass, time, current, resistance, and voltage,
- use a stopwatch or light gates for timing,
- use methods to increase accuracy of measurements, such as repeating and averaging timings or using a fiduciary marker, set square, or plumb line,
- correctly construct circuits from own or given circuit diagrams using dc power supplies, cells, and a range of circuit components, including those where polarity is important,
- generate and measure waves, using a microphone and loudspeaker, or ripple tank, or vibration transducer, or microwave / radio wave source,
- use a suitable light source to investigate characteristics of light, including interference and diffraction,
- use a data logger with a variety of sensors to collect data, or use software to process data.

During A Level Year 2, as outlined in Topic 29.1, in addition to being able to use instruments and techniques that you used in A Level Year 1, you are also expected to be able to use instruments as described below that are more complex. Such instruments include the oscilloscope, the Geiger–Müller tube with a scaler counter or ratemeter, and data loggers and/or light gates in some Year 2 investigations (e.g., oscillations, capacitor discharge). You should also know how to time multiple oscillations and how to avoid parallax errors when reading a scale. Two further instruments, the travelling microscope (see later) and the spectrometer, are also described in this book – although you do *not* need to know how to use these in your A Level Physics course, they give more experience of using very accurate instruments.

An oscilloscope

An oscilloscope is used to display waveforms and to measure pds and time intervals. You will have probably used an oscilloscope in the Year 1 part of your A Level course when you studied sound waves. In Year 2 topics such as capacitor discharge, you will use an oscilloscope to measure the pd across a discharging capacitor. When you use an oscilloscope, you should assume that the control dials for its time base and voltage gain are calibrated accurately. However,

- always check if the oscilloscope has a variable control for either the time base or the voltage gain in addition to the fixed settings of each control dial. If so, you need to ensure that the variable control is at the correct setting (e.g., fully clockwise) for the calibration figure for each of the fixed settings to apply. See Using an oscilloscope in Topic 4.6 if you need to revise the use of an oscilloscope.

Synoptic link

If you are unsure about how to use any of the listed instruments, see Topic 14.3, Everyday physics instruments, and consult your teacher.

Synoptic link

The spectrometer was covered in Topic 5.7, The diffraction grating.

▲ **Figure 1** *The oscilloscope*

- make sure if you are measuring direct pds that the oscilloscope input is set for direct pd measurements rather than for ac measurements. Likewise, if you are measuring an ac waveform, you should check the input is set for ac measurements rather than dc measurements.

In addition, when measuring

- an ac waveform, ensure the y-gain is adjusted so the vertical height of the waveform is as large as possible with the full waveform from top to bottom on the screen. When measuring a time period, ensure several cycles are displayed across the screen and that you measure across as many cycles as possible to reduce experimental uncertainty.

- dc potentials, for example in a capacitor discharge experiment, ensure the zero reading is correct for zero input pd and check it has not drifted during the investigation.

A data logger

A data logger enables routine or remote measurements to be made as well as measurements over very long or short time scales. Electronic sensors connected to a data logger are necessary to record the variation of a physical property such as temperature. Ammeter and voltage sensors are necessary to measure currents and potential differences.

Data loggers vary considerably in complexity and ease of use. Assuming the data logger and sensors are set up, before using a data logger, you may need to choose:

- the most appropriate time scale for the recording,
- the time interval between successive recordings (or the number of recordings per second/minute/hour),
- the most appropriate range of each sensor.

If a recording is too fast or too long or the sensors are out of range, the recording should be repeated if possible.

Most data loggers will be linked to computers which are loaded with appropriate software for recording, processing and/or plotting graphs of the results. You may need to print a graph out if you intend to use it to measure, for example, the gradient if it is a straight-line graph. However, the computer software may do such measurements for you.

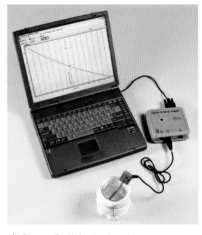

▲ **Figure 2** *Using a data logger*

Light gates

Light gates are used with a computer or a data logger or timer to remove some of the random errors associated with personal judgements when a moving object passes a certain position, for example, if you have to time an object to move from rest through a certain distance.

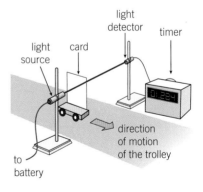

▲ **Figure 3** *Using a light gate*

The effect of using light gates should be to reduce the range of the readings for a given measurement. However, light gates may not be suitable for every experiment in which a moving object has to be timed. For example, the time period of an oscillating object that repeatedly moves backwards and forwards through a light gate could only be timed for one half-cycle of the object's motion, corresponding to the object moving through the light gate in one direction to start the timing and then in the opposite direction to stop the light

Study tip

After starting the counter and the timer simultaneously, observe the timer and stop the counter exactly when the time interval has elapsed. Don't try to stop the timer at the same time as stopping the counter and watching the timer.

+ A travelling microscope

This is a microscope on an adjustable platform that can be moved vertically or horizontally by turning an adjustment screw. The platform is fitted with a horizontal and a vertical vernier scale, so its horizontal or vertical position can be measured to within ±0.1 mm. A travelling microscope would be used, for example, to measure the internal diameter of a glass tube (e.g., 1 mm bore), because a micrometer or a conventional vernier could not access the internal surface of the tube. Before use, a travelling microscope should be levelled using a spirit level in two perpendicular directions so that its platform is horizontal. Otherwise, there could be a systematic error in the measurements.

gate. The light gate would need to be exactly at the centre of the oscillations, otherwise the timing would not be exactly one half-cycle. Repeated measurements of one half-cycle could be made to give a more reliable mean value, and this might give better results than using a stopwatch if the oscillations are too fast to time manually.

An ionising radiation detector

The ionising radiation detector that you use will probably be a Geiger–Müller tube. This may be used with a scaler counter, which counts the number of ionising particles that enter the tube or it may be used with a ratemeter which gives a read-out of the count rate (i.e., number of counts per unit time) of the particles entering the tube. The tube pd must be set at its operating pd which is normally in excess of 300 V. The number of counts in a certain time interval is measured by setting the counter to zero, then starting the counter and stopping it after a certain time.

Figure 4b shows how the count rate varies with the tube pd. The operating pd corresponds to the plateau of the graph sufficiently far from the minimum pd necessary for the tube to operate (i.e., the threshold pd) as to be unaffected by random fluctuations in the tube pd.

When using the tube with a scaler counter, the number of counts in a given time (e.g., 100 s) should be measured several times to give a mean value of the count rate (i.e., counts per second). The bigger the total number of counts, the smaller the uncertainty in the measurement. If the time interval is too short, random errors that may occur in starting and stopping the counter could be more significant than if a longer time interval were used. If the time interval is too long, it would be difficult to tell if the activity of the source is decreasing or if an error in starting or stopping the timer has occurred.

When using the tube with a ratemeter, ensure the ratemeter is set on the range which gives the largest reading. For example, if the range dial has three positions, 1, 10, and 100 counts per second, the range may need to be set at 1 count per second if the reading is very small on the '10' and '100' positions.

When using either a scaler counter or a ratemeter, remember to measure the background count rate and subtract it from the measurements made when the source is present.

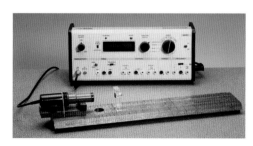

a *a Geiger–Müller tube connected to a scaler counter*

▲ **Figure 4** *Using a Geiger–Müller tube*

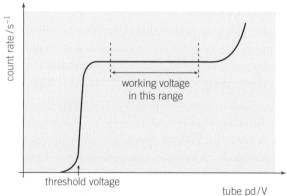

b *a graph of count rate against tube pd*

30 Mathematical skills in Year 2 Physics
30.1 Trigonometry

In this chapter, you will consider only what you need to know at A Level Physics Year 2. If you need to check any of the mathematical skills that you need for your Year 1 course, you can use Chapter 16 More on mathematical skills, and/or use the Kerboodle exercises online.

Angles and arcs

The radian
The radian (rad) is a unit used to express or measure angles. It is defined such that 2π radians = $360°$.

When using a calculator to work out sines, cosines, tangents, or the corresponding inverse functions, always check that the calculator is in the correct 'angle' mode. This is usually indicated on the display by 'deg' for degrees or 'rad' for radians. Your calculations will be incorrect if you work in one mode when you should be working in the other mode. For example, check for yourself that sin $30° = 0.5$, whereas sin $(30$ rad$) = -0.988$. You also need to know how to change from one mode to the other; read your calculator manual or ask you teacher if you can't do this.

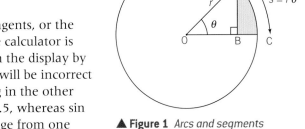
▲ **Figure 1** *Arcs and segments*

Arcs and segments
Consider an arc of length s on the circumference of a circle of radius r, as shown in Figure 1. The angle θ, in degrees, subtended by the arc to the centre of the circle is given by the equation

$$\theta / \text{degrees} = \frac{s}{2\pi r} \times 360$$

Because 2π radians = $360°$, applying this conversion factor to the above equation for θ gives

$$\theta / \text{radians} = \frac{s}{r}$$

Rearranging this equation gives

$$\textbf{arc length } s = r\theta$$

where θ is the angle subtended in radians.

> **Note**
>
> Note that for $s = r$, $\theta = 1$ rad
>
> $$\left(= \frac{360}{2\pi} = 57.3°\right)$$

The small angle approximation
For angle θ less than about $10°$,

$$\sin \theta \approx \tan \theta \approx \theta \text{ in radians, and}$$

$$\cos \theta \approx 1$$

To explain these approximations, consider Figure 1 again. If angle θ is sufficiently small, then the segment OAC will be almost the same as triangle OAB, as shown in Figure 2.

- AB $\approx$ arc length s so $\sin \theta = \dfrac{\text{AB}}{\text{OA}} \approx \dfrac{s}{r} = \theta$ in radians, and therefore, $\sin \theta \approx \theta$ in radians.

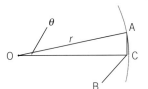
▲ **Figure 2** *The small angle approximation*

- OB ≈ radius r, so $\tan\theta = \dfrac{AB}{OB} \approx \dfrac{s}{r} = \theta$ in radians, and therefore, $\tan\theta \approx \theta$ in radians.

- Also, $\cos\theta = \dfrac{OB}{OA} \approx \dfrac{r}{r} = 1$, and therefore, $\cos\theta \approx 1$

Use a calculator to prove for yourself that for $\sin 10° = 0.1736$, $\tan 10° = 0.1763$ and $10° = 0.1745$ rad. Also, $\cos 10° = 0.9848$. So the small angle approximation is almost 99% accurate up to 10°.

The small angle approximation is used to show that the time period of a simple pendulum of length L is given by the formula $T = 2\pi\sqrt{\dfrac{L}{g}}$, provided the maximum angular displacement of the pendulum from equilibrium is less than about 10°. See Topic 18.4 for more about this equation.

Sine and cosine curves

Figure 3 shows how $\sin\theta$ and $\cos\theta$ change as θ increases. Notice that $\sin\theta \approx \theta$ and $\cos\theta \approx 1$ up to about 10°. The general shape of a cosine wave is the same that as that of a sine wave so we refer to them both as 'sinusoidal' waveforms. In addition, notice that:

- the sine wave starts at zero and rises to a maximum from $\theta = 0$ to $\theta = \frac{1}{2}\pi$ rad (= 90°) whereas the cosine wave starts at +1 and falls to zero from $\theta = 0$ to $\theta = \frac{1}{2}\pi$ rad (= 90°).

- the gradient of the sine wave is zero where the cosine wave is zero (i.e. where it crosses the horizontal axis) and the gradient of the cosine wave is zero where the sine wave is zero.

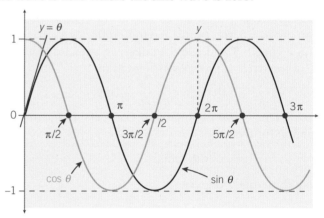

▲ **Figure 3** *Sine and cosine curves*

Equations that describe sine waves and cosine waves are often used to calculate, for example, the displacement of an oscillating particle at a certain time or of a wave at a particular position along the wave.

For example, consider the displacement x of a particle on a spring oscillating vertically in simple harmonic motion, as shown in Figure 4a. Let f represent its frequency and A its amplitude. Its displacement varies sinusoidally between a maximum value $+A$ and a minimum value $-A$. Also, suppose $x = +A$ at time $t = 0$. In other words, the object is held above the equilibrium position at displacement $x = +A$ and released at time $t = 0$.

- Its displacement–time curve will therefore be a cosine wave, as shown in Figure 4b where its time period $T = \dfrac{1}{f}$

- At time t later, the particle will have gone through ft cycles of oscillation, corresponding to $\theta = 2\pi ft$ radians. Hence its displacement at time t is given by $x = A \cos 2\pi ft$

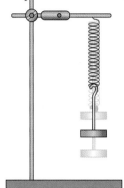

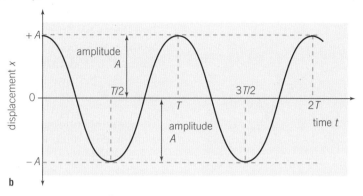

▲ **Figure 4 a** *An object oscillating on a spring*   **b** *Displacement–time curve for x = A cos 2πft*

Summary questions

1 a Convert the following angles from degrees into radians and express your answer to one further significant figure than in each question:

   i 30°,

   ii 50°,

   iii 120°,

   iv 230°,

   v 300°.

b Convert the following angles from radians into degrees and express your answer to one further significant figure than in each question:

   i 0.10 rad,

   ii 0.50 rad,

   iii 1.20 rad,

   iv 2.50 rad,

   v 6.00 rad.

2 a Measure the diameter of a 1p coin to the nearest millimetre. Calculate the angle subtended at your eye, in degrees, by a 1p coin held at a distance of 50 cm from your eye.

b i Estimate the angular width of the Moon, in degrees, at your eye by holding a millimetre scale at 50 cm from your eye and measuring the distance on the scale covered by the lunar disc.

   ii The diameter of the Moon is 3500 km. The average distance to the Moon from the Earth is 380 000 km. Calculate the angular width of the Moon as seen from the Earth and compare the calculated value with your estimate in **b i**.

3 a Use the small angle approximation to calculate $\sin \theta$ for $\theta =$

   i 2.0°,

   ii 8.0°.

b Show that the small angle approximation for $\sin \theta$ is more than 99% accurate for $\theta = 10°$.

4 Use your calculator to find

   i $\sin \theta$,

   ii $\cos \theta$ for the following values of θ:

a 0.1 rad,

b 10°,

c 45°,

d 0.25π rad.

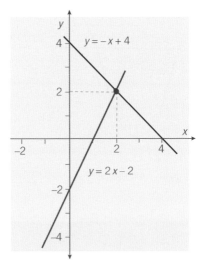

▲ **Figure 1** *A graphical solution*

Linear simultaneous equations

Two linear equations with two variable quantities, x and y, in each can be solved to find the values of x and y. Such a pair of equations are referred to as **simultaneous equations** because they have the same solution. They are described as **linear** because they contain terms in x and y and do not contain any higher order terms such as x^2 or y^2.

The general equation for a straight-line graph is $y = mx + c$, as explained in Topic 16.4. Two straight lines on a graph can be represented by two such equations. Provided the two lines are not parallel to one another, they cross each other at a single point. The coordinates of this point are the values of x and y that fit both equations. In other words, these coordinates are the solution of a pair of simultaneous equations representing the two straight lines..

The graph approach to finding the solution of a pair of simultaneous equations is shown in Figure 1 and is described in Topic 16.4. However, plotting graphs takes time and is not as accurate as a systematic algebraic method. This method can best be explained by considering an example, as follows

$$2x - y = 2 \qquad \text{(equation 1)}$$

$$x + y = 4 \qquad \text{(equation 2)}$$

Make the coefficient of x the same in both equations by multiplying one or both equations by a suitable number. In the above equation, this is most easily achieved by multiplying equation 2 throughout by 2 to give $2x + 2y = 8$.

The two equations to be solved are now

$$2x - y = 2 \qquad \text{(equation 1)}$$

$$2x + 2y = 8 \qquad \text{(modified equation 2)}$$

Subtracting modified equation 2 from equation 1 gives

$$(2x - y) - (2x + 2y) = 2 - 8$$

$$\text{therefore, } -y - 2y = -6$$

$$-3y = -6$$

$$y = \frac{-6}{-3} = 2$$

Substituting this value into equation 1 or equation 2 enables the value of x to be determined. Using equation 2 for this purpose gives $x + 2 = 4$, hence $x = 4 - 2 = 2$.

The solution of the two equations is therefore $x = \mathbf{2}$, $y = \mathbf{2}$.

Linear simultaneous equations with two unknown quantities can arise in several parts of the A level physics course, for example

- $v = u + at$ in kinematics (see Topic 7.3),
- $V = \varepsilon - Ir$ in electricity (see Topic 13.3),
- $E_{kmax} = hf - \phi$ (see Topic 3.2).

Hint

You can check this using the equations (e.g. $x = \mathbf{2}$, $y = \mathbf{2}$ is also a solution for $x + y = \mathbf{4}$).

The quadratic equation

Any quadratic equation can be written in the form $ax^2 + bx + c = 0$, where a, b and c are constants. The general solution of the quadratic equation $ax^2 + bx + c = 0$ is

$$x = \frac{-b \pm \sqrt{(b^2 - 4ac)}}{2a}$$

Note that every quadratic equation has two solutions, one given by the + sign before the square root sign in the above expression, and the other given by the − sign. For example, consider the solution of the equation $2x^2 + 5x - 3 = 0$.

As $a = 2$, $b = 5$ and $c = -3$, then the solution is

$$x = \frac{-5 \pm \sqrt{\left(5^2 - \left(4 \times 2 \times -3\right)\right)}}{2 \times 2} = \frac{-5 \pm \sqrt{49}}{4} = \frac{-5 \pm 7}{4} = +0.5 \text{ or } -3$$

A graph of $y = 2x^2 + 5x - 3$ is shown in Figure 2. Note that the two solutions above are the values of the x-intercepts, which is where $y = 0$.

Quadratic equations occur in A level physics where a formula contains the square of a variable. The equation $s = ut + \frac{1}{2}at^2$ for displacement at constant acceleration is a direct example. Other examples can arise indirectly. For example, suppose the pd across a certain type of component varies with current I according to the equation $V = kI^2$. In a circuit with a battery of negligible internal resistance and a resistor of resistance R, the battery pd, $V_0 = IR + kI^2$. Given values of R, k and V_0, the current could be calculated using the solution for the quadratic equation with $a = k$, $b = R$ and $c = -V_0$.

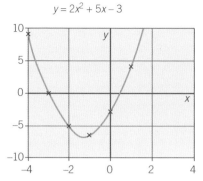

▲ **Figure 2** $y = 2x^2 + 5x - 3$

Summary questions

1 Solve each of the following pairs of simultaneous equations.

  a $3x + y = 6$; $2y = 5x + 1$

  b $3a - 2b = 8$; $a + b = 2$

  c $5p + 2q = 18$; $q = 2p$

2 Use the data and the given equation to write down a pair of simultaneous equations and so determine the unknown quantities in each case:

  a For $v = u + at$, when $t = 3.0$ s, $v = 8.0$ m s^{-1} and when $t = 6.0$ s, $v = 2.0$ m s^{-1}. Determine the values of u and a.

  b For $\varepsilon = IR + Ir$, when $R = 5.0\ \Omega$, $I = 1.5$ A and when $R = 9.0\ \Omega$, $I = 0.9$ A. Determine the values of ε and r.

3 Solve each of the following quadratic equations.

  a $2x^2 + 5x - 3 = 0$

  b $x^2 - 7x + 8 = 0$

  c $3x^2 + 2x - 5 = 0$

4 Use the data and the given equation to write down a quadratic equation and so determine the unknown quantity in each case:

  a $s = ut + \frac{1}{2}at^2$, where $s = 20$ m, $u = 4$ m s^{-1} and $a = 6$ m s^{-2}; find t.

  b $P = V^2 \dfrac{R}{(R + r)^2}$, where $P = 16$ W, $V = 12$ V, $r = 2.0\ \Omega$; find R.

Logarithms and powers

Any number can be expressed as any other number raised to a particular power. You can use the y^x key on a calculator to show, for example that $8 = 2^3$ and $9 = 2^{3.17}$. In these examples, 2 is referred to as the base number and is raised to a different power in each case to generate 8 or 9. The power is defined as the **logarithm** of the number generated.

In general, for a number $n = \mathbf{b}^p$ where b is the base number, then $p = \mathbf{log_b}\, n$ where $\log_b$ means a logarithm using b as the base number.

Note:
$\log_b(b^p) = p$ as $b^p = n$ and $\log_b n = p$.

Applying the general definition above gives the following rules to remember when working with logs:

1 **For any two numbers *m* and *n*,**

$$\log_b (nm) = \log_b n + \log_b m$$

Let $p = \log_b n$ and let $q = \log_b m$ so $n = b^p$ and $m = b^q$

therefore, $nm = b^p b^q = b^{p+q}$ so $\log_b(nm) = p + q = \log_b m + \log_b n$

2 **For any two numbers *m* and *n*,**

$$\log_b \frac{n}{m} = \log_b n - \log_b m$$

Let $p = \log_b n$ and let $q = \log_b m$ so $n = b^p$ and $m = b^q$.

Therefore $\dfrac{1}{m} = \dfrac{1}{b^q} = b^{-q}$

so, $\dfrac{n}{m} = b^p b^{-q} = b^{p-q}$ so $\log_b \left(\dfrac{n}{m}\right) = p - q = \log_b n - \log_b m$

3 **For any number *m* raised to a power *p*,**

$$\log_b (m^p) = p \log_b m$$

This is because $m^p = m$ multiplied by itself p times.

$$\xleftarrow{\hspace{2cm}} p \text{ terms} \xrightarrow{\hspace{2cm}}$$

Therefore $\log_b m^p = \{\log_b m + \log_b m + \dots + \log_b m\} = p \log_b m$

Base 10 logarithms and natural logarithms are used extensively in physics and are explained below.

Base 10 logarithms

Base 10 logs are written as $\log_{10}$ or lg (or sometimes incorrectly as log).

For example,
- $100 = 10^2$ so $\log_{10} 100 = 2$
- $50 = 10^{1.699}$ so $\log_{10} 50 = 1.699$
- $10 = 10^1$ so $\log_{10} 10 = 1$
- $5 = 10^{0.699}$ so $\log_{10} 5 = 0.699$

The base 10 examples illustrate the product rule for logs

(i.e. $\log_b nm = \log_b n + \log_b m$)

since $\log_{10} 50 = \log_{10} 5 + \log_{10} 10 = 0.699 + 1 = 1.699$.

Uses of base 10 logs

In graphs where a **logarithmic scale** is necessary to show the full range of a variable that covers a very wide range, as shown in Figure 2. Notice in Figure 2 that the frequency increases by ×10 in equal intervals along the horizontal axis.

In data analysis where a relationship between two variables is of the form $y = kx^n$ and k and n are unknown constants. As explained on the previous page, for an equation of the form $y = kx^n$, then

$$\log_{10} y = \log_{10} k + \log_{10} x^n = \log_{10} k + n \log_{10} x$$

The graph of $\log_{10} y$ (on the vertical axis) against $\log_{10} x$ is therefore a straight line of gradient n with an intercept equal to $\log_{10} k$. Hence n and k can be determined.

In certain formulae where a ×10 scale is used. For example, the gain of an amplifier in decibels (dB) is a ×10 scale defined by the formula

$$\text{voltage gain / dB} = 10 \log_{10} \frac{V_{out}}{V_{in}}$$

where V_{out} and V_{in} are the output and input voltages, respectively. If $V_{out} = 50V_{in}$, the gain of the amplifier is 17 dB (= $10 \log_{10} 50$).

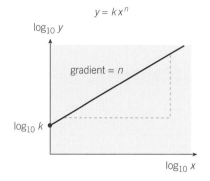

▲ **Figure 1** Using logs to test $y = kx^n$

> **Hint**
>
> Note that natural logs could be used in Figure 1 instead of base 10 logs; the gradient would still be n but the y-intercept would be $\ln k$.

> **Hint**
>
> The Richter scale for earthquakes is another example. A 'Richter scale 8' earthquake is ten times as powerful as a 'Richter scale 7' earthquake.

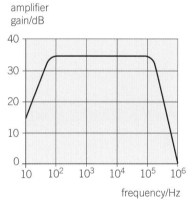

▲ **Figure 2** Logarithmic scales

Natural logarithms

Natural logs are written as $\log_e$ or $\ln$, where e is the exponential number used as the base of natural logarithms and is equal to 2.718. For example,

- $2.718 = e^1$ so $\ln 2.718 = 1$
- $7.389 = e^2$ so $\ln 7.389 = 2$
- $20.009 = e^3$ so $\ln 20.009 = 3$
- In general, for any number n, if p is such that $n = e^p$, then $\ln n = p$

Uses of natural logarithms

Natural logs are used in the equations for radioactive decay (Topic 26.5) and capacitor discharge (Topic 23.3) or any other process where the rate

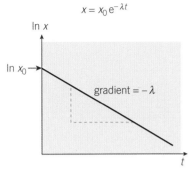

$x = x_0 e^{-\lambda t}$

▲ **Figure 3** *Using logs to test* $x = x_0 e^{-\lambda t}$

of change of a quantity is proportional to the quantity itself. For example, the rate of decrease of pd across a capacitor discharging through a resistor is proportional to the pd across the capacitor. This type of change is described as an exponential decrease because the quantity decreases by the same factor in equal intervals of time.

Applying the general rule (that if p is such that $n = e^p$, then $p = \ln n$) to the equation $x = x_0 e^{-\lambda t}$ gives $\ln x = \ln x_0 - \lambda t$.

Therefore, a graph of $\ln x$ (on the vertical axis) against t (on the horizontal axis) is a straight line with a gradient equal to $-\lambda$ and a y-intercept equal to $\ln x_0$.

Comparing the equation for capacitor discharge $Q = Q_0 e^{\frac{-t}{RC}}$ with the radioactive decay equation $N = N_0 e^{-\lambda t}$,

- for capacitor discharge, $\ln Q = \ln Q_0 - \dfrac{t}{RC}$ so a graph of $\ln Q$ (on the vertical axis) against t is a straight line which has a gradient $\dfrac{-1}{RC}$ and $\ln Q_0$ as its y-intercept,

- for radioactive decay, $\ln N = \ln N_0 - \lambda t$ so a graph of $\ln N$ (on the vertical axis) against t is a straight line which has a gradient $-\lambda$ and $\ln N_0$ as its y-intercept.

Summary questions

1 a Use your calculator to work out

    i $\log_{10} 3$

    ii $\log_{10} 15$

  b Use your answers in **a** to work out

    i $\log_{10} 45$

    ii $\log_{10} 5$

2 The gain of an amplifier, in decibels, is given by the formula $10 \log_{10} \dfrac{V_{out}}{V_{in}}$.

  a Calculate the gain, in decibels (dB), for

    i $V_{out} = 12V_{in}$

    ii $V_{out} = 5V_{in}$

  b Show that the gain, in decibels, of an amplifier for which $V_{out} = 60V_{in}$ is equal to the sum of the gain in **a i** and the gain in **a ii** above.

3 Write down the gradient and the y-intercept of a line on a graph representing the equation $\log_{10} y = n \log_{10} x + \log_{10} k$ for

  a $y = 3x^5$

  b $y = \frac{1}{2}x^3$

  c $y = x^2$

4 a Use your calculator to work out

    i $\ln 3$

    ii $\ln 15$

  b Use your answers in **a** to work out

    i $\ln 45$

    ii $\ln 5$

Rates of change

Consider a variable quantity y that changes with respect to a second quantity x as shown in Figure 1. The gradient of the curve at any point is the rate of change of y with respect to x at that point. This can be worked out from the graph by drawing a tangent to the curve at that point and measuring the gradient of the tangent. The rate of change of y with respect to x at point P is equal to the gradient of the tangent to the curve at P which is $\frac{\Delta y}{\Delta x}$.

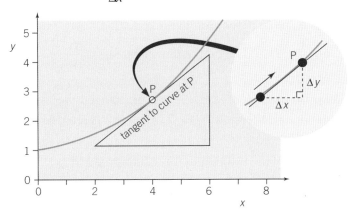

▲ **Figure 1** *Tangents and curves*

The rate of change of y with respect to x can be worked out algebraically if the equation relating y and x is known. This process is known as **differentiation**. For example,

- for $y = x^2$, then increasing x to $x + \Delta x$ increases y to $y + \Delta y$ where $y + \Delta y = (x + \Delta x)^2$

  Multiplying out $(x + \Delta x)^2$ gives $y + \Delta y = x^2 + 2x\Delta x + \Delta x^2$

  Subtracting $y = x^2$ from this equation gives $\Delta y = 2x\Delta x + \Delta x^2$

  Dividing by Δx therefore gives $\frac{\Delta y}{\Delta x} = \frac{2x\Delta x + \Delta x^2}{\Delta x} = 2x + \Delta x$

  Therefore, as $\Delta x \longrightarrow 0$, $\frac{\Delta y}{\Delta x} \longrightarrow 2x$ which is therefore the formula for the gradient at x.

  This is written $\frac{dy}{dx} = 2x$, where $\frac{dy}{dx}$ is the mathematical expression for the rate of change of y with respect to x.

- for the general expression $y = x^n$, it can be shown that $\frac{dy}{dx} = nx^{n-1}$

For example, if $y = 3x^5$, then $\frac{dy}{dx} = 15x^4$

> **Hint**
>
> Note that differentiation of functions is not required in the AS or the A2 specification. The information on differentiation is provided to help you develop your understanding of exponential change in the next section. But from your studies of simple harmonic motion (Chapter 3), you should know that the gradient of $\sin x$ is $\cos x$ and the gradient of $\cos x$ is $-\sin x$. In other words, differentiating $\sin x$ gives $\cos x$ and differentiating $\cos x$ gives $-\sin x$.

Exponential change

Exponential change happens when the change of a quantity is proportional to the quantity itself. Such a change can be an increase (i.e. exponential growth) or a decrease (i.e. exponential decay).

In both cases, the quantity changes by a fixed proportion in equal intervals of time. The Year 2 specification requires knowledge and understanding of the equations and graphs for exponential decrease in radioactive decay and capacitor discharging and for exponential growth in capacitor charging. The notes below concentrate on the mathematical basis of exponential decrease as the equations for capacitor charging are developed on p383 from the preceding exponential decrease equations for capacitor discharge.

In your studies of capacitor discharge (Topic 22.3) and of radioactive decay (Topic 26.5), you will have met and used the equation $\frac{dx}{dt} = -\lambda x$ and the solution of this equation $x = x_0 e^{-\lambda t}$.

Let's consider why the equation $\frac{dx}{dt} = -\lambda x$ represents an exponential decrease which is a change where the variable quantity x decreases with time at a rate in proportion to the quantity.

If x decreases by Δx in time Δt, the rate of change is $\frac{\Delta x}{\Delta t}$. This is written as $\frac{dx}{dt}$ in the limit $\Delta t \longrightarrow 0$.

For an exponential decrease, the rate of change is negative and is proportional to x, therefore $\frac{dx}{dt} = -\lambda x$, where λ is referred to as the decay constant.

Now consider why the solution of this equation is $x = x_0 e^{-\lambda t}$, where x_0 is a constant.

Look at the function:

$$x = x_0 \left[1 + t + \frac{t^2}{2 \times 1} + \frac{t^3}{3 \times 2 \times 1} + \frac{t^4}{4 \times 3 \times 2 \times 1} \right.$$
$$\left. + \text{ similar higher order terms} \right] \ldots$$

Applying the rules of **differentiation** to it gives:

$$\frac{dx}{dt} = x_0 \left[0 + 1 + t + \frac{t^2}{2 \times 1} + \frac{t^3}{3 \times 2 \times 1} \right.$$
$$\left. + \text{ similar higher order terms} \right] \ldots \text{ which is the same as } x.$$

So $\frac{dx}{dt} = x$ if x is the above function.

It can be shown that the function in brackets above may be written as n^t, where n is a specific number which is referred to as the exponential number e.

Therefore, $e^t = 1 + t + \frac{t^2}{2 \times 1} + \frac{t^3}{3 \times 2 \times 1} + \frac{t^4}{4 \times 3 \times 2 \times 1}$
$+ \text{ similar higher order terms}$

To show that the solution of the equation $\frac{dx}{dt} = -\lambda x$ is $x = x_0 e^{-\lambda t}$, divide both sides of the equation by $-\lambda$ to give $\frac{dx}{-\lambda dt} = x$

Substituting z for $-\lambda t$ therefore gives $\frac{dx}{dz} = x$ which has the solution $x = x_0 e^z = x_0 e^{-\lambda t}$

The half-life $T_{1/2}$ of an exponential decrease is the time taken for x to decrease from x_0 to $\frac{1}{2} x_0$.

Note

The value of e, the exponential number, can be worked out by substituting $t = 1$ in the above expression for e^t, giving $e = 1 + 1 + \frac{1}{2} + \frac{1}{6} + \text{etc.} = 2.718$ to four significant figures.

Substituting $x = \frac{1}{2}x_0$ and $t = T_{\frac{1}{2}}$ into $x = x_0 e^{-\lambda t}$ gives $\frac{x_0}{2} = x_0 e^{-\lambda T_{\frac{1}{2}}}$

Applying logs to both sides gives $\ln x_0 - \ln 2 = \ln x_0 - \lambda T_{\frac{1}{2}}$ which simplifies to $\lambda T_{\frac{1}{2}} = \ln 2$

therefore, $T_{\frac{1}{2}} = \dfrac{\ln 2}{\lambda} = \dfrac{0.693}{\lambda}$

The time constant τ of an exponential decrease is the time taken for x to decrease from x_0 to $\dfrac{x_0}{e}$ ($= 0.368\, x_0$ as $\dfrac{1}{e} = 0.368$).

Substituting $x = \dfrac{x_0}{e}$ and $t = \tau$ into $x = x_0 e^{-\lambda t}$ gives $\dfrac{x_0}{e} = x_0 e^{-\lambda \tau}$

Applying natural logs to both sides gives $\ln x_0 - \ln e = \ln x_0 - \lambda \tau$

which simplifies to $\tau = \dfrac{1}{\lambda}$ as $\ln e = 1$

For capacitor discharge, $\lambda = \dfrac{1}{CR}$ therefore $\tau = \dfrac{1}{\lambda} = CR$ (see Topic 22.3).

Testing exponential decrease

As explained on page 248, $\ln(e^{-\lambda t}) = -\lambda t$

Therefore, $\ln x = \ln(x_0 e^{-\lambda t}) = \ln x_0 + \ln(e^{-\lambda t}) = \ln x_0 - \lambda t$

Suppose two physical variables x and t are thought to relate to each other through an equation of the form $x = x_0 e^{-\lambda t}$. If so, a graph of $\ln x$ on the y-axis against t on the x-axis would be a straight line in accordance with the equation $\ln x = \ln x_0 - \lambda t$ where the gradient is $-\lambda$ and the y-intercept is $\ln x_0$.

> **Note**
>
> Note that knowledge of the equation for the exponential function $e^t = 1 + t + \dfrac{t^2}{2 \times 1} + \text{etc.}$, is not required in the AS or the A2 specification. The information is provided to help you develop your understanding of the exponential number e and the exponential function e^{-x}.

Summary questions

1 **a** For each exponential decrease equation, write down the initial value at $t = 0$ and the decay constant:

     i $x = 2e^{-3t}$

     ii $x = 12e^{-t/5}$

     iii $x = 4e^{-0.02t}$

   b For each exponential decrease equation above, work out the half–life.

2 A radioactive isotope has a half-life of 720 s and it decays to form a stable product. A sample of the isotope is prepared with an initial activity of 12.0 kBq. Calculate the activity of the sample after:

   a 1 min,

   b 5 min,

   c 1 h.

3 A capacitor of capacitance 22 µF discharged from a pd of 12.0 V through a 100 kΩ resistor.

   a Calculate:

     i the time constant of the discharge circuit,

     ii the half-life of the exponential decrease.

   b Calculate the capacitor pd

     i 2.0 s, and

     ii 5.0 s after the discharge started.

4 A certain exponential decrease process is represented by the equation

     $x = 1000e^{-5t}$

   a **i** Calculate the half-life of the process.

     ii Calculate x when $t = 0.5$ s.

   b Show that the above equation can be rearranged as an equation of the form $\ln x = a + bt$ and determine the values of a and b.

From the Year 1 course, you should recall that the area under a line on a graph can give useful information if the product of the y-variable and the x-variable represents another physical variable. For example, the tension against extension graph for a spring is a straight line through the origin and the area under the line represents the work done to stretch the spring. See Topic 10.1.

Table 1 gives some further examples where the area under a graph has physical significance

▼ **Table 1** *Areas in graphs*

examples (y-variable first)	area between the line and the x-axis	equation	units
power against time	energy transferred	energy transferred = power × time	$1\,W = 1\,J\,s^{-1}$
potential difference against charge (or stored)	electrical energy transferred	electrical energy transferred = pd × charge	$1\,V = 1\,J\,C^{-1}$
force against time	change of momentum (or impulse)	change of momentum = force × time	$1\,kg\,m\,s^{-1}$ $= 1\,N\,s$
force against distance	work done	work done = force × distance	$1\,J = 1\,N\,m$

To find the area under the line, we can either:

- count the squares of the grid under the line and multiply the number of squares by the amount of the physical variable that 1 square of area represents, or

- use the mathematical process known as **integration** as outlined below. The notes below are intended to give a deeper understanding of how areas under curves can be calculated. They are not part of the A2 Physics specification.

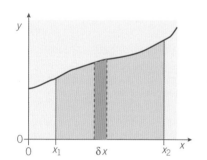

▲ **Figure 1** *Integration*

Consider Figure 1 which shows a y-variable that changes as the x-variable changes. A small increase of the x-variable, δx (δ for small) gives little or no change of the y-variable. The area under that section of the curve, $\delta A = y\,\delta x$ as it is a strip of width δx and height y. Note that rearranging $\delta A = y\,\delta x$ gives $y = \dfrac{\delta A}{\delta x}$.

Hence the total area under the line from x_1 to x_2 in Figure 1 is equal to the area of all the strips, each of width δx, from x_1 to x_2. The process of adding the individual strip areas together to give the total area is called **integration.**

In mathematical terms,

Total area $A = \displaystyle\int_{x_1}^{x_2} \delta A = \int_{x_1}^{x_2} y\,\delta x$ where $\int$ is the mathematical symbol for integration.

Note

Note that integration of functions is not required in the specification. The information on integration is provided to help you develop your understanding of how areas under graphs can be found precisely if the equation of the line in known.

As $y = \dfrac{\delta A}{\delta x}$, then differentiating A in terms of x gives y. If we know the formula for y in terms of x, we can find the formula for A in terms of x by using the differentiation formula in Topic 30.4 in reverse.

For example, if $y = 2x$, then using the process of reverse differentiation gives $A = x^2$

Force-field curves, such as the inverse square law of force between two point charges, give areas that represent potential energy. We can use the ideas outlined above to obtain an exact formula for the potential energy of two point charges at a certain distance apart.

Consider the two point charges q_1 and q_2 at distance apart r. The force F between the charges is given by Coulomb's law

$$F = \frac{q_1 q_2}{4\pi\varepsilon_0} \times \frac{1}{r^2}$$

If the charges move so their distance apart changes by δr, the work they do in this movement $= F\delta r$. This is represented on Figure 2 by the narrow strip of width δr.

Since the work they do reduces their potential energy, their change of potential energy $\delta E_P = -F\delta r$

When the distance apart decreases to r_1 from infinite separation, the potential energy changes from zero at infinite separation to E_P at distance apart r_1. Since E_P is represented by the total area under the line from $r =$ infinity to $r = r_1$

$$E_P = \int_{\text{infinity}}^{r_1} \delta E_P = \int_{\text{infinity}}^{r_1} -F\delta r$$

Because the force is given by the inverse square law above,

$$E_P = \int_{\text{infinity}}^{r_1} \frac{-k}{r^2} \delta r$$

where $k = \dfrac{q_1 q_2}{4\pi\varepsilon_0}$

Therefore, $E_P = \dfrac{k}{r}$ because differentiating $\dfrac{1}{r}$ gives $\dfrac{-1}{r^2}$.

If you need to, look again at Topic 30.4.

Hence their potential energy at distance r_1 apart, $E_P = \dfrac{q_1 q_2}{4\pi\varepsilon_0 r_1}$

as the potential energy at infinity is zero.

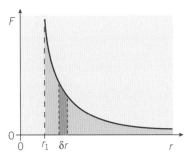

▲ **Figure 2** *The inverse square law of force*

Note

The inverse square law also applies to the gravitational force $F = \dfrac{GMm}{r^2}$ between two point masses M and m. The constant k is written as $-GMm$.

(The minus sign represents the attractive nature of the force.)

Therefore, for a small mass m at distance r from the centre of a spherical planet of mass M at or beyond its surface, the gravitational potential energy $E_P = -\dfrac{GMm}{r}$

Summary questions

1 For a velocity–time graph, what physical variable is represented by:

  a the gradient,   **b** the area under the line?

2 What physical variable is represented by:

  a the area under a graph of acceleration against time,

  b the area under a graph of current against time,

  c What physical variable is represented by the area under a graph of pressure against volume for a gas?

  d State the unit of

    i pressure,

    ii volume,

    iii pressure × volume.

3 For the electric field near a point charge, what physical variable is represented by

  a the area under the graph of electric field strength E against distance r in Figure 3a?

  b the gradient of the graph of electric potential V against distance r in Figure 3b?

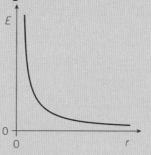

▲ Figure 3a

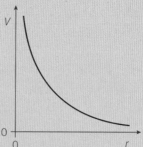

▲ Figure 3b

4 **a** For the gravitational field strength near a spherical object, what physical variable is represented by

    i the area under the graph of gravitational field strength g against distance r in Figure 4a,

    ii the gradient of the graph of gravitational potential V_{grav} against distance r in Figure 4b?

  b Which of the graphs shown in Figures 3 and 4 are inverse square curves?

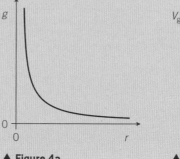

▲ Figure 4a

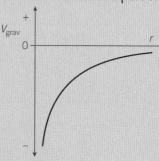

▲ Figure 4b

Graphical methods

Useful physical quantities in equations can be worked out directly and indirectly by measuring rates of change from graph gradients. For example, the acceleration of an object can be determined from its displacement–time graph by

- measuring the gradient at two different points on the graph to give the velocity at each point,

- dividing the change of velocity between the two points by the time between them.

Figure 1 is a graph of height (h) against time (t) graph for a projectile. The graph can be used to find g, the acceleration of free fall, by measuring the gradient of the curve at two different times. The gradient at A is zero so in this example, only the gradient at a second point B needs to be found.

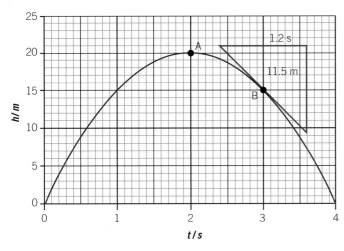

▲ **Figure 1** *Projectile motion*

- The velocity at A = 0.
- The velocity 1.0 s later at B = gradient of the *blue* triangle = $-\dfrac{11.5\,\text{m}}{1.2\,\text{s}}$ = $-9.6\,\text{m s}^{-1}$ to two significant figures.

Therefore, the vertical acceleration = $g = -\dfrac{9.6\,\text{m s}^{-1}}{1.0\,\text{s}} = -9.6\,\text{m s}^{-2}$.

Modelling a physical system

Physicists often try to understand the equations for a physical system by modelling the system. For example, consider the 'dice' model of radioactive decay, which you studied in Topic 26.5. In this model, a large number of dice (1000 to start with) are considered, each one representing a radioactive nucleus. Each time the dice are thrown, all the dice that show '1' uppermost are removed before the remaining dice are thrown. Each throw should result in a sixth of the dice being removed (because each of the six numbers on a dice are equally probable as the uppermost number). Table 2 shows how the number of dice decreases after each of nine throws.

▼ Table 1

	number of dice at the start of each throw, N	number of dice removed	number of dice remaining
1st throw	1000	167	833
2nd throw	833	139	694
3rd throw	694	116	578
4th throw	578	96	482
5th throw	482	80	402
6th throw	402	67	335
7th throw	335	56	279
8th throw	279	47	232
9th throw	232	39	193

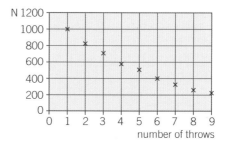

▲ **Figure 2** *Number of dice, N, against number of throws*

Figure 2 shows how the number of dice, N, decreases with the number of throws. A curve drawn through the points would be similar to a radioactive decay curve.

* After about four throws, the number of dice remaining is about half the initial number.
* After about seven to eight throws, the number of dice remaining is about a quarter the initial number.

The curve is like a radioactive decay curve, so the dice model seems to work – but it isn't good enough to tell us with confidence.

To improve the dice model, you could use a *spreadsheet* to carry out the calculations and to display the results. The use of a spreadsheet also has the advantage that we can change the half-life and use much bigger numbers.

Before setting up the spreadsheet, you need to recall that the rate of change of the number of nuclei N of a radioactive isotope decreases with time t according to the equation

$$\frac{\Delta N}{\Delta t} = -\lambda N$$

where λ is the decay constant, which is the probability per second of a nucleus decaying. See Topic 26.6.

Rearranging this equation gives the change of the number of nuclei of the isotope:

$$\Delta N = -\lambda N \Delta t$$

Comparing this with the dice model, you can see that the product $\lambda \Delta t$ tells us the fraction of the nuclei that decay in time Δt. So $\lambda \Delta t$ is like the fraction $\frac{1}{6}$ in the dice model because this tells us that $\frac{1}{6}$th of the dice are removed after each throw.

By choosing suitable values for λ, Δt, and the initial number of nuclei N_0, you can use a spreadsheet to calculate the number of nuclei remaining after any number of time intervals.

You can set up your own radioactive decay spreadsheet by following the instructions below.

Step 1 Choose values of λ, N_0, and Δt. To compare the results with the dice experiment, choose λ and Δt such that $\lambda \Delta t = \frac{1}{6}$ (= 0.1667 to four significant figures).

- Insert λ = in cell A1, Δt = in cell C1 and N_0 = in cell E1.
- Insert a value of 16.67 (s^{-1}) for λ in cell B1, a value of 0.01 s for Δt in cell D1, and a value of 1000 for N_0 in cell F1.
- Insert column headings for time t, the initial number of nuclei N_i, ΔN, and the final number of nuclei N into cells A3 to D3.

Step 2 Calculate ΔN and N after the first time interval.

- Insert 0 into cell A4, +F1 into cell B4 and the expression +(B4)*(D1)*(B1) for ΔN into cell C4.
- Insert the expression +B4 − C4 into cell D4 to calculate the final number of nuclei N.

The values of t, N_i, ΔN and N after the first interval should now be in cells A4 and D4.

Step 3 Repeat the calculation for the second time interval 0.01 s to 0.02 s.

- Insert +A4+D1 into cell A5 to calculate t.
- Insert +D4 into cell B5, and copy cells C4 and D4 into cells C5 and D5.

The values of t, N_i, ΔN, and N after the second interval should now be in in cells C5 and D5.

Step 4 Copy cells A5 to D5 into the rows below down to row 49.

- The values of t, N_i, ΔN, and N for successive intervals should now appear in cells C6 to C49 and D6 to D49.

To display the results as a graph, select the block of cells from A4 to B24 at the same time, and select a graph option from the *Insert* menu across the top of the spreadsheet. A graph showing N against t should then appear on the screen as shown in Figure 3.

Note

Don't include the unit symbol when you enter a value into a cell

Study tip

You *don't* need to know how to design your own spreadsheets in your A Level Physics course.

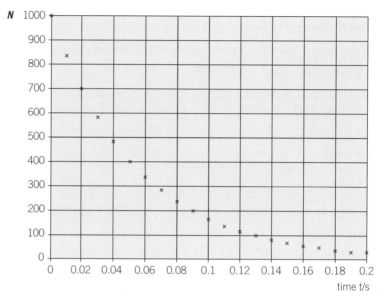

▲ **Figure 3** *A radioactive decay spreadsheet graph*

+ About spreadsheets

A spreadsheet consists of a matrix of cells labelled alphabetically across the top and numerically down the left-hand side. Data entered in a cell can be processed mathematically and displayed in another cell.

The basic rules for designing a spreadsheet are:

- A number entered into a cell can be processed mathematically and displayed in another cell. For example, if a number is entered into cell A1, the square of the number in cell A1 can be displayed in cell B1 by entering +A1*A1 (or +A1^2) into cell B1.

- The contents of a cell can be copied into other cells. For example, suppose the content of cell B1 is copied into cells B2, B3, B4, and B5. Cells B2 to B5 will then display the square of the numbers entered into cells A2 to A5. See Figure 4.

- Data from a specific cell (e.g., A5) can be copied into other cells by prefixing the cell address with the dollar symbol $. For example, if +((A1) − (A1)) is entered into cell C1 and copied into cells C2 to C5, cell C1 will display 0, and cells C2 to C5 will display the number in cells A2 to A5 minus the number in cell A1, as shown in Figure 4.

	A	B	C
1	0.5	0.25	0
2	1.5	2.25	1
3	2.5	6.25	2
4	3.5	12.25	3
5	4.5	20.25	4

▲ **Figure 4** *A simple spreadsheet*

Summary questions

1 Determine a value of g from Figure 1 by using the gradient given at B and measuring the gradient at 0.6 s.

2 a Estimate from Figure 3 the half-life $T_{\frac{1}{2}}$ of the isotope.

   b Describe and explain what difference it makes to the graph in Figure 3 if the decay constant is halved.

Useful data for A Level Physics

Data

Fundamental constants and values

quantity	symbol	value	units
speed of light in vacuo	c	3.00×10^8	$m\,s^{-1}$
permeability of free space	μ_0	$4\pi \times 10^{-7}$	$H\,m^{-1}$
permittivity of free space	ε_0	8.85×10^{-12}	$F\,m^{-1}$
magnitude of charge of electron	e	1.60×10^{-19}	C
Planck constant	h	6.63×10^{-34}	J s
gravitational constant	G	6.67×10^{-11}	$N\,m^2\,kg^{-2}$
Avogadro constant	N_A	6.02×10^{23}	mol^{-1}
molar gas constant	R	8.31	$J\,K^{-1}\,mol^{-1}$
Boltzmann constant	k	1.38×10^{-23}	$J\,K^{-1}$
Stefan constant	σ	5.67×10^{-8}	$W\,m^{-2}\,K^{-4}$
Wien constant	α	2.90×10^{-3}	m K
electron rest mass (equivalent to 5.5×10^{-4} u)	m_e	9.11×10^{-31}	kg
electron charge/mass ratio	$\dfrac{e}{m_e}$	1.76×10^{11}	$C\,kg^{-1}$
proton rest mass (equivalent to 1.007 28 u)	m_p	$1.67(3) \times 10^{-27}$	kg
proton charge/mass ratio	$\dfrac{e}{m_p}$	9.58×10^7	$C\,kg^{-1}$
neutron rest mass (equivalent to 1.008 67 u)	m_n	$1.67(5) \times 10^{-27}$	kg
gravitational field strength	g	9.81	$N\,kg^{-1}$
acceleration due to gravity	g	9.81	$m\,s^{-2}$
atomic mass unit (1 u is equivalent to 931.3 MeV)	u	1.661×10^{-27}	kg

Astronomical data

body	mass/kg	mean radius/m
Sun	1.99×10^{30}	6.96×10^8
Earth	5.98×10^{24}	6.37×10^6

Geometrical equations

arc length = $r\,\theta$

circumference of circle = $2\pi r$

area of circle = πr^2

surface area of cylinder = $2\pi rh$

volume of cylinder = $\pi r^2 h$

area of sphere = $4\pi r^2$

volume of sphere = $\frac{4}{3}\pi r^3$

Momentum

force $\qquad F = \dfrac{\Delta(mv)}{t}$

impulse $\qquad F\Delta t = \Delta(mv)$

Circular motion

magnitude of angular speed $\omega = \dfrac{v}{r}$

$\qquad\qquad\qquad\qquad\qquad \omega = 2\pi f$

centripetal acceleration $\qquad a = \dfrac{v^2}{r} = \omega^2 r$

centripetal force $\qquad F = \dfrac{mv^2}{r} = m\omega^2 r$

Simple harmonic motion

acceleration $\qquad a = -\omega^2 x$

displacement $\qquad x = A\cos(\omega t)$

speed $\qquad v = \pm\, \omega\sqrt{A^2 - x^2}$

maximum speed $\qquad v_{\text{max}} = \omega A$

maximum acceleration $\qquad a_{\text{max}} = \omega^2 A$

for a mass-spring system $\qquad T = 2\pi\sqrt{\dfrac{m}{k}}$

for a simple pendulum $\qquad T = 2\pi\sqrt{\dfrac{l}{g}}$

Gases and thermal physics

gas law $\qquad pV = nRT$

$\qquad pV = NkT$

kinetic theory model $\qquad pV = \dfrac{1}{3} Nm(c_{\text{rms}})^2$

kinetic energy of a gas molecule $\qquad \dfrac{1}{2}m(c_{\text{rms}})^2 = \dfrac{3}{2}kT$

$\qquad = \dfrac{3RT}{2N_A}$

energy to change temperature $\qquad Q = mc\Delta T$

energy to change state $\qquad Q = ml$

Gravitational fields

force between two masses $\qquad F = \dfrac{Gm_1 m_2}{r^2}$

magnitude of gravitational field strength $\qquad g = \dfrac{F}{m}$

$\qquad g = \dfrac{GM}{r^2}$

gravitational potential $\qquad \Delta W = m\Delta V$

$\qquad V = -\dfrac{GM}{r}$

$\qquad g = -\dfrac{\Delta V}{\Delta r}$

Electric fields and capacitors

force between two point charges $\qquad F = \dfrac{Q_1 Q_2}{4\pi\varepsilon_0 r^2}$

force on a charge $\qquad F = EQ$

field strength for a uniform field $\qquad E = \dfrac{V}{d}$

field strength for a radial field $\qquad E = \dfrac{Q}{4\pi\varepsilon_0 r^2}$

work done $\qquad \Delta W = Q\Delta V$

electric potential $\qquad V = \dfrac{Q}{4\pi\varepsilon_0 r}$

capacitance $\qquad C = \dfrac{Q}{V}$

$\qquad C = \dfrac{A\varepsilon_0 \varepsilon_r}{d}$

decay of charge $\qquad Q = Q_0 e^{\frac{-t}{RC}}$

capacitor charge $\qquad V = V_0 (1 - e^{\frac{-t}{RC}})$

time constant $\qquad RC$

time to halve $\qquad T_{1/2} = 0.69RC$

capacitor energy stored $\qquad E = \dfrac{1}{2}QV = \dfrac{1}{2}CV^2$

$\qquad = \dfrac{1}{2}\dfrac{Q^2}{C}$

Magnetic fields

force on a current $\qquad F = BIl$

force on a moving charge $\qquad F = BQv$

magnetic flux $\qquad \phi = BA$

magnetic flux linkage $\qquad N\phi = BAN\cos\theta$

induced emf $\qquad \varepsilon = N\dfrac{\Delta\phi}{\Delta t}$

emf induced in a rotating coil $\qquad N\phi = BAN\cos\theta$

$\qquad \varepsilon = BAN\,\omega\,\sin\,\omega t$

alternating current $\qquad I_{\text{rms}} = \dfrac{I_0}{\sqrt{2}} \qquad V_{\text{rms}} = \dfrac{V_0}{\sqrt{2}}$

transformer equations $\qquad \dfrac{N_s}{N_p} = \dfrac{V_s}{V_p}$

efficiency $\qquad = \dfrac{I_s V_s}{I_p V_p}$

Nuclear physics

the inverse-square law for γ variation $\qquad I = \dfrac{k}{x^2}$

radioactive decay $\qquad \dfrac{\Delta N}{\Delta t} = -\lambda N \qquad N = N_0 e^{-\lambda t}$

activity $\qquad A = \lambda N$

half-life $\qquad T_{1/2} = \dfrac{\ln 2}{\lambda}$

nuclear radius $\qquad R = R_0 A^{\frac{1}{3}}$

energy–mass equation $\qquad E = mc^2$

Particle physics
Rest energy values

class	name	symbol	rest energy / MeV
photon	photon	γ	0
lepton	neutrino	ν_e	0
		ν_μ	0
	electron	$e^\pm$	0.510 99
	muon	$\mu^\pm$	105.659
mesons	pion	$\pi^\pm$	139.576
		π^0	134.972
	kaon	$K^\pm$	493.821
		K^0	497.762
baryons	proton	p	938.257
	neutron	n	939.551

Properties of quarks
Antiparticles have opposite signs

type	charge	baryon number	strangeness
u	$+\frac{2}{3}e$	$+\frac{1}{3}$	0
d	$-\frac{1}{3}e$	$+\frac{1}{3}$	0
s	$-\frac{1}{3}e$	$+\frac{1}{3}$	-1

Properties of leptons

lepton	lepton number
particles: e^-, ν_e; μ^-, ν_μ	$+1$
antiparticles: e^+, $\bar{\nu}_e$; μ^+, $\bar{\nu}_\mu$	-1

Photons and energy levels

photon energy $\qquad E = hf = \dfrac{hc}{\lambda}$

photoelectric effect $\qquad hf = \phi + E_{K(max)}$

energy levels $\qquad hf = E_1 - E_2$

de Broglie wavelength $\qquad \lambda = \dfrac{h}{p} = \dfrac{h}{mv}$

Waves

wave speed $\qquad c = f\lambda$

period $\qquad f = \dfrac{1}{T}$

first harmonic $\qquad f = \dfrac{1}{2l}\sqrt{\dfrac{T}{\mu}}$

fringe spacing $\qquad w = \dfrac{\lambda D}{s}$

diffraction grating $\qquad d\sin\theta = n\lambda$

refractive index of a substance s $\quad n_s = \dfrac{c}{c_s}$

For two different substances of refractive indices n_1 and n_2,

law of refraction $\qquad n_1 \sin\theta_1 = n_2 \sin\theta_2$

critical angle $\qquad \sin\theta_c = \dfrac{n_2}{n_1}$ for $n_1 > n_2$

Mechanics

Moments $\qquad \text{moment} = Fd$

velocity and acceleration $\qquad v = \dfrac{\Delta s}{\Delta t}; \quad a = \dfrac{\Delta v}{\Delta t}$

equations of motion $\qquad v = u + at$

$$s = \dfrac{(u+v)t}{2}$$

$$v^2 = u^2 + 2as$$

$$s = ut + \dfrac{1}{2}at^2$$

force $\qquad F = ma$

work, energy, and power $\qquad W = Fs\cos\theta$

$$E_K = \dfrac{1}{2}mv^2$$

$$\Delta E_p = mg\Delta h$$

$$P = \dfrac{\Delta W}{\Delta t} \quad P = Fv$$

$$\text{efficiency} = \dfrac{\text{useful output power}}{\text{input power}}$$

Materials

density $\qquad \rho = \dfrac{m}{V}$

Hooke's law $\qquad F = k\Delta L$

$$\text{Young modulus} = \dfrac{\text{tensile stress}}{\text{tensile strain}}$$

tensile stress $\quad = \dfrac{F}{A}$

tensile strain $\quad = \dfrac{\Delta L}{L}$

energy stored $\quad E = \dfrac{1}{2}F\Delta L$

Electricity

current and pd $\qquad I = \dfrac{\Delta Q}{t}$

$$V = \dfrac{W}{Q}$$

$$R = \dfrac{V}{I}$$

emf $\qquad \varepsilon = \dfrac{E}{Q}$

$$\varepsilon = I(R + r)$$

resistors in series $\qquad R = R_1 + R_2 + R_3 + \ldots$

resistors in parallel $\qquad \dfrac{1}{R} = \dfrac{1}{R_1} + \dfrac{1}{R_2} + \dfrac{1}{R_3} + \ldots$

resistivity $\qquad \rho = \dfrac{RA}{L}$

power $\qquad P = VI = I^2R = \dfrac{V^2}{R}$

Glossary

A glossary of practical terms follows this glossary.

A

absolute scale: temperature scale in kelvins (K) defined in terms of *absolute zero*, 0 K, and the triple point of water, 273.16 K, which is the temperature at which ice, water and water vapour are in thermal equilbrium.

absolute zero: the lowest possible temperature, the temperature at which an object has minimum internal energy.

absolute temperature T: in kelvin = temperature in °C + 273(.15).

activity A: of a radioactive isotope, the number of nuclei of the isotope that disintegrate per second. The unit of activity is the becquerel (Bq), equal to 1 disintegration per second.

alpha (α) decay: change in an unstable nucleus when it emits an α particle which is a particle consisting of two protons and two neutrons.

alpha radiation: particles that are each composed of two protons and two neutrons. An alpha (α) particle is emitted by a heavy unstable nucleus which is then less unstable as a result. Alpha radiation is easily absorbed by paper, has a range in air of no more than a few centimetres and is more ionising than beta (β) or gamma (γ) radiation.

angular displacement: the angle an object in circular motion turns through. If its time period is T and its frequency is f, its angular displacement in time t, in radians = $2\pi ft = 2\pi t/T$

angular speed ω: the rate of change of angular displacement of an object in circular (or orbital or spinning) motion.

angular frequency ω: for an object oscillating at frequency f in simple harmonic motion, its angular frequency = $2\pi f$.

antiquark: antiparticle of a quark.

atomic mass unit u: correctly referred to as the unified atomic mass constant; $\frac{1}{12}$th of the mass of an atom of the carbon isotope $^{12}_{6}C$, equal to 1.661×10^{-27} kg.

atomic number Z: of an atom of an element is the number of protons in the nucleus of the atom. It is also the order number of the element in the Periodic Table.

Avogadro constant N_A: the number of atoms in 12 g of the carbon isotope $^{12}_{6}C$. N_A is used to define the mole. Its value is 6.02×10^{23} mol^{-1}.

B

back emf: emf induced in the spinning coil of an electric motor or in any coil in which the current is changing (e.g., the primary coil of a transformer). A back emf acts against the change of applied pd.

background radiation: radiation due to naturally occurring radioactive substances in the environment (e.g., in the ground or in building materials or elsewhere in the environment). Background radiation is also caused by cosmic radiation.

beta (β) decay: change in a nucleus when a neutron changes into a proton and a β^- particle and an antineutrino are emitted if the nucleus is neutron-rich or a proton changes to a neutron and a β^+ particle and a neutrino are emitted if the nucleus is proton-rich.

beta-minus (β^-) radiation: electrons (β^-) emitted by unstable neutron-rich nuclei (i.e., nuclei with a neutron/proton ratio greater than for stable nuclei). β^- radiation is stopped by about 5 mm of aluminium, has a range in air of up to a metre and is less ionising than alpha (α) radiation and more ionising than gamma (γ) radiation.

beta-plus (β^+) radiation: *positrons* (β^+) emitted by unstable proton-rich nuclei (i.e., nuclei with a neutron/proton ratio smaller than for stable nuclei). Positrons emitted in solids or liquids travel no further than about 2 mm before they are annihilated.

binding energy of a nucleus: the work that must be done to separate a nucleus into its constituent neutrons and protons. Binding energy = mass defect $\times c^2$. Binding energy in MeV = mass defect in u $\times$ 931.3.

binding energy per nucleon: the average work done per nucleon to separate a nucleus into its constituent parts. The binding energy per nucleon of a nucleus = the binding energy of a nucleus/ mass number A. The binding energy per nucleon is greatest for iron nuclei of mass number about 56. The binding energy curve is a graph of binding energy per nucleon against mass number A.

boiling point: the temperature at which a pure liquid at atmospheric pressure boils.

Boyle's law: for a fixed mass of gas at constant temperature, its pressure $\times$ its volume is constant. A gas that obeys Boyle's law is said to be an *ideal gas*.

Boltzmann constant k: the molar gas constant divided by the Avogadro number $\left(\text{i.e., } \frac{R}{N_A}\right)$. See *kinetic energy of the molecules of an ideal gas*.

Brownian motion: the random and unpredictable motion of a particle such as a smoke particle caused by molecules of the surrounding substance colliding at random with the particle. Its discovery provided evidence for the existence of atoms.

C

capacitance: the charge stored per unit pd of a capacitor. The unit of capacitance is the farad (F), equal to 1 coulomb per volt. For a capacitor of capacitance C at pd V, the charge stored, $Q = CV$.

capacitor energy: energy stored by the capacitor,
$$E = \frac{1}{2}QV = \frac{1}{2}CV^2 = \frac{1}{2}\frac{Q^2}{C}$$

capacitor discharge: through a fixed resistor of resistance R; time constant = RC; exponential decrease equation for current or charge or pd;
$$x = x_0 e^{\frac{-t}{RC}}$$

Celsius scale: temperature, in degrees Celsius or °C, is defined as absolute temperature in kelvins −273.15. This definition means that the temperature of pure melting ice (ice point) is 0 °C, and the temperature of steam at standard atmospheric pressure (steam point) is 100 °C.

centripetal acceleration: 1. For an object moving at speed v (or angular speed ω) in uniform circular motion, its centripetal acceleration $a = \frac{v^2}{r} = \omega^2 r$ towards the centre of the circle. 2. For a satellite in a circular orbit, its centripetal acceleration $\frac{v^2}{r} = g$

centripetal force: the resultant force on an object that moves along a circular path. For an object of mass m moving at speed v along a circular path of radius r, the centripetal force $= \frac{mv^2}{r}$ towards the centre of the circle.

chain reaction: a series of reactions in which each reaction causes a further reaction. In a nuclear reactor, each fission event is due to a neutron colliding with a $^{235}_{92}$U nucleus which splits and releases two or three further neutrons that can go on to produce further fission. A steady chain reaction occurs when one fission neutron on average from each fission event produces a further fission event.

Charles' law: for a fixed mass of an ideal gas at constant pressure, its volume is directly proportional to its *absolute temperature*.

collisions: see *elastic collision*.

conservation of momentum: for a system of interacting objects is the total momentum of the objects remains constant provided no external resultant force acts on the system.

control rods: rods made of a neutron-absorbing substance such as cadmium or boron that are moved in or out of the core of a nuclear reactor to control the rate of fission events in the reactor.

coolant: a fluid that is used to prevent a machine or device from becoming dangerously hot. The coolant of a nuclear reactor is pumped through the core of the reactor to transfer thermal energy from the core to a heat exchanger.

Coulomb's law of force: for two point charges Q_1 and Q_2 at distance apart r, the force F between the two charges is given by the equation
$F = \frac{Q_1 Q_2}{4\pi\varepsilon_0 r^2}$, where ε_0 is the permittivity of free space.

count rate: the number of counts per unit time detected by a Geiger Müller tube. Count rates should always be corrected by measuring and subtracting the background count rate (i.e., the count rate with no radioactive source present).

critical mass: the minimum mass of the fissile *isotope* (e.g., the uranium *isotope* $^{235}_{92}$U) in a nuclear reactor necessary to produce a *chain reaction*. If the mass of the fissile isotope in the reactor is less than the critical mass, a chain reaction does not occur because too many fission neutrons escape from the reactor or are absorbed without fission.

D

damped oscillations: oscillations that reduce in *amplitude* due to the presence of resistive forces such as friction and drag.

1. For a lightly damped system, the amplitude of oscillations decreases gradually.

2. For a heavily damped system displaced from equilibrium then released, the system slowly returns to equilibrium without oscillating.

3. For a critically damped system, the system returns to equilibrium in the least possible time without oscillating.

de Broglie wavelength: a particle of matter has a wave-like nature which means that it can behave as a wave. For example, electrons directed at a thin crystal are diffracted by the crystal. The de Broglie wavelength, λ, of a matter particle depends on its momentum, p, in accordance with de Broglie's equation $\lambda = \frac{h}{p} = \frac{h}{mv}$, where h is the Planck constant.

decay constant λ: the probability of an individual nucleus decaying per second.

decay curve: an exponential decrease curve showing how the mass or activity of a radioactive isotope decreases with time.

dielectric: material that increases the capacity of a parallel-plate capacitor to store charge when placed between the plates of the capacitor. Polythene and waxed paper are examples of dielectrics.

differentiation: mathematical process of finding the gradient of a line from its equation.

diffraction: the spreading of waves when they pass through a gap or round an obstacle. X-ray diffraction is used to determine the structure of crystals, metals and long molecules. Electron diffraction is used to probe the structure of materials. High-energy electron scattering is used to determine the diameter of the nucleus.

dissipative forces: forces that transfer energy which is wasted.

dose equivalent: a comparative measure of the effect of each type of *ionising radiation*, defined as the energy that would need to be absorbed per unit mass of matter from 250 k of X-radiation to have the same effect as a certain 'dose' of the *ionising radiation*. The unit of dose equivalent is the sievert (Sv).

E

eddy currents: induced currents in the metal parts of ac machines.

elastic collision: an elastic collision is one in which the total kinetic energy after the collision is equal to the total kinetic energy before the collision.

electrical conductor: an object that can conduct electricity.

electrically insulating materials: an electrical insulator is a material that cannot conduct electricity; a thermal insulator is a material that is a poor conductor of heat.

electric field strength *E*: at a point in an electric field, is the force per unit charge on a small positively charged object at that point in the field.

electric potential *V*: at a point in an electric field is the work done per unit charge on a small positively charged object to move it from infinity to that point in the field.

electromagnetic induction: the generation of an emf when the *magnetic flux linkage* through a coil changes or a conductor cuts across magnetic field lines.

electromagnetic wave: an electric and magnetic wavepacket or *photon* that can travel through free space.

electromotive force (emf): the amount of electrical energy per unit charge produced inside a source of electrical energy.

electron: a *lepton* of rest mass 9.11×10^{-31} kg and electric charge -1.60×10^{-19} C (to 3 significant figures).

electron capture: a proton-rich *nucleus* captures an inner-shell electron to cause a *proton* in the nucleus to change into a *neutron*. An electron *neutrino* is emitted by the nucleus. An X-ray *photon* is subsequently emitted by the atom when the inner shell vacancy is filled.

equipotential: a line or surface in a field along which the electric or gravitational potential is constant.

escape velocity: the minimum velocity an object must be given to escape from the planet when projected vertically from the surface.

excited state: an atom which is not in its ground state (i.e., its lowest energy state).

explosion: when two objects fly apart, the two objects carry away equal and opposite momentum.

exponential change: exponential change happens when the change of a quantity is proportional to the quantity itself. For an exponential decrease of a quantity x, $\frac{dx}{dt} = -\lambda x$, where λ is referred to as the decay constant. The solution of this equation is $x = x_0 e^{-\lambda t}$ where x_0 is an initial value of x.

F

Faraday's law of electromagnetic induction: the induced emf in a circuit is equal to the rate of change of *magnetic flux linkage* through the circuit. For a changing magnetic field in a fixed coil of area A and N turns, the induced emf $= -NA\frac{\Delta B}{\Delta t}$

field line: see *line of force*.

fission: the splitting of a $^{235}_{92}$U *nucleus* or a $^{235}_{94}$Pu nucleus into two approximately equal fragments. Induced fission is fission caused by an incoming *neutron* colliding with a $^{235}_{92}$U nucleus or a $^{235}_{94}$Pu nucleus.

fission neutrons: neutrons released when a nucleus undergoes fission and which may collide with nuclei to cause further fission.

Fleming's left-hand rule: rule that relates the directions of the force, magnetic field and current on a current-carrying conductor in a magnetic field.

Fleming's right-hand rule: rule that relates the directions of the induced current, magnetic field and velocity of the conductor when the conductor cuts across magnetic field lines and an emf is induced in it.

force: = rate of change of *momentum*
$$= \frac{\text{change of momentum}}{\text{time taken}}$$
(= mass × acceleration for fixed mass).

forced vibrations: vibrations (oscillations) of a system subjected to an external periodic force.

free electrons: electrons in a conductor that move about freely inside the metal because they are not attached to a particular atom.

free vibrations: vibrations (oscillations) where there is no damping and no periodic force acting on the system, so the amplitude of the oscillations is constant.

fusion (nuclear): the fusing together of light nuclei to form a heavier nucleus.

fusion (thermal): the fusing together of metals by melting them together.

G

geostationary satellite: a satellite that stays above the same point on the Earth's equator as it orbits the Earth because its orbit is in the same plane as the equator, its period is exactly 24 h and it orbits in the same direction as the Earth's direction of rotation.

gold leaf electroscope: a device used to detect electric charge.

gravitational constant G: the constant of proportionality in *Newton's law of gravitation*.

gravitational field: the region surrounding an object in which it exerts a gravitational force on any other object.

gravitational field strength g: the force per unit mass on a small mass placed in the field.

1. $g = \dfrac{F}{m}$, where F is the gravitational force on a small mass m.

2. At distance r from a point mass M, $g = \dfrac{GM}{r^2}$

3. At or beyond the surface of a sphere of mass M, $g = \dfrac{GM}{r^2}$ where r is the distance to the centre.

4. At the surface of a sphere of mass M and radius R, $g_s = \dfrac{GM}{R^2}$

gravitational force: an attractive force that acts equally on any two objects due to their mass.

gravitational potential V: at a point in a gravitational field is the work done per unit mass to move a small object from infinity to that point. At distance r from the centre of a spherical object of mass M, $V = -\dfrac{GM}{r}$

gravitational potential energy: at a point in a gravitational field is the work done to move a small object from infinity to that point. The change of gravitational potential energy of a mass m moved through height h near the Earth's surface, $\Delta E_p = mg\,\Delta h$

grid system: the network of transformers and cables that is used to distribute electrical power from power stations to users.

H

half-life $T_{1/2}$: the time taken for the mass of a radioactive isotope to decrease to half the initial mass or for its activity to halve. This is the same as the time taken for the number of nuclei of the isotope to decrease to half the initial number.

Hall probe: a device used to measure *magnetic flux density*.

heat Q: energy transfer due to a difference of temperature.

heat capacity: the energy needed to raise the temperature of an object by 1 K

heat exchanger: a steel vessel containing pipes through which hot coolant in a sealed circuit is pumped, causing water passing through the steel vessel in separate pipes to turn to steam which is used to drive turbines.

I

ideal gas: a gas under conditions such that it obeys *Boyle's law*.

ideal gas equation: $pV = nRT$, where p is the gas pressure, V is the gas volume, n is the number of moles of gas, T is the absolute temperature and R is the *molar gas constant*.

impulse: of a force acting on an object, force × time for which the force acts.

integration: mathematical process of finding the area under a curve from its mathematical equation.

intensity of radiation: at a surface is the radiation energy per second per unit area at normal incidence to the surface. The unit of intensity is J s^{-1} m^{-2} or W m^{-2}.

internal energy: of an object is the sum of the random distribution of the kinetic and potential energies of its molecules.

ionising radiation: radiation that produces ions in the substances it passes through. It destroys cell membranes and damages vital molecules such as DNA directly or indirectly by creating 'free radical' ions which react with vital molecules.

inverse-square laws: 1. Force: *Newton's law of gravitation* and *Coulomb's law* of force between electric charges are inverse-square laws because the force between two point objects (masses in the case of gravitation and charge in the case of charges) is inversely proportional to the square of the distance between the two objects. Because these two laws are inverse-square laws, the field strength due to a point mass or a point charge varies with distance according to the inverse of the square of the distance to the point object.

2. Intensity: the intensity of γ radiation from a point source varies with the inverse of the square of the distance from the source. The same rule applies to radiation from any point source that spreads out equally in all directions and is not absorbed.

K

Kepler's third law: for any planet, the cube of its mean radius of orbit r is directly proportional to the square of its time period T. Using *Newton's law of gravitation*, it can be shown that $\dfrac{r^3}{T^2} = \dfrac{GM}{4\pi^2}$

kinetic energy: the energy of a moving object due to its motion. For an object of mass m moving at speed v, its kinetic energy $E_K = \frac{1}{2}mv^2$, provided $v \ll c$ (the speed of light in free space).

kinetic energy of the molecules of an ideal gas:
1. Mean kinetic energy of a molecule of an *ideal gas* = $\frac{3}{2}kT$, where the Boltzmann constant $k = \dfrac{R}{N_A}$
2. Total kinetic energy of n moles of an ideal gas = $\frac{3}{2}nRT$

kinetic theory of a gas:
1. Assumptions; a gas consists of identical point molecules which do not attract one another. The molecules are in continual random motion colliding elastically with each other and with the container.
2. The pressure p of N molecules of such a gas in a container of volume V is given by the equation $pV = \frac{1}{3}Nm(c_{rms})^2$, where m is the mass of each molecule and $(c_{rms})^2$ is the mean square speed of the gas molecules.
3. Assuming that the mean kinetic energy of a gas molecule $\frac{1}{2}m(c_{rms})^2 = \frac{3}{2}kT$, where $k = \dfrac{R}{N_A}$, it can be shown from $pV = \frac{1}{3}Nm(c_{rms})^2$ that $pV = nRT$, which is the *ideal gas law*.

kinetic theory of gases equation: $pV = \frac{1}{3}Nm(c_{rms})^2$

L

latent heat of fusion: the energy needed to change the state of a solid to a liquid without change of temperature. See *specific latent heat of fusion*.

latent heat of vaporisation: the energy needed to change the state of a liquid to a vapour without change of temperature. See *specific latent heat of vaporisation*.

Lenz's law: when a current is induced by electromagnetic induction, the direction of the induced current is always such as to oppose the change that causes the current.

line of force or a field line: the direction of a line of force or a field line indicates the direction of the force. An electric field line is the path followed by a free positive test charge. The gravitational field lines of a single mass point towards that mass.

logarithms: for a number $n = b^p$ where b is the base number, then $p = \log_b n$
$\log (nm) = \log n + \log m$,
$\log \left(\dfrac{n}{m}\right) = \log n - \log m$,
$\log (m^p) = p \log m$
natural logs: for $n = e^p$, then $\ln n = p$
base 10 logs: for $n = 10^p$, then $\log_{10} n = p$

logarithmic scale: a scale such that equal intervals correspond to a change by a constant factor or multiple (e.g., ×10).

log graphs: 1. For $y = kx^n$, $\log_{10} y = \log_{10} k + n \log_{10} x$; the graph of $\log_{10} y$ (on the vertical axis) against $\log_{10} x$ is therefore a straight line of gradient n with an intercept equal to $\log_{10} k$;
2. For $x = x_0 e^{-\lambda t}$, $\ln x = \ln x_0 - \lambda t$; the graph of $\ln x$ (on the vertical axis) against t is a straight line with a gradient equal to $-\lambda$ and a y-intercept equal to $\ln x_0$

M

magnetic flux ϕ: $\phi = BA$ for a uniform magnetic field of flux density B that is perpendicular to an area A. The unit of magnetic flux is the weber (Wb)

magnetic flux density B: the magnetic force per unit length per unit current on a current carrying conductor at right angles to the field lines. The unit of magnetic flux density is the tesla (T). B is sometimes referred to as the magnetic field strength.

magnetic flux linkage $N\phi$: through a coil of N turns, $= N\phi = NBA$ where B is the magnetic flux density perpendicular to area A. The unit of magnetic flux and of flux linkage is the weber (Wb), equal to 1 T m^2 or 1 V s.

magnetic force: 1. $F = BIl \sin \theta$ gives the force F on a current-carrying wire of length l in a uniform magnetic field B at angle θ to the field lines, where I is the current. The direction of the force is given by *Fleming's left-hand rule* where the field direction is the direction of the field component perpendicular to the wire.
2. $F = BQv \sin \theta$ gives the force F on a particle of charge Q moving through a uniform magnetic field B at speed v in a direction at angle θ to the field. If the velocity of the charged particle is perpendicular to the field, $F = BQv$. The direction of the force is given by Fleming's left-hand rule, provided the current is in the direction that positive charge would move in.
3. $BQv = \dfrac{mv^2}{r}$ gives the radius of the orbit of a charge moving in a direction at right angles to the lines of a magnetic field.

mass defect: of a nucleus is the difference between the mass of the separated nucleons (i.e., protons and neutrons from which the nucleus is composed) and the nucleus.

mean kinetic energy: for a molecule in a gas at absolute temperature T, its mean kinetic energy = $\frac{3}{2}kT$, where k is the Boltzmann constant $\left(= \frac{R}{N_A}\right)$

melting point: the temperature at which a pure substance melts.

metastable state: an excited state of the nuclei of an isotope that lasts long enough after α or β emission for the isotope to be separated from the parent isotope (e.g., technetium $^{99}_{43}$Tc).

moderator: substance in a thermal nuclear reactor that slows the fission neutrons down so they can go on to produce further fission.

mole: one mole of a substance consisting of identical particles is the quantity of substance that contains N_A particles of the substance.

molarity: the number of moles in a certain quantity of a substance. The unit of molarity is the mol.

molar mass: the mass of one mole of a substance.

motor effect: the force on a current-carrying conductor due to a magnetic field.

N

natural frequency: the frequency of free oscillations of an oscillating system.

neutron: an uncharged particle that has a rest mass of 1.674×10^{-27} kg. Neutrons are in every atomic nucleus except that of hydrogen $^1_1$H.

Newton's law of gravitation: the gravitational force F between two point masses m_1 and m_2 at distance r apart is given by $F = \frac{Gm_1m_2}{r^2}$.

nucleus: the relatively small part of an atom where all the atom's positive charge and most of its mass is concentrated.

nuclide of an isotope ^A_ZX: a *nucleus* composed of Z *protons* and $(A - Z)$ *neutrons*, where Z is the proton number (and also the atomic number of element X) and A is the mass number (or nucleon number, i.e., the number of protons and neutrons in a nucleus).

P

period of a wave: time for one complete cycle of a wave to pass a point.

periodic force: a *force* that varies regularly in magnitude with a definite time period.

permittivity of free space ε_0: the charge per unit area in coulombs per square metre on oppositely charged parallel plates in a vacuum when the *electric field strength* between the plates is 1 volt per metre. See *Coulomb's law of force*.

phase difference: in radians, for two objects oscillating with the same time period, T_p, the phase difference = $\frac{2\pi\,\Delta t}{T_p}$, where Δt is the time between successive instants when the two objects are at maximum displacement in the same direction.

photon: electromagnetic radiation consists of photons. Each photon is a wave packet of electromagnetic radiation. The energy of a photon, $E = hf$, where f is the frequency of the radiation and h is the Planck constant.

polarised: the positive charge and the negative charge of a polarised molecule are displaced in opposite directions.

potential gradient: at a point in a field is the change of potential per unit change of distance along the field line at that point. The potential gradient = − the field strength at any point.

pressure law: for a fixed mass of an ideal gas at constant volume, its pressure is directly proportional to its absolute temperature.

principle of conservation of momentum: when two or more bodies interact, the total *momentum* is unchanged, provided no external forces act on the bodies.

proton: a particle that has equal and opposite charge to the electron and has a rest mass of 1.673×10^{-27} kg which is about 1836 times that of the electron. Protons are in every atomic nucleus. The nucleus of hydrogen $^1_1$H is a single proton. The proton is the only stable baryon.

R

radial field: a field in which the *field lines* are straight and converge or diverge as if from a single point.

radian: 1 radian = $\frac{360}{2\pi}$ degrees. 2π radians = $360°$.

reactor core: the fuel rods and the control rods together with the moderator substance are in a steel vessel through which the coolant (which is also the moderator in 'pressurised water reactor') is pumped.

relative permittivity: ratio of the charge stored by a parallel-plate capacitor with dielectric filling the space between its plates to the charge stored without the dielectric for the same pd.

renewable energy: energy from a source that is continually renewed. Examples include hydroelectricity, tidal power, geothermal power, solar power, wave power, and wind power.

resonance: the amplitude of vibration of an oscillating system subjected to a periodic force is largest when the periodic force has the same frequency as the resonant frequency of the system.

For a lightly damped system, the frequency of the periodic force = natural frequency of the oscillating system. At resonance, the system vibrates such that its velocity is in phase with the periodic force.

resonant frequency: the frequency of an oscillating system in resonance.

root mean square speed, c_{rms}: square root of the mean value of the square of the molecular speeds of the molecules of a gas.

$$c_{rms} = \left(\frac{c_1^2 + c_2^2 + \dots + c_N^2}{N}\right)^{\frac{1}{2}}$$

where $c_1, c_2, c_3 \dots c_N$ represent the speeds of the individual molecules and N is the number of molecules in the gas.

Rutherford's α-particle scattering experiment: demonstrated that every atom contains a positively charged nucleus which is much smaller than the atom and where all the positive charge and most of the mass of the atom is located.

S

satellite: a small object in orbit round a larger object.

satellite motion: for a satellite moving at speed v in a circular orbit of radius r round a planet, its centripetal acceleration, $\frac{v^2}{r} = g$. Substituting $v = \frac{2\pi r}{T}$, where T is its time period, and $g = \frac{GM}{r^2}$, where M is the mass of the planet, $T^2 = \left(\frac{4\pi^2}{GM}\right)r^3$. See *geostationary satellite*.

simple electric motor: an electric motor with an armature consisting of a single coil of insulated wire.

simple harmonic motion: motion of an object if its acceleration is proportional to the displacement of the object from equilibrium and is always directed towards the equilibrium position.

1. The acceleration, a, of an object oscillating in simple harmonic motion is given by $a = -(2\pi f)^2 x = -\omega^2 x$, where x = displacement from equilibrium, and f = frequency of oscillations and ω = the angular frequency = $2\pi f$.

2. The solution of this equation depends on the initial conditions. If $x = 0$ and the object is moving in the + direction at time $t = 0$, then $x = A\sin(2\pi ft)$. If the object is at maximum displacement, $+A$, at time $t = 0$, then $x = A\cos(2\pi ft)$

simple harmonic motion applications: 1. For a simple pendulum of length L, its time period

$$T = 2\pi \left(\frac{L}{g}\right)^{\frac{1}{2}}$$

2. For an oscillating mass m on the end of a vertical spring, its time period $T = 2\pi \left(\frac{m}{k}\right)^{\frac{1}{2}}$, where k is the spring constant.

sinusoidal curves: Any curve with the same shape as a sine wave (e.g., a cosine curve).

specific heat capacity c: of a substance is the energy needed to raise the temperature of 1 kg of the substance by 1 K without change of state. To raise the temperature of mass m of a substance from T_1 to T_2, the energy needed, $Q = mc(T_2 - T_1)$, where c is the specific heat capacity of the substance.

specific latent heat of fusion: of a substance is the energy needed to change the state of unit mass of a solid to a liquid without change of temperature.

specific latent heat of vaporisation: for a substance is the energy needed to change the state of unit mass of a liquid to a vapour without change of temperature.
To change the state of mass m of a substance without change of temperature, the energy needed $Q = ml$, where l is the specific latent heat of fusion or vaporisation of the substance.

sublimation: the change of state when a solid changes to a vapour directly.

T

temperature: the degree of hotness of an object. Defined in terms of 'fixed points' (e.g., the triple point of water = 273.16 K).

thermal energy: the internal energy of an object due to temperature.

thermal equilibrium: when no overall heat transfer occurs between two objects at the same temperature.

thermal nuclear reactor: nuclear reactor which has a moderator in the core.

time constant: the time taken for a quantity that decreases exponentially to decrease to 0.37 $\left(= \frac{1}{e}\right)$ of its initial value. For the discharge of a capacitor through a fixed resistor, the time constant = resistance × capacitance.

time period or period: time taken for one complete cycle of oscillations.

transformer: converts the amplitude of an alternating pd to a different value. It consists of two insulated coils, the primary coil and the secondary coil, wound round a soft iron laminated core; **step-down transformer:** a transformer in which the rms pd across the secondary coil is less than the rms pd applied to the primary coil; **step-up transformer:** a transformer in which the rms pd across the secondary coil is greater than the rms pd applied to the primary coil.

transformer rule: the ratio of the secondary voltage to the primary voltage is equal to the ratio of the number of secondary turns to the number of primary turns.

transformer efficiency: for an ideal transformer (i.e., one that is 100% efficient), the output power (= secondary voltage × secondary current) = the input power (= primary voltage × primary current). Transformer inefficiency is due to: resistance heating of the current in each coil; the heating effect of eddy currents (i.e., unwanted induced currents) in the core; and heating caused by repeated magnetisation and demagnetisation of the core.

U

uniform circular motion: motion of an object moving at constant speed along a circular path.

uniform field: a region where the field strength is the same in magnitude and direction at every point in the field.

1. The **electric field** between two oppositely charged parallel plates is uniform. The electric field strength $E = \dfrac{V}{d}$, where V is the pd between the plates and d is the perpendicular distance between the plates.

2. The **gravitational field** of the Earth is uniform over a region which is small compared to the scale of the Earth.

3. The **magnetic field** inside a solenoid carrying a constant current is uniform along and near the axis.

X

X-rays: electromagnetic radiation of wavelength less than about 1 nm. X-rays are emitted from an X-ray tube as a result of fast-moving electrons from a heated filament as the cathode being stopped on impact with the metal anode. X-rays are ionising and they penetrate matter. Thick lead plates are needed to absorb a beam of X-rays.

Glossary of practical terms

accepted value: value of the most accurate measurement available, sometimes referred to as the 'true value'.

accuracy: The closeness of a measurement to the true value (if known).

dependent variable: the variable of which the value is measured for each and every change in the independent variable.

error bar: representation of an uncertainty on a graph.

errors: 1. **Random** errors cause readings to be spread about the true value, due to the results varying in an unpredictable way from one result to the next.

2. **Systematic** errors cause readings to differ from the true value by a consistent amount each time a measurement is made.

independent variable: physical quantities whose values are selected or controlled by the experimenter.

linearity: an instrument that gives readings that are directly proportional to the magnitude of the quantity being measured.

mean value of a set of readings: sum of the readings divided by the number of readings.

percentage uncertainty: $= \dfrac{\text{uncertainty}}{\text{mean value}} \times 100\%$

precision of a measurement: precise measurements are ones in which there is very little spread about the mean value. Precision depends only on the extent of random error and gives no indication of how close the results are to the true value.

precision of an instrument: the smallest non-zero reading that can be measured using the instrument, also sometimes referred to as the instrument sensitivity or resolution.

range of a set of readings: the maximum and minimum values of the readings.

range of an instrument: the maximum and minimum values that can be obtained using the instrument.

reliability: an experiment or measurement is reliable if a consistent value is obtained each time it is repeated under identical conditions. The reliability of an experiment is increased if random and systematic errors have been considered and eliminated and, where approriate, a more precise best fit line has been obtained.

repeatable: an experiment or measurement that gives the same results when it is repeated by the original experimenter using the same method and equipment.

reproducible: an experiment or measurement that gives the same results when it is repeated by another person or by using different equipment or techniques.

sensitivity of an instrument: output response per unit input quantity.

uncertainty of a measurement: the interval within which the true value can be expected to lie, with a given level of confidence or probability.

valid measurement: measurements that give the required information by an acceptable method.

zero error any indication that a measuring system gives a false reading when the true value of a measured quantity is zero. A zero error may result in a systematic uncertainty.

Answers to summary questions

17.1

1 a 1.75×10^{-3} rad
   b 0.105 rad
   c 6.28 rad
2 a 20 ms
   b i 0.31 rad
      ii 310 rad
3 a 465 m s^{-1}
   b i 0.0042°
      ii 7.3×10^{-5} rad
4 a 7.0 km s^{-1}
   b i 0.050°
      ii 8.7×10^{-4} rad

17.2

1 a 0.23 m s^{-1}
   b i 7.9×10^{-4} m s^{-2}
      ii 5.1×10^{-2} N
2 a 0.53 m s^{-1}, 0.66 m s^{-2}
   b 9.9×10^{-2} N
3 a i 3.0×10^{4} m s^{-1}
      ii 6.0×10^{-3} m s^{-2}
   b i 7.9×10^{3} m s^{-1}
      ii 5.1×10^{3} s
4 a 8.4 m s^{-1}
   b 88 m s^{-2}
   c 175 N

17.3

1 a 6.7 m s^{-2}
   b 3.8 kN
2 a 4.1 m s^{-2}
   b 3.0 kN
3 a 40 m s^{-1}

17.4

1 a 30 m s^{-1}
   b i 11.3 m s^{-2}
      ii 690 N
2 a 25 m s^{-1}
   b 20 m s^{-2}
   c 2000 N

3 a 13 m s^{-1}
   b 13 m s^{-2}
   c 240 N
4 −0.04 N

18.1

1 a 0.48 s
   b 2.1 Hz
2 a $\frac{\pi}{2}$ radian
   b π radian

18.2

1 a +25 mm, changing direction from up to down
   b 0, moving down
   c −25 mm, changing direction from down to up
   d 0, moving up
2 a 0.5 Hz
   b i −0.25 m s^{-2}
      ii 0
      iii 0.25 m s^{-2}
3 a 0.5 Hz
   b −0.32 m s^{-2}
4 a −32 mm   0.32 m s^{-2}
   b 0, 0

18.3

1 a 0.33 Hz
   b 0.25 m s^{-2}
2 a i 12 mm
      ii 0.63 s
   b 6.5 mm
3 a 2.1 Hz
   b 0.057 m
4 a 3.7 Hz
   b i −8.2 mm towards maximum negative displacement
      ii −0.7 mm towards maximum positive displacement.

18.4

1 a i 0.33 s
      ii 3.1 Hz

b   i   0

   ii $-3.7\,\text{m s}^{-2}$

   iii $-7.5\,\text{m s}^{-2}$

2 a   i   3.0 Hz

   ii   0.33 s

 b $f_2 < f_1 \therefore m_2 > m_1$

3 a   i   70 mm

   ii $21\,\text{N m}^{-1}$

 b   ii   0.53 s

4 a   i   1.25 N

   ii $2.5\,\text{m s}^{-2}$

 b   ii   1.1 Hz, +47 mm

5 a   i   2.0 s

   ii   1.0 s

 b   5.0 s

18.5

1 a   i   1.50 s

   ii   0.56 m

   iii   0.029 J

 b   See Figure 2

2 a   i $60\,\text{N m}^{-1}$

   ii   0.54 s

 b   i   75 mJ

   ii   75 mJ

   iii $0.50\,\text{m s}^{-1}$

3 a   i   light

   ii   heavy

4 b   82 mm

 d   44 mm

18.6

1 a $27\,\text{N m}^{-1}$

 b   1.7 Hz

4 b $2.8\,\text{m s}^{-1}$

19.1

4 a   i   273 K

   ii   293 K

   iii   77 K

 b   i   328 K

   ii   137 kPa

19.2

1 a   23 kJ

 b   535 kJ

2 a   280 s

 b   10.3 MJ

3 a   320 J

 b $130\,\text{J kg}^{-1}\,\text{K}^{-1}$

4 3.2 kW (3.15 kW to 3 sig. figs)

19.3

2 0.16 kg

3 a $4.2\,\text{J s}^{-1}$

 b   6400 s

4 a $22\,\text{J s}^{-1}$

 b   6.5 kJ

20.1

1 126 kPa

2 79 kPa

3 $0.097\,\text{m}^3$

4 b $2.3 \times 10^{-5}\,\text{m}^3$, $1600\,\text{kg m}^{-3}$

20.2

1 a   1.1 moles

 b   109 kPa

2 a 9.3×10^{-4} moles

 b $2.1 \times 10^{-5}\,\text{m}^3$

3 b $1.3\,\text{kg m}^{-3}$

4 a $1.2\,\text{kg m}^{-3}$

 b 2.5×10^{22}

20.3

1 b   i   0.97 mol

   ii   4.5 kJ

2 c 5.7×10^{-21} J

 d $1.8 \times 10^3\,\text{m s}^{-1}$

4 a   1.48 mol

 b $4.2 \times 10^{-2}\,\text{kg}$

 c $5.2 \times 10^2\,\text{m s}^{-1}$

21.1

2 a i 33 N
  ii 160 N
 b i 16 N kg^{-1}
  ii 4.0 N kg^{-1}

21.2

1 a 235 J
2 a 2.0 MJ kg^{-1}
 b i −61 MJ kg^{-1}
  ii 2.2 × 10^9 J
3 a i −250 J
  ii −200 J
  iii −200 J
 b i 50 J
  ii 0
4 b 5 N kg^{-1}
 c 25 MJ

21.3

1 a 1.3 × 10^{-6} N
 b 5.4 mm
2 a 780 N
 b 6.0 × 10^{24} kg
3 a 54 N
 b 0.24 N
4 a i 16.6 N
  ii 0.2 N
 b 16.4 N towards the centre of the Earth

21.4

1 a 7.35 × 10^{22} kg
2 a i 272 N kg^{-1}
  ii 5.9 × 10^{-3} N kg^{-1}
3 a 0.028 N kg^{-1}
 c 7.1 × 10^6 J
4 2.8 MJ kg^{-1}, 1410 MJ

21.5

2 a 3.4 × 10^6 m
 b 3.0 m s^{-2}
 c 5.2 × 10^{23} kg

3 b i 9.5 N kg^{-1}
  ii 7.9 km s^{-1}
  iii 5200 s
4 b ii 7100 s

22.1

2 b i 75 nA
  ii 1.9 × 10^{11}

22.2

1 a 1.4 × 10^{-3} N
 b 4.0 × 10^4 V m^{-1}
2 a i negative
  ii 1.3 × 10^{-7} C
 b i 7.3 × 10^{-3} N
  ii towards the metal surface
3 a i 9.0 × 10^4 V m^{-1}
  ii 7.2 × 10^{-14} N
 b 80 mm
4 a 2.9 kV
 b i 5.6 × 10^{-15} N

22.3

1 a i −8.0 × 10^{-18} J
  ii +7.2 × 10^{-17} J
 b +8.0 × 10^{-17} J
2 a −1.8 × 10^{-3} J
 b +1.2 × 10^{-3} J
3 a i 250 V m^{-1}
  ii 8.0 × 10^{-17} N (towards the negative plate)
 b −8.0 × 10^{-19} J
4 b i 3000 V m^{-1}

22.4

1 a 3.7 × 10^{-11} N
 b 2.6 × 10^{-10} N
2 a i 69 mm
  ii 3.6 × 10^{-6} N
 b 2.5 × 10^{-5} N repulsion
3 a 6.1 nC, negative
 b 2.2 × 10^{-2} N
4 a 2.7 nC, attract
 b 6.2 × 10^{-2} m, repel

22.5

1 **a** 5.3×10^6 V m^{-1}

  b 10 mm

2 **a** **i** 3.7×10^8 V m^{-1}

    ii 5.6×10^{-3} N towards Q_1

3 **a** **i** 4.5×10^8 V m^{-1} towards Q_2

    ii 2.6×10^8 V m^{-1} away from Q_1

  b **ii** 11 mm from Q_1, 9 mm from Q_2

4 **a** -9.0×10^6 V

  b **ii** 2.0×10^9 V m^{-1} directly towards Q_2

23.1

1 **a** 5.0 µF

  b 2.2 V

  c 9.9 mC

  d 1.4 µF

2 **a** 264 µC

  b 106 s

3 **a** 27.5 µC

  b 5.5 µF

4 **a** 0.91 µC

  b 0.22 µF

  c 700 µC

  d 7.4 V

23.2

1 **a** 30 µC, 45 µJ

  b 60 µC, 180 µJ

2 **a** 0.45 C, 2.0 J

  b 10 W

3 **a** **i** 6.6 µC, 9.9 µJ

    ii 6.6 µC, 19.8 µJ

4 **a** 56 µC, 338 µJ

  b **i** 2.2 µF; 18 µC, 4.7 µF; 38 µC

    ii 8.2 V

c 2.2 µF; 73 µJ, 4.7 µF; 157 µJ

23.3

1 **a** **i** 300 µC

    ii 5.0 s

  b **i** 5 s approx

    ii 20 kΩ

2 **a** **i** 0.61 mC

    ii 0.45 mA

  b 0.23 V, 11 µA

3 **a** 13 µC, 40 µJ

  b 0.62 V, 0.42 µJ

4 **a** **i** 60 mA

    ii 0.34 mJ

  b 1.4 s

  c 0.32 mJ

23.4

1 **a** **i** C increases

    ii Q increases

  b The energy stored $= \frac{1}{2}QV$, so the energy stored increases because Q increases and V is unchanged.

2 The capacitor is isolated from the battery, so the charge stored does not change. The capacitance decreases when the dielectric is removed. The energy stored $= \frac{1}{2}\frac{Q^2}{C}$, so it increases because C decreases and Q is unchanged. The increase of energy is due to the work done to overcome the electrostatic attraction between the charge on each dielectric surface and the opposite type of charge on the adjacent plate.

3 **a** 9.8 pF

  b 1100 pJ

4 **a** 0.14 µm

  b 2.2 MJ m^{-3}

24.1

1 **a** 2.4×10^{-2} N; west

  b 4.5 A; east to west

  c 0.20 T; vertically down

  d 8.0×10^{-3} N; due south

2 **a** 22 mN due east

  b 4.0 A west to east

3 Short sides; zero, long sides; 2.72 N vertically up on one side and vertically down on the other side

4 **a** 58 µT

  b 6.5×10^{-5} N due east

24.2

1 b i 1.9×10^{-13} N

    ii 0

2 3.8×10^{-23} N horizontal due East

4 b i 4.4 m s^{-1}

    ii 8.5×10^{-20} N

24.3

1 a ii 21 mm

  b 2.8 mT

2 a 4.7 mT

  b 17.5 mm

3 b 1.2 MeV

4 a 8.0×10^{6} C kg^{-1}

  b 1.4×10^{7} C kg^{-1}

25.2

1 a i 1.1 mWb

    ii 2.0 s

    iii 0.54 mV

2 a 1.4 mWb

  b 23 mV

3 a i 4.5×10^{-4} m^2

    ii 1.5(4) mWb

  b i 3.1 mWb

    ii 33 mV

4 a 8.0 µWb

  b 40 µV

25.3

2 flux linkage 0, 0, $-BAN$; induced emf 0, $-\varepsilon_0$, 0

3 a 26 mWb

25.4

1 a 5 ms

  b 0.0707 A

2 a i 2.12 A

    ii 8.49 V

  b i 36 W

    ii 18 W

3 a i 4.35 A

    ii 6.15 A

    iii 2.0 kW

b The fuse melts only if the rms current exceeds 5 A. The rms current in this case is less than 5 A, so the fuse does not melt.

4 a i 4.0 A

    ii 5.66 A

    iii 17.0 V

  b i $V_{rms} = I_{rms} R = 4.0 \times 0.5 = 2.0$ V

    ii 19.8 V

25.5

1 a 11.5 V

  b i 0.26 A

    ii 5.2 A

2 b i 17 A (16.7 A to 3 sig. fig.)

    ii 56 kW

26.1

4 b 3×10^{-14} m

  c 4.5×10^{-14} m

26.2

1 a 42 %

  b 58 %

26.3

1 a $^{235}_{92}\text{U} \rightarrow {}^{234}_{90}\text{Th} + {}^{4}_{2}\alpha$

  b $^{228}_{90}\text{Th} \rightarrow {}^{224}_{88}\text{Ra} + {}^{4}_{2}\alpha$

2 a $^{64}_{29}\text{Cu} \rightarrow {}^{64}_{30}\text{Zn} + {}^{0}_{-1}\beta \ (+ \bar{\nu})$

  b $^{32}_{15}\text{P} \rightarrow {}^{32}_{16}\text{S} + {}^{0}_{-1}\beta \ (+ \bar{\nu})$

3 a $^{213}_{84}\text{Po}, \ {}^{290}_{82}\text{Pb}, \ {}^{209}_{83}\text{Bi}$

  b i 83 p + 130 n

    ii 83 p + 126 n

4 a 3.2 counts per second

  b 160 mm

26.5

1 a 40 s

  b 6 mg

2 a i 9.0×10^{14}

    ii $2.2(5) \times 10^{14}$

  b i 9.0×10^{14}

    ii 15.8×10^{14}

  c 1.3×10^{3} J

3 a 19 kBq

  b 2.4 kBq

4 a 4.4×10^{21}

  b 1.1×10^{21} atoms

26.6

1 a $1.0 \times 10^{-6} \text{ s}^{-1}$

  b 3.9×10^{16}

2 a $6.3 \times 10^{-10} \text{ s}^{-1}$

  b 20.5 kBq

3 a 2.7×10^{24}

  b i 0.65 kg

    ii 1.7×10^{24}

4 a $1.3 \times 10^{-6} \text{ s}^{-1}$

  b 149 hours

26.7

1 b 1340 years

2 a i 730 kBq

    ii 870 kBq

    iii 1.9×10^{-13} kg

4 b i 5.2×10^{13} Bq

    ii 3.3 J s^{-1}

26.8

2 a 82 p, 126 n

4 a

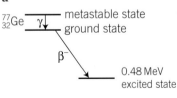

  b 0.21 MeV, 0.27 MeV

26.9

2 a 3.2 fm

  b 5.2 fm

3 b 6.5 fm, $1.2 \times 10^{-42} \text{ m}^3$

4 b 2.5 fm, $3.4 \times 10^{17} \text{ kg m}^{-3}$

27.1

1 a 2.18×10^{-15} kg

  b i 8.89×10^{-33} kg

    ii 8.89×10^{-30} kg

2 6.2 MeV

3 0.56 MeV

4 1.68 MeV

27.2

1 a 7.4 MeV

  b 8.8 MeV

2 a i 7.1 MeV

    ii 2.6 MeV

3 a $\text{p} + \text{p} \rightarrow {}_{1}^{2}\text{H} + {}_{+1}^{0}\beta + \nu$

  b 1.12 MeV

27.3

2 b i $a = 56$, $b = 98$

    ii 206 MeV

3 b ii 0.43 MeV, 5.5 MeV

4 b 12.9 MeV

30.1

1 a i 0.524 rad

    ii 0.873 rad

    iii 2.094 rad

    iv 4.014 rad

    v 5.236 rad

  b i 5.73°

    ii 28.7°

    iii 68.8°

    iv 143.2°

    v 343.8°

2 a 20 mm, 2.3°

  b ii 0.5°

3 a i 0.035

    ii 0.140

4 a i 0.0998

    ii 0.995

  b i 0.1736

    ii 0.9848

  c i 0.7071

    ii 0.7071

  d i 0.7071

    ii 0.7071

30.2

1 a $x = 1, y = 3$
  b $a = 2.4, b = -0.4$
  c $p = 2, q = 4$
2 a $u = 14$ m s^{-1}, $a = -2.0$ m s^{-2}
  b $r = 1.0\ \Omega$, $\varepsilon = 9.0$ V
3 a 0.5 or −3
  b 1.4 or 5.6
  c $-\dfrac{10}{6} = -1.67$ (to 3 sig. figs) or 1
4 a $t = -\dfrac{20}{6} = -3.33$ (to 3 sig. figs) or 2 s
  b $R = 1$ or $4\ \Omega$

30.3

1 a i 0.477
    ii 1.176
  b i 1.653
    ii 0.699
2 a i 10.8 dB
    ii 7.0 dB
  b 17.8 dB
3 a $n = 5, k = 3$
  b $n = 3, k = \dfrac{1}{2}$
  c $n = 2, k = 1$
4 a i 1.10
    ii 2.71
  b i 3.81
    ii 1.61

30.4

1 a i 2, 3
    ii 12, 0.2
    iii 4, 0.02
  b i 0.23 s
    ii 3.5 s
    iii 35 s
2 a 11.3 kBq
  b 9.0 kBq
  c 0.38 kBq
3 a i 2.2 s
    ii 1.52 s
  b i 4.83 V
    ii 1.24 V
4 a i 0.14 s
    ii 82
  b $a = 6.9, b = -5$

30.5

2 d i Pa or N m^{-2}
    ii m^3
    iii J

30.6

1 The gradient at 0.6 s = 14.4 m s^{-1}. This gives a value of g equal to 10.0 m s^{-2}.
2 a 0.038 s^{-1}
  b N decreases at a slower rate, and the half-life is doubled.

19.1 The coldest places in the world

The thermometer would not respond at such low temperatures, because the liquid would be frozen.

20.3 Making assumptions

It would be greater than if the molecules were point molecules.

21.3 Cavendish's measurement of *G*

The forces on each small ball from the nearest massive ball act in opposite directions, but not along the same line, so they act as a couple, which causes the balance to turn.

26.1 What made Rutherford originate the nuclear model of the atom?

The electrostatic force of repulsion between an α particle and a nucleus.

26.4 Radiation dose limits

Any two from the ground, cosmic rays, nuclear weapons, food and drink.

27.2 α particle tunnelling

Tunnelling is more likely higher up the barrier as it gets narrower with height.

27.4 Moderators at work

42. Note that $0.72^n = 10^{-6}$ gives $n = 41.4$, so 42 collisions are necessary (because 41 collisions reduces the energy to 1.4 eV).

27.4 A nuclear future?

Many of the radioactive isotopes from the reactor that fell on the surrounding area have long half-lives and are still active.

Index

Acknowledgements

The author wishes to acknowledge the active support, advice and contributions he has received from Marie Breithaupt, and Patrick Organ, and from Sze-Kie Ho, Alison Schrecker, Matteo Orsini Jones and their colleagues at OUP.

I would like to thank my family for their support in the preparation of this book, particularly my wife, Marie, for secretarial support and cheerful encouragement. I am also grateful to the publishing team at Nelson Thornes, in particular Eleanor O'Byrne and Carol Usher who initiated the project. I am grateful to Ian Holt and Patrick Organ for their contributions and advice and I wish also to thank AQA for their support.

Jim Breithaupt

The authors and publisher are grateful to the following for permission to reproduce photographs and other copyright material in this book.

Cover: ferrantraite / iStock; **p2-3**: ZargonDesign/iStockphoto; **p4**: Cycreation/Shutterstock; **p11**: Liewluck/Shutterstock; **p31**(T): Library of Congress/Science Photo Library; **p31**(B): Adrian Bicker/Science Photo Library; **p36**: Stephen Harrison/Alamy; **p39**: Pascal Goetgheluck/Science Photo Library; **p61**: ZargonDesign/iStockphoto; **p66-67**: Digi_guru/iStockphoto; **p68**: NASA; **p70**: NASA; **p75**: Science Museum/Science & Society Picture Library; **p80**: NASA; **p81**: NASA; **p82**: George Clerk/iStockphoto; **p84**: Viktar/iStockphoto; **p89**: Charistoone-stock/Alamy; **p93**: Piotr Krzeslak/Shutterstock; **p99**: Corbis; **p128**: Valerii Ivashchenko/Shutterstock; **p130**: Goronwy Tudor Jones, University Of Birmingham/Science Photo Library; **p131**: Andrew Lambert Photography/Science Photo Library; **p141**: Wynnter/iStockphoto; **p144**: 360b/Shutterstock; **p161**: Digi_guru/iStockphoto; **p166-167**: RelaxFoto.de/iStockphoto; **p169**: Science Photo Library; **p171**: Georgios Kollidas/Shutterstock; **p172**(L): Science Photo Library; **p172**(R): Science Photo Library; **p179**: Martyn F. Chillmaid/Science Photo Library; **p189**(T): Ria Novosti/Science Photo Library; **p189**(B): Fertnig/iStockphoto; **p210**: EFDA-Jet/Science Photo Library; **p213**: Digital Globe/Science Photo Library; **p214**: Paul Shambroom/Science Photo Library; **p219**: RelaxFoto.de/iStockphoto; **p232-233**: Stefano Garau/Shutterstock; **p235**: Stefano Garau/Shutterstock; **p236-237**: Pedrosala/Shutterstock; **p239**: Pedrosala/Shutterstock; **p240-241**: Stason4ik/Shutterstock; **p243**: Stason4ik/Shutterstock; **p244-245**: Giphotostock/Science Photo Library; **p247**: Giphotostock/Science Photo Library; **p248-249**: Tushchakorn/Shutterstock; **p251**: Tushchakorn/Shutterstock; **p252**: Slobo/iStockphoto; **p256**: Gina Randall/U.S. Air Force; **p257**: Martyn F. Chillmaid/Science Photo Library; **p258**: Martyn F. Chillmaid/Science Photo Library;

Artwork by Q2A Media

Although we have made every effort to trace and contact all copyright holders before publication this has not been possible in all cases. If notified, the publisher will rectify any errors or omissions at the earliest opportunity.

Links to third party websites are provided by Oxford in good faith and for information only. Oxford disclaims any responsibility for the materials contained in any third party website referenced in this work.